# Hydrogen-Bonded Interpolymer Complexes

## Formation, Structure and Applications

# Hydrogen-Bonded
# Interpolymer
# Complexes
## Formation, Structure and Applications

Editors

## Vitaliy V Khutoryanskiy
*University of Reading, UK*

## Georgios Staikos
*University of Patras, Greece*

**W⊖ World Scientific**

NEW JERSEY · LONDON · SINGAPORE · BEIJING · SHANGHAI · HONG KONG · TAIPEI · CHENNAI

*Published by*

World Scientific Publishing Co. Pte. Ltd.

5 Toh Tuck Link, Singapore 596224

*USA office:* 27 Warren Street, Suite 401-402, Hackensack, NJ 07601

*UK office:* 57 Shelton Street, Covent Garden, London WC2H 9HE

QD
381
·H86
2009

**British Library Cataloguing-in-Publication Data**
A catalogue record for this book is available from the British Library.

ISBN-13 978-981-270-785-7
ISBN-10 981-270-785-9

Printed by Fuisland Offset Printing (S) Pte Ltd. Singapore

# ABOUT THE EDITORS

**Dr Vitaliy V. Khutoryanskiy** is a lecturer in pharmaceutics in Reading School of Pharmacy, University of Reading. His current research interests include polymers for pharmaceutical applications, water-soluble polymers and polymeric amphiphiles, polymeric complexes and blends, muco-adhesion and mucoadhesive drug delivery systems, hydrogels and biomaterials. He has published 65 original research papers and four reviews in peer reviewed journals.

**Dr Georgios Staikos** is a professor in the Department of Chemical Engineering of the University of Patras (Greece). His current research interests include water-soluble polymers, thermoresponsive polymers, inter-polymer complexes and synthesis of graft copolymers, properly designed to form colloidal nanoparticles through hydrogen-bonding or Coulombic interactions with neutral polymers, polyelectrolytes and proteins. He has published 54 original research papers.

# PREFACE

Noncovalent interactions play key roles in many natural processes leading to self-assembly of molecules with formation of supramolecular structures. One of the most significant forces responsible for self-assembly is hydrogen-bonding, which also plays an important role in self-assembly of synthetic polymers in aqueous solutions. Proton-accepting polymers can associate with proton-donating polymers via hydrogen-bonding in aqueous solutions and form polymer-polymer or interpolymer complexes. There has been an increased interest of researchers in the hydrogen-bonded interpolymer complexes since the first pioneering papers published in the early 1960s. Several hundreds of research papers have been published on various aspects of the complex formation reactions in solutions and interfaces, the properties of interpolymer complexes and their potential applications.

The book is focused on the latest developments in the area of interpolymer complexation via hydrogen-bonding. It represents a collection of original and review articles written by recognized experts in this field.

*V.V. Khutoryanskiy*
*G. Staikos*

# CONTENTS

# CHAPTER 1

## pH- AND IONIC STRENGTH EFFECTS ON INTERPOLYMER COMPLEXATION VIA HYDROGEN-BONDING

Vitaliy V. Khutoryanskiy[1]*, Artem V. Dubolazov[2], Grigoriy A. Mun[2]

[1]*Reading School of Pharmacy, University of Reading, Whiteknights, PO Box 224, RG6 6AD, Reading, United Kingdom*
*\*E-mail: v.khutoryanskiy@reading.ac.uk*
[2]*Department of Chemical Physics and Macromolecular Chemistry, Kazakh National University, 95 Karasai Batyra Street, 050012 Almaty, Kazakhstan*

When poly(carboxylic acids) are mixed with non-ionic polymers in aqueous solutions, a phase separation can be observed which is often accompanied by formation of precipitates. These precipitates are in fact novel individual polymeric compounds termed interpolymer complexes (IPCs) or polymer-polymer complexes. The precipitation in the mixtures of poly(carboxylic acids) with non-ionic polymers was reported for the first time in 1959 by Smith and co-workers,[1] when two water-soluble polymers (poly(acrylic acid) (PAA) and poly(ethylene oxide)) were mixed together in aqueous solutions resulting in immediate formation of insoluble IPCs. These precipitates were separated, dried and their physicochemical properties were examined by a number of techniques, including the study of mechanical properties at different temperatures and polymer ratios, X-ray patterns and heat stability evaluation. It was concluded that the unique properties of precipitates result from hydrogen-bonding between ether and carboxyl groups of the polymers. It was also pointed out that the complexes may be formed not only between PAA and poly(ethylene oxide) (PEO) but also by many other non-ionic polymers such as poly(vinyl ethers), cellulose ethers and a large variety of hydroxyl-containing polymers.

Few years later, Bailey and co-workers[2] published the second paper on the complexation between polymers via hydrogen-bonding. This study was also focused on the interaction between PAA and PEO with the aim to elucidate the effects of pH and composition on the solution viscosity of polymer/polymer mixture. They demonstrated that the product of interaction between PAA and PEO forms a precipitate at pH below 3.8, whereas at higher pH the polymers co-exist in aqueous solution. Depending on solution pH and PAA/PEO ratio the rheological properties of the polymers mixture displayed a varying degree of association between carboxyl and ether groups. Working with the relatively concentrated solutions of PAA and PEO (0.05–2 wt %) the authors[2] have demonstrated that their mixtures exhibit a higher viscosity compared to the pure components with a maximum corresponding to the IPC stoichiometry. It was demonstrated that the interaction between these polymers is driven by hydrogen-bonding and the IPC stoichiometry approaches 1:1. Whilst at low pH (pH < 3.8), the interaction results in phase separation, at higher pH, the polycomplex exists in solution. In the neutral pH region, they also observed some interaction between the two polymers.

At present, more than 50 poly(carboxylic acid) – non-ionic polymer pairs are known to form hydrogen-bonded complexes. The most commonly used water-soluble poly(carboxylic acids) include PAA, poly(methacrylic acid) (PMAA), styrene-maleic acid copolymer (PSMA) as well as poly(itaconic acid) and its derivatives:

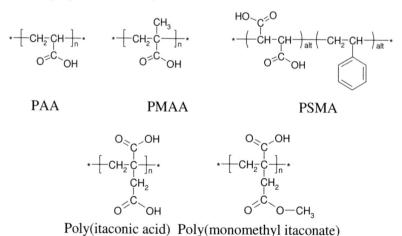

PAA                    PMAA                    PSMA

Poly(itaconic acid)  Poly(monomethyl itaconate)

The list of non-ionic polymers forming IPC with poly(carboxylic acids) is more broad and includes several classes[3]:

*1. Polymers, containing lactame groups.*

Poly(N-vinyl pyrrolidone) (PVP)      Poly(N-vinyl caprolactame)

*2. Polymers containing ether groups either in the backbone (PEO, poly(propylene glycol) (PPG)) or as pendants (poly(vinyl methyl ether (PVME).*

PEO                PPG                PVME

*3. Acrylic type polymers, such as polyacrylamide (PAAM), poly(N-isopropylacrylamide) (PNIPAAM) and poly(N,N-dimethylacrylamide) (PDMAA).*

PAAM              PNIPAAM              PDMAA

4.   *Polymeric alcohols such as poly(vinyl alcohol) (PVA), poly(2-hydroxyethylacrylate) (PHEA) and poly(2-hydroxyethyl vinyl ether) (PHEVE).*

         PVA                    PHEA                   PHEVE

5.   *Other synthetic polymers such as poly(2-ethyl-2-oxazoline) (PEOX) and poly(N-acetyliminoethylene) (PAIE).*

         PEOX                              PAIE

6.   *Water-soluble non-ionic polysaccharides.*

Methylcellulose (MC)

Hydroxyethylcellulose (HEC)

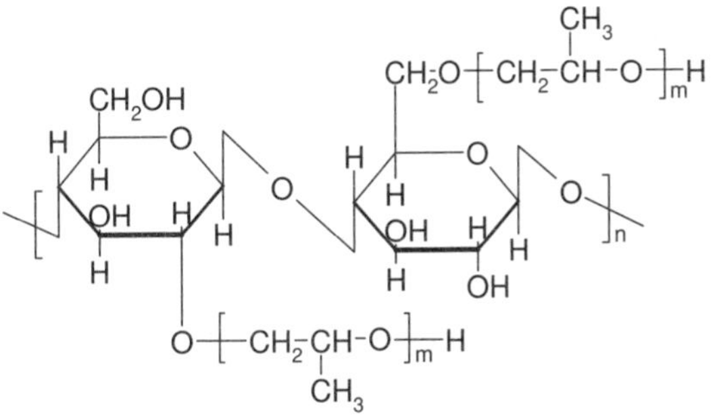

Hydroxypropylcellulose (HPC)

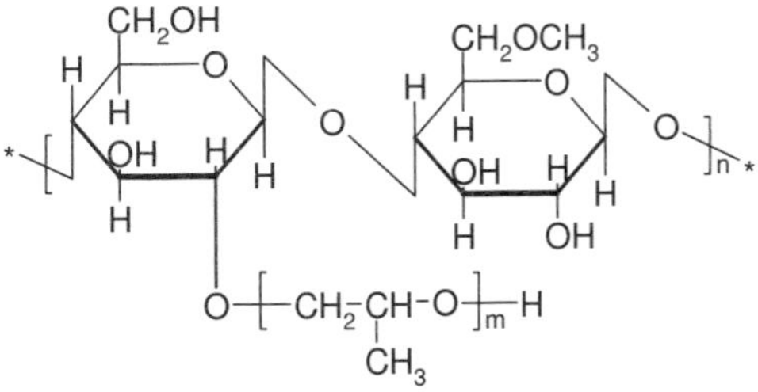

Hydroxypropylmethylcellulose (HPMC)

All poly(carboxylic acids) are weak polyelectrolytes, which ionisation is pH dependent. At low pH, they exist in non-ionised form and at high pH the carboxylic groups are fully ionised. Only unionised carboxylic groups can form hydrogen bonds with proton-accepting non-ionic polymers. By this reason, insoluble IPC may be formed under acidic conditions and dissociate upon increase in pH:

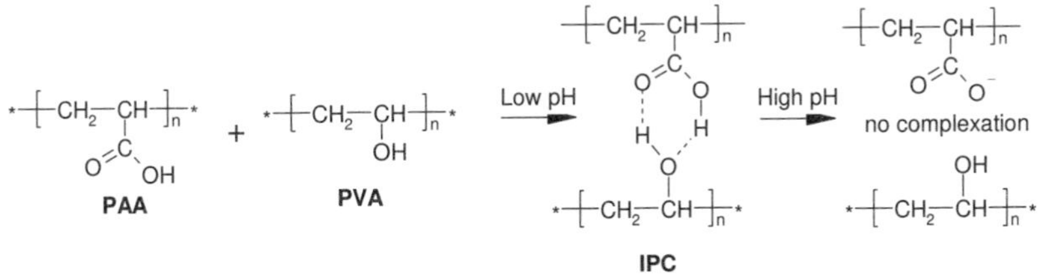

The complexation between poly(carboxylic acids) and non-ionic polymers occurs instantaneously upon mixing of solutions. This very fast process cannot be studied by conventional physicochemical techniques. However, the fast complexation can be followed by the aggregation of primary complexes, which rate is strongly dependent on many factors including temperature, concentration and pH. Usaitis and co-workers[4] have studied the aggregation of IPC formed by PMAA and PVP at different pHs using dynamic light scattering (DLS). The most pronounced increase in the IPC particle diameter was observed for the complexation at pH 3.2, where the particle size increased from 100 to 550 nm within 100 min. At lower pHs (3.4 and 3.6), the aggregation was less intensive resulting in smaller IPC particles (**Fig. 1**).

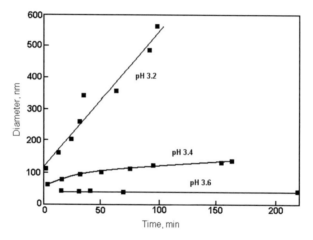

**Fig. 1.** Average diameter of PMAA-PVP complex particles as a function of time at different pH. Reproduced from Usaitis A., Maunu L.S., Tenhu H. Aggregation of the interpolymer complex of poly(methacrylic acid) and poly(vinyl pyrrolidone) in aqueous solutions, *European Polymer Journal*, 33, 219-223 ©1997 (Ref. 4) with permission from Elsevier.

Ohno and co-workers[5] as well as Hemker and Frank[6] have studied the aggregation of complexes formed by PMAA and PEO by dynamic light scattering. It was demonstrated[5] that lower pH and higher temperature make the aggregation faster and spherical shape of aggregates was confirmed using scanning electron microscopy. A dramatic growth in light scattering was observed at pH < 3.0, whereas no significant changes in scattering-light intensity were detected at pH within 4.0–6.0. Similar observations were reported by Hemker and Frank,[6] however, in this study a significant increase in the IPC radius (from 50 to more than

200 nm) was observed at pH < 1.9. In the pH range 2.1–2.7, the radius remained unchanged at 48 nm. The difference of this result from the data reported by Ohno and co-workers[5] was explained by lower molecular weight of PMAA. The kinetics of IPC aggregation was also studied and it was demonstrated that the increase in size can be described by a power law relationship:

$$<R> = R' \cdot t^b,$$

where <R> is IPC radius (nm), t is time (min), R' and b are constants.

The effect of pH on the complexation of PEO with different poly(carboxylic acids) such as PAA, PMAA and PSMA in aqueous solutions has extensively been studied by Ikawa and co-workers.[7] They demonstrated that the yield of IPC increases drastically below a certain pH value, which was called critical pH of complexation ($pH_{crit}$). It was shown that the $pH_{crit}$ depends on the nature of poly(carboxylic acid) and increases with increase in dissociation constant ($pK_a$). The composition and the structure of IPC were also found to be pH-dependent. The existence of a certain level of unionised carboxylic groups is necessary for PMAA and PEO to form stable IPC through hydrogen-bonding. A stable IPC with 1:1 molar stoichiometry is formed at pH < $pH_{crit}$ (Structure 1), whereas at pH slightly higher than $pH_{crit}$ the deviation from the 1:1 stoichiometry is observed (Structure 2). When the pH is high and carboxylic groups are fully ionised, the IPC is not formed (Structure 3):

Structure 1        Structure 2        Structure 3

Based on the studies discussed above, it can be concluded that the existence of a critical pH is a fundamental property of hydrogen-bonded IPCs, which determines the possibility of their formation, stoichiometry and structure. In the following sections we will consider the physical methods used to determine pH$_{crit}$ and factors affecting this value.

## 1. Methods used for determination of critical pH values

At present a number of methods are commonly used for determination of pH$_{crit}$ values. All these methods are based on a measurement of a complexation parameter as a function of environmental pH.

### *a)  Yield of IPC precipitate as a function of pH*

The gravimetric method suggested by Ikawa and co-workers[7] involves a mixing of the interacting polymers in solutions with different pH, separation of precipitate formed, drying and weighing of the final IPC. pH at which a drastic increase in the IPC yield is observed corresponds to the pH$_{crit}$. However, this method is time-consuming and centrifugation must be applied to ensure a complete isolation of insoluble complexes.

### *b)  Turbidity of IPC solution as a function of pH*

We suggested use of turbidimetry as a simple and fast alternative to the gravimetric method for determination of critical pH.[8-11] In this method, solutions of interacting polymers are mixed at pH 5–6 to make sure that insoluble complex is not formed. pH is then gradually decreased upon addition of small portions of 0.1 M HCl and solution turbidity is measured in parallel with pH control. The wavelength for measurement of turbidity can be in the range of 400-600 nm. The typical dependence of solution turbidity on pH is shown in **Fig. 2.** When pH is high enough and an insoluble complex is not formed, the turbidity readings remain low. However, when solution pH reaches pH$_{crit}$, the turbidity increases drastically, which can often be seen even by the naked eye. The drastic increase in solution turbidity within a narrow pH range is an indication of the cooperativity of the complexation process.

When pH is below pH$_{crit}$, the IPC continues to aggregate and forms larger particles. The aggregation process reaches saturation at some pH and then the turbidity does not change until the IPC particles start to precipitate, which may result in a slight decrease in turbidity.

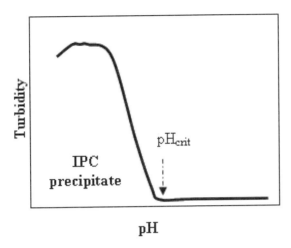

**Fig. 2.** Typical dependence of solution turbidity on pH for the complexation between poly(carboxylic acid) and non-ionic polymer.

### c) Viscosity of IPC solution as a function of pH

This method is based on the possibility to follow the conformational changes happening with the macromolecules upon complexation.[12,13] When the interacting macromolecules are mixed at relatively high pH, the solution viscosity is quite high. However, a gradual decrease in pH will lead to compaction of macromolecules, which is most pronounced when IPC is formed.

### d) Fluorescence spectra as a function of pH

The fluorescent methods can be used for the study of interpolymer interactions when fluorescent molecules are either covalently linked with one of the interacting polymers or simply dispersed in the IPC solution. These methods will be considered in detail in Chapter 4. Here, we will briefly discuss the use of fluorescent methods for the study of pH effects in complexation.

The ability of the fluorescent probe pyrene to migrate into a more hydrophobic environment and change the intensity ratio of the third (383 nm) to the first (373 nm) vibronic peaks $I_3/I_1$ in its emission spectra can be used to study pH-effects in the complexation between poly(carboxylic acids) and non-ionic polymers and to determine the critical pH values. The $I_3/I_1$ value for pyrene solubilised in distilled water is around 0.60-0.64. In more hydrophobic environment, this ratio increases, which may help to follow formation of non-polar domains upon complexation. In a series of studies,[13-15] we have demonstrated the possibility of using pyrene probe for determination of the critical pH. In these experiments free pyrene molecules were introduced to the IPC aqueous dispersion and pH of solution was varied to monitor how it affects the $I_3/I_1$ ratio. **Figure 3** shows the dependence of pyrene $I_3/I_1$ ratio and solution viscosity for PAA-PVA complexes as a function of pH.

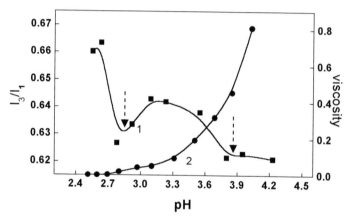

**Fig. 3.** Dependence of pyrene $I_{383}/I_{373}$ ratio (1) and specific viscosity (2) of PAA-PVA equimolar 0.01 base-mol/L solutions on pH. Reproduced from Nurkeeva Z.S., Mun G.A., Dubolazov A.V., Khutoryanskiy V.V. pH effects on the complexation, miscibility and radiation-induced crosslinking in poly(acrylic acid)-poly(vinyl alcohol) blends, *Macromolecular Bioscience*, 2005, 5, 424-432 (Ref. 13). Copyright Wiley-VCH Verlag GmbH & Co. KGaA. Reproduced with permission.

When pH is high (pH 3.75–4.25) and the IPC are not formed the $I_{383}/I_{373}$ ratio approaches 0.62, the value typical for a polar, aqueous environment. A decrease in pH within 3.75–3.20 is accompanied by an increase in $I_3/I_1$, confirming the formation of more hydrophobic environment in the system. A further decrease in pH results in the lowering of $I_3/I_1$ until pH 2.75, then this ratio increases again. We explained this complicated dependence of $I_3/I_1$ by the existence of two critical pH values (pH$_{crit1}$ and pH$_{crit2}$), which are shown by the arrows. Below pH$_{crit1}$, the hydrophobic IPC aggregate is formed; above pH$_{crit2}$, there is no interaction between macromolecules. At pH in between pH$_{crit1}$ and pH$_{crit2}$, the product of interpolymer interaction exists in a more hydrophilic form. It should be noted that the values of pH$_{crit1}$ determined by this fluorescent method are in good agreement with the values determined by turbidimetric and viscometric approaches.

Numerous attempts have also been made to study the pH effects on the complexation between polymers, when pyrene is covalently linked to macromolecules of one type.[16] However, in this case the complexation ability of modified macromolecules is enhanced greatly due to the contribution of non-polar pyrene groups to the stabilisation of IPC by hydrophobic effects. For example, according to Sivadasan and co-workers[16] the IPC formed by PAA and PAAM labeled with pyrene is characterised by pH$_{crit}$ 4–5, whereas the turbidimetric determination of pH$_{crit1}$ carried out for PAA and unmodified PAAM by using the turbidimetric method gives pH$_{crit1}$ 2.5–3.5.[17]

## 2. Factors affecting the critical pH values

### a)  Concentration of polymers in solution

In a number of studies, it has been shown that an increase in polymer concentration shifts the critical pH values to the higher region.[9-11,17-19] It is believed that an increase in the polymer concentration suppresses the ionisation of PAA, which favors the formation of intermacromolecular hydrogen bonds.

### b)  Molecular weight of polymers

It is well known that the molecular weight of polymers affects the complexation. Kabanov and Papisov[20] have demonstrated  the existence of a lower critical molecular weight of polymers, below which IPC are not formed. It was related to the importance of cooperative effects in complexation, which can be achieved only when the interacting macromolecules are long enough.

Critical pH values of complexation were also found to be affected by the molecular weight of the interactive polymers. **Fig. 4** shows the dependence of solution turbidity of 1:1 molar mixtures of PAA and HEC on pH for PAA samples with different molecular weight. An increase in PAA molecular weight within 250–1250 kDa leads to increase in the $pH_{crit1}$ values indicating an enhancement in the stability of IPC. However, the mixture of HEC with PAA 2 kDa remains fully transparent even at very low pH (pH < 1.0). This behavior of the binary system can be due to the absence of the complexation because the chain length of PAA is lower than the minimal critical molecular weight.

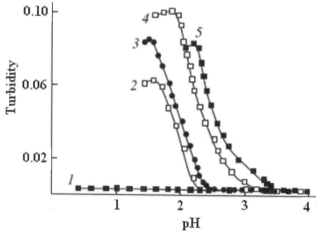

**Fig. 4.** Dependence of PAA-HEC (1:1 molar ratio) solution turbidity on pH. Molecular weights of PAA ($M_w$): 2 (1), 250 (2), 450 (3), 750 (4) and 1250 kDa (5). The initial concentration of polymers is 0.01 base-mol/L. Reproduced with permission from Ref. 21.

## c)   *Hydrophobicity of interacting polymers*

In addition to hydrogen-bonding, formation of IPC in aqueous solutions is stabilised by hydrophobic effects. As a result, the hydrophobic/hydrophilic properties of interacting polymers can affect the stability of IPC as well as the critical pH values. For example, PMAA, as a more hydrophobic polymer compared to PAA, forms stronger complexes, which $pH_{crit}$ is generally higher.[9, 11]

An introduction of hydrophobic groups into macromolecules can also shift the $pH_{crit}$ to higher pH values. In the study of the complexation of PAA with a series of vinyl ether copolymers it was found that an introduction of hydrophobic vinyl butyl ether (VBE) into the structure of PHEVE can enhance their complexation ability with respect to PAA in aqueous solutions.[10, 22] The copolymers containing more VBE exhibited higher critical pH values (**Fig. 5**).

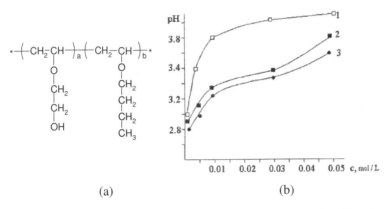

(a)                 (b)

**Fig. 5.** Structure of HEVE-VBE copolymers (a) and dependence of the $pH_{crit1}$ for the complexes of PAA and HEVE-VBE on the polymer concentration for the copolymers containing 22 (1), 12 (2) and 6 mol. % VBE (3). Reproduced from Mun G.A., Nurkeeva Z.S., Khutoryanskiy V.V., Bitekenova A.B.: Effect of copolymer composition on interpolymer complex formation of (co)polyvinyl ethers with polyacrylic acid in aqueous and organic solutions, Macromolecular Rapid Communications, 2000, 21, 381-384 (Ref. 10). Copyright Wiley-VCH Verlag GmbH & Co. KGaA. Reproduced with permission.

Unlike PHEVE homopolymer, the HEVE-VBE copolymers exhibit lower critical solution temperature (LCST) behaviour in aqueous solutions. It means that they can undergo phase separation upon heating. Staikos and co-workers[23] have previously demonstrated that non-ionic polymers with LCST are characterised by higher complexation ability with respect to poly(carboxylic acids) due to more effective stabilisation of IPC via hydrophobic effects. Indeed, the non-ionic polymers with a LCST form complexes with higher $pH_{crit1}$ as compared to their analogues without a LCST. For example, PVME,[18] PNIPAAM,[19] MC,[24] HPC[15] and

HPMC[25] have LCST in aqueous solutions and form stronger complexes with higher $pH_{crit}$ compared to PHEVE,[9] PAAM[17] and HEC[26] as LCST is not typical for the later group of polymers.

### d) Nature of non-ionic polymers

The critical pH values ($pH_{crit1}$), determined for complexes of PAA with different non-ionic polymers are summarised in **Table 1**.

**Table 1\***. Critical pH values ($pH_{crit1}$) of some non-ionic polymers in 1:1 complexes with PAA ($M_w$ 450 kDa). The concentrations of both polymers in solution were 0.01 base-mol/L.

| Non-ionic polymer | $M_w$, kDa | $pH_{crit1}$ | Comment |
|---|---|---|---|
| PHEVE | 40 | $2.45 \pm 0.05$ | - |
| PVA | 205 | $2.67 \pm 0.05$ | - |
| PHEA** | 420 | $2.75 \pm 0.05$ | data taken from Ref. 27 |
| HEC | 100 | $2.85 \pm 0.05$ | - |
| PEO | 20 | $2.88 \pm 0.05$ | - |
| PAAM | 6000 | $3.00 \pm 0.05$ | - |
| HPC | 100 | $3.66 \pm 0.05$ | LCST*** |
| PNIPAAM | 450 | $4.60 \pm 0.05$ | LCST |
| PVME | 60 | $4.85 \pm 0.05$ | LCST |
| PVP | 24 | $4.85 \pm 0.05$ | - |

\*Reproduced from Ref. 19. Copyright Society of Chemical Industry. Reproduced with permission. Permission granted by John Wiley & Sons Ltd on behalf of the SCI.
\*\*Data listed for complexes with PAA ($M_w$ 250 kDa)
\*\*\*Non-ionic polymer exhibits LCST in aqueous solution

Although, a strict comparison of different polymers in this case is rather difficult as they have significantly different molecular weights, there is still a possibility to draw several conclusions. The lowest $pH_{crit1}$ values are observed for the complexes of PAA with polymeric alcohols such as PHEVE, PVA, PHEA and HEC, which means that polymers bearing hydroxyl-groups have lower complexation ability. However, if hydroxyl-containing polymers exhibit LCST in aqueous solutions their complexation ability is greatly enhanced and these systems show significantly higher $pH_{crit1}$ values. This phenomenon is observed for complexes of PAA with HPC. The polymers having either ether-groups such as PEO or amide-groups such as PAAM exhibit intermediate complexation ability. Again, when the presence of ether-groups or amide-groups is additionally backed by LCST the complexation ability of the polymers is enhanced significantly; this is observed for PVME and PNIPAAM. The highest complexation ability is observed for polymers with lactame-groups such as PVP.

*e) Presence of inorganic salts*

Several authors reported different phenomena observed upon the addition of inorganic salts to the polymer solutions forming IPC. Bel'nikevich and co-workers[28] reported that increasing ionic strength suppresses the complexation of PAA with PVA and HEC. Chen and Morawetz[29] demonstrated that the addition of NaCl reduces the intensity of interactions of PAA with PVP and PEO. Sivadasan *et al.*[16] found that an increase in ionic strength leads to a decrease in pH at which the interpolymer interaction begins between PAA and pyrene-modified PAAM. However, Staikos *et al.*[30] showed that the presence of sodium bromide in solution favours the formation of a compact IPC between PAA and PEO. Iliopoulos and Audebert[31] studied the effect of tetramethylammonium chloride on the complexation between PAA and PEO by the potentiometric technique. They showed that the complexation degree is higher in salt solutions than in pure water, which was related to the change in the thermodynamic quality of the solvent. Indeed, deterioration in the thermodynamic quality of the solvent leads to more effective stabilisation of IPC through hydrophobic effects. Prevysh *et al.*[32] studied the effect of added salts on the stability of IPCs composed of PAA and PEO, PVP and HPC. It was demonstrated that the addition of NaCl results in IPC aggregation.

In fact, the influence of inorganic salts on the complexation between poly(carboxylic acids) and non-ionic polymers in aqueous solutions is associated with two opposite effects. The presence of inorganic salts in solution favours the ionisation of carboxylic groups of poly(carboxylic acids), which is unfavourable for hydrogen-bonding. On the other hand, the solvation of macromolecules is reduced in the presence of inorganic salts, which weakens polymer-solvent interactions and strengthens the intermacromolecular interactions between polymers. A prevalence of one of these effects will lead either to stabilisation or destabilisation of IPC upon addition of inorganic salts.

When analysing the effect of salt addition on the critical pH values ($pH_{crit1}$), we have found that the IPC that have a relatively low $pH_{crit1}$ (2.45–2.88) exhibit a tendency to have an increased $pH_{crit}$ upon the addition of inorganic salts.[19] This trend is observed for complexes of PAA with PHEVE, PVA, HEC and PEO (see **Table 1**). The stronger complexing systems with $pH_{crit1}$ within 3.66–4.85 range show the opposite trend. The complexes of PAA with HPC, PNIPAAM, PVME and PVP decrease their $pH_{crit1}$ upon addition of NaCl. Hence, the effect of ionic strength on $pH_{crit1}$ depends on the pH at which the complexation is occured.

Interesting phenomena were observed when transition metal ions such as $Cu^{2+}$ and $Fe^{3+}$ are added to poly(carboxylic acid) – non-ionic polymer mixtures.[17,33,34] A significant enhancement in the complexation ability of the polymers in the presence of these ions was detected and related to the formation of triple poly(carboxylic acid) – metal ions – non-ionic polymer complexes. Since the enhancement of the complexation in the presence of transition metal ions was reported for complexes of PAA with three different non-ionic polymers such as PVP,[33] PEO[34] and PAAM[17] it can be anticipated that this effect is common for all hydrogen-bonded IPC. However, further studies are required to clarify it.

*f)   Presence of organic molecules in solution*

Only few studies reporting the effects of organic molecules on the complexation between poly(carboxylic acids) and non-ionic polymers in aqueous solutions have been published. In the study of the complexation between PAA and poly(vinyl ether of diethylene glycol) (PVEDEG), it was found[35] that the addition of phenol and hydroquinone shifts the $pH_{crit1}$ to the lower pH region even at relatively low concentrations of the additives (**Fig. 6**). Phenol and hydroquinone are known to form complexes with a number of water-soluble polymers such as PEO, PVP and PVA.[36,37] Moreover, this complexation occurs more readily with polyphenols.[38] It was assumed that the strong influence of these molecules on the complexation can also be related to the competitive interactions between phenols and individual polymers via hydrogen-bonding; this may affect the complexation between the polymers.

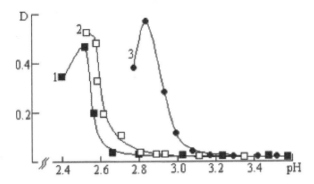

**Fig. 6.** Dependence of turbidity of PAA–PVEDEG (2:1 mol./mol.) solutions on pH in the presence of hydroquinone (1), phenol (2) and without additives (3).The concentration of added phenol and hydroquinone is 0.01 mol/L. The initial concentration of polymers is 0.01 base-mol/l. Reproduced from Nurkeeva Z.S., Mun G.A., Khutoryanskiy V.V., Sergaziev A.D. Complex formation between poly(vinyl ether of diethyleneglycol) and polyacrylic acid: I. Effect of low molecular salts and phenols additives, *European Polymer Journal*, 37, 1233-1237, ©2000, with permission from Elsevier (Ref. 35).

Drug molecules such as lidocaine hydrochloride (Lid·HCl) can also affect the complexation between PAA and PHEVE.[39] The presence of Lid·HCl in solutions of PAA and PHEVE shifts the $pH_{crit1}$ to the higher pH region when the drug concentration is within 0.005-0.095 mol/L. It was found that this drug can interact with PAA via ionic contacts and with PHEVE via hydrogen-bonding. The mixing of PAA, Lid·HCl and PHEVE results in the formation of triple complexes with reduced solubility.

Cationic surfactants such as cetyl pyridinium bromide (CPB) can also affect the complexation due to the specific binding to PAA via ionic contacts.[27] Depending on the ratio between PAA and CPB the polycomplexes can form clear solutions, stable colloidal dispersions or precipitates. In order to study the effect of CPB addition on the $pH_{crit1}$ for the complexation between PAA and PHEA the concentration of the surfactant was kept at very low level ($1.5 \cdot 10^{-4}$ mol/L). It was found that the addition of CPB shifts the $pH_{crit1}$ to the higher pH region and this shift is more pronounced compared to the addition of NaCl of the same concentration.

## 3. pH-induced complexation – miscibility – immiscibility transitions

Formation of insoluble IPC between poly(carboxylic acids) and non-ionic polymers in aqueous solutions can be quite undesirable for the preparation of polymeric films. These films can be used for design of novel drug delivery systems (see Chapter 9), as packaging materials and as membranes in separation technologies. From a practical point of view, it is important to develop polymeric films with good mechanical properties and uniformity; however, the precipitation of IPC leads to the formation of uneven materials.

In the attempt to prepare the polymeric films based on PAA and HPC, we have tried to avoid the complexation by neutralising PAA with addition of NaOH.[40] A complete neutralisation of PAA helped to avoid the complexation and mixing the polymers has not resulted in the formation of a turbid IPC solution. However, after evaporation of water, the film formed had poor mechanical properties, was not fully transparent and had inhomogeneous morphology confirming a complete immiscibility between poly(sodium acrylate) and HPC (**Fig. 7**).

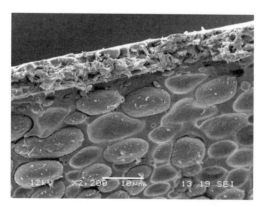

**Fig. 7.** Scanning electron microscopy image of the film based on the blend of HPC with poly(sodium acrylate) (33:67 mol. % ratio).

The immiscibility in this binary system was due to the lack of intermacromolecular hydrogen-bonding, which is known to facilitate the formation of miscible blends.[41] In a later study[15] we have attempted to cast the polymeric films from PAA and HPC mixtures at varied pH and found that a miscibility window is observed at a casting solution pH higher than $pH_{crit1}$ but lower than $pH_{crit2}$. Similar results were obtained for the blends of PAA with PEO[14] and PVA.[13] **Figure 8** shows the scanning electron microscopy images of the film cross-sections obtained by casting PAA-PVA mixtures at different pH.

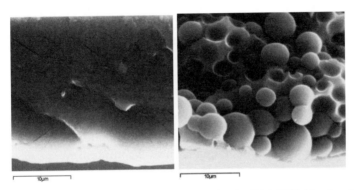

**Fig. 8.** Scanning electron microscopy image of an equimolar PAA-PVA blend cast from a solution with pH 3.32 (a) and 4.6 (b). Reproduced from Nurkeeva Z.S., Mun G.A., Dubolazov A.V., Khutoryanskiy V.V. pH effects on the complexation, miscibility and radiation-induced crosslinking in poly(acrylic acid)-poly(vinyl alcohol) blends, *Macromolecular Bioscience*, 2005, 5, 424-432 (Ref. 13). Copyright Wiley-VCH Verlag GmbH & Co. KGaA. Reproduced with permission.

## 4. Concluding remarks

pH plays an important role in the complexation between poly(carboxylic acid) and non-ionic polymers in aqueous solutions. The structure of IPCs, as well as their physicochemical properties, depends on the solution pH. Two critical pH values exist in binary poly(carboxylic acid)-non-ionic polymer mixtures. The formation of colloidal dispersions of hydrophobic interpolymer complexes and their further precipitation is observed below the first critical pH of complexation ($pH_{crit1}$). Above this value, but below the second critical pH of complexation ($pH_{crit2}$), the products of interaction are more hydrophilic. Above $pH_{crit2}$, the polymers do not interact with one another leading to homogeneous solution mixtures.

The $pH_{crit1}$ depends on the concentration of polymers in solution, their molecular weight, the hydrophobic-hydrophilic balance, the nature of the functional groups in the non-ionic polymers, and the presence of inorganic salts or organic molecules in solution. The determination of $pH_{crit1}$ allows the comparison of the complexation ability of different polymeric pairs. Higher $pH_{crit1}$ corresponds to a stronger complexation ability between poly(carboxylic acid)s and non-ionic polymers.

A careful control of solution pH is also important for preparing polymeric films based on blends of poly(carboxylic acids) and non-ionic polymers. The films cast from solutions below $pH_{crit1}$ are not uniform because of IPC precipitation. The films cast from solutions above $pH_{crit2}$ are cloudy and have poor mechanical properties due to lack of intermacromolecular hydrogen-bonding and immiscibility between the polymers. The formation of miscible films is possible within the $pH_{crit1}$-$pH_{crit2}$ range, where the polymers form a hydrophilic associate and weak interpolymer hydrogen-bonding ensures the miscibility.

## References

1.  K.L. Smith, A.E. Winslow, and D.E. Petersen, *Ind. Eng. Chem.* 51, 1361 (1959).
2.  F.E. Bailey, R.D. Lundberg, and R.W. Callard, *J. Polym. Sci. Part A* 2, 845 (1964).
3.  Z.S. Nurkeeva, G.A. Mun, and V.V. Khutoryanskiy, *Macromol. Biosci.* 3, 283 (2003).
4.  A. Usaitis, S.L. Maunu, and H. Tenhu, *Eur. Polym. J.* 33, 219 (1997).
5.  H. Ohno, H. Matsuda, and E. Tsuchida, *Makromol. Chem.* 182, 2267 (1981).
6.  D.J. Hemker and C.W. Frank, *Macromolecules* 23, 4404 (1990).
7.  T. Ikawa, K. Abe, K. Honda, and E. Tsuchida, *J. Polym. Sci. Polym. Chem. Ed.* 13, 1505 (1975).
8.  G.A. Mun, Z.S. Nurkeeva, and V.V. Khutoryanskiy, *Macromol. Chem. Phys.* 200, 2136 (1999).
9.  Z.S. Nurkeeva, G.A. Mun, V.V. Khutoryanskiy, A.A. Zotov, and R.A. Mangazbaeva, *Polymer* 41, 7647 (2000).
10. G.A. Mun, Z.S. Nurkeeva, V.V. Khutoryanskiy, and A.B. Bitekenova, *Macromol. Rapid Commun.* 21, 381 (2000).
11. Z.S. Nurkeeva, G.A. Mun, and V.V. Khutoryanskiy, *Polym. Sci. Ser. B*, 43, 148 (2001).
12. V. Baranovsky, T. Petrova, and I. Rashkov, *Eur. Polym. J.* 27, 1045 (1991).
13. Z.S. Nurkeeva, G.A. Mun, A.V. Dubolazov, and V.V. Khutoryanskiy, *Macromol. Biosci.* 5, 424 (2005).
14. V.V. Khutoryanskiy, A.V. Dubolazov, Z.S. Nurkeeva, and G.A. Mun, *Langmuir* 20, 3785 (2004).
15. A.V. Dubolazov, Z.S. Nurkeeva, G.A. Mun, and V.V. Khutoryanskiy, *Biomacromolecules* 7, 1637 (2006).

16. K. Sivadasan, P. Somasundaran, and N.J. Turro, *Colloid Polym. Sci.* 269, 131 (1991).
17. G.A. Mun, Z.S. Nurkeeva, V.V. Khutoryanskiy, G.S. Sarybayeva, and A.V. Dubolazov, *Eur. Polym. J.* 39, 1687 (2003).
18. Z.S. Nurkeeva, V.V. Khutoryanskiy, G.A. Mun, and A.B. Bitekenova, *Polym. Sci. Ser. B*, 45, 365 (2003).
19. V.V. Khutoryanskiy, G.A. Mun, Z.S. Nurkeeva, and A.V. Dubolazov, *Polym. Int.* 53, 1382 (2004).
20. V.A. Kabanov, I.M. Papisov, *Polym. Sci. U.S.S.R.* 21, 261 (1979).
21. Y.J. Bo, V.V. Khutoryanskiy, G.A. Mun, and Z.S. Nurkeeva, *Polym. Sci. Ser. A*, 44, 1094 (2002).
22. S.E. Kudaibergenov, Z.S. Nurkeeva, G.A. Mun, B.B. Ermukhambetova, and A.T. Akbauova, *Macromol. Chem. Phys.* 196, 2203 (1995).
23. M. Koussathana, P. Lianos, and G. Staikos, *Macromolecules* 30, 7798 (1997).
24. V.V. Khutoryanskiy, M.G. Cascone, L. Lazzeri, Z.S. Nurkeeva, G.A. Mun, and R.A. Mangazbaeva, *Polym. Int.* 52, 62 (2003).
25. R.A. Mangazbaeva, G.A. Mun, Z.S. Nurkeeva, and V.V. Khutoryanskiy, *Polym. Int.* 55, 668 (2006).
26. G.A. Mun, Z.S. Nurkeeva, V.V. Khutoryanskiy, and A.V. Dubolazov, *Polym. Sci. Ser. B*, 45, 361 (2003).
27. G.A. Mun, V.V. Khutoryanskiy, G.T. Akhmetkalieva, S.N. Shmakov, A.V. Dubolazov, Z.S. Nurkeeva, and K. Park, *Colloid Polym. Sci.* 283, 174 (2004).
28. N.G. Belnikevich, T.V. Budtova, N.P. Ivanova, Y.F. Panarin, Y.N. Panov, and S.Y. Frenkel, *Vysokomolek. Soed. Ser. A*, 31, 1691 (1989).
29. H.L. Chen and H. Morawetz, *Eur. Polym. J.* 19, 923 (1983).
30. G. Staikos, P. Antonopoulou, and E. Christou, *Polym. Bull.* 21, 209 (1989).
31. I. Iliopoulos and R. Audebert, *Eur. Polym. J.* 24, 171 (1988).
32. V.A. Prevysh, B.C. Wang, and R.J. Spontak, *Colloid Polym. Sci.* 274, 532 (1996).
33. R. Subramanian and P. Natarajan, *J. Polym. Sci. Polym. Chem.* 22, 437 (1984).
34. N. Angelova, N. Manolova, and I. Rashkov, *Eur. Polym. J.* 31, 741 (1995).
35. Z.S. Nurkeeva, G.A. Mun, V.V. Khutoryanskiy, and A.D. Sergaziev, *Eur. Polym. J.* 37, 1233 (2001).
36. B.N. Kabadi, R. Hammarlund, *J. Pharm. Sci.* 55, 1069 (1966).
37. B.N. Kabadi, R. Hammarlund, *J. Pharm. Sci.* 55, 1072 (1966).
38. P. Molyneux, "Water-soluble synthetic polymers: properties and behavior." Boca Raton: CRC Press, 1984.
39. Z.S. Nurkeeva, G.A. Mun, V.V. Khutoryanskiy, A.B. Bitekenova, and A.B. Dzhusupbekova, *J. Biomat. Sci. Polym. Edn.* 13, 759 (2002).
40. V.V. Khutoryanskiy, M.G. Cascone, L. Lazzeri, N. Barbani, Z.S. Nurkeeva, G.A. Mun, and A.V. Dubolazov, *Polym. Int.* 53, 307 (2004).
41. M. Jiang, M. Li, M.L. Xiang, and H. Zhou, *Adv. Polym. Sci.*, 146, 121 (1999).

# CHAPTER 2

# HYDROGEN-BONDED INTERPOLYMER COMPLEXES SOLUBLE AT LOW pH

Georgios Staikos[1*], Maria Sotiropoulou[1], Georgios Bokias[2], Frederic Bossard[3], Julian Oberdisse[4], Eric Balnois[5]

[1]*Department of Chemical Engineering, University of Patras, GR-26504 Patras, Greece*
*[*]E-mail: staikos@chemeng.upatras.gr*
[2]*Department of Chemistry, University of Patras, GR-26504 Patras, Greece*
[3]*Laboratoire de Rheologie, UMR 5520, Université Joseph Fournier, 1301, rue de la piscine, BP 53, 38041 Grenoble Cedex, France*
[4]*Laboratoire des Colloides, Verres et Nanomateriaux, UMR CNRS/UM2, Université Montpellier, France*
[5]*Laboratoire Polyméres, Propriétés aux Interfaces et Composites (L2PIC), Université de Bretagne Sud, Rue de Saint Maudé, BP 92116, 56231 Lorient, France*

## 1. Introduction

The formation of hydrogen-bonded interpolymer complexes (IPCs) between proton donor and proton acceptor polymers in aqueous solution has been widely studied during the past four decades.[1-4] The proton donors are usually weak polycarboxylic acids, such as poly(acrylic acid) (PAA) and poly(methacrylic acid) (PMAA), whereas the proton acceptors are nonionic polybases, such as poly(ethylene glycole) (PEG) or polyethyleneoxide (PEO),[5-9] polyacrylamides,[10-14] poly(vinyl ethers),[15-17] etc. The research interest in this field has been further stimulated by potential applications of these hydrogen-bonded interpolymer complexes to various fields, such as drug delivery formulations,[18,19] biomaterials,[20] emulsifiers,[21] and membrane and separation technology.[22,23]

In general, in aqueous mixtures of such complementary polymers hydrogen-bonding association occurs between the carboxylic groups of the polyacid and the nonionic polybase, leading to the formation of compact hydrogen-bonded IPCs. An important limitation is that they are usually soluble only within a narrow pH window.[24,25] At pH values higher than 4–5, dissociation of these hydrogen-bonded IPCs occurs, due to the increase of the ionized sites (carboxylate groups) in the polyacid chain. These non-complexable, negatively charged, groups are structural defects, not allowing also to the polyacid chain to adopt the appropriate conformation in order to associate with the non-ionic polybase chain.[26-29] On the other hand, at pH values lower than 3–3.5, the hydrogen-bonded IPC formed precipitates, because the fraction of the carboxylate anions in the polyacid chain, mainly responsible for the solubility of the complex, decreases considerably.[5, 6,10,11,30] In summary, when pH is higher than 4–5 interpolymer complexation via hydrogen-bonding is not possible, while at pH lower than 3–3.5, associative phase separation takes place.

Extension of the solubility of the hydrogen-bonded IPCs in the low pH region is of importance, as this would allow original properties to be observed and could enlarge the spectrum of the potential applications of such complexes. To achieve this aim, the non-ionic polybase poly (*N, N*-dimethylacrylamide) (PDMAM) was grafted onto a highly charged anionic backbone, consisting of 2-acrylamido-2-methyl-1- propane-sulphonic acid (AMPSA) and acrylic acid (AA) monomer units, P(AA-*co*-AMPSA). It was shown that, the graft copolymers synthesized, P(AA-*co*-AMPSA)-*g*-PDMAM, could form soluble hydrogen-bonded IPCs with PAA at low pH.[31] Alternatively, it has been shown that PEG also forms soluble hydrogen-bonded IPCs with PAA grafted with the negatively charged poly(2-acrylamido-2-methyl-1-propanesulphonic acid) (PAMPSA) chains.[32]

## 2. Structure of the graft copolymers

**Scheme 1** presents the two kinds of graft copolymers suitable to form soluble hydrogen-bonded IPCs. The product shown in **Scheme 1(a)** was synthesized by grafting amine-terminated PDMAM chains onto an anionic backbone comprised by a statistical copolymer of AMPSA and

AA monomer units. AA, in a minor proportion, was used just for making possible the grafting of the amine-terminated PDMAM chains onto the negatively charged backbone, through an amide formation reaction, using 1-(3-(dimethylamino)propyl)-3-ethyl-carbodiimide hydrochloride (EDC) as a condensing agent.[31]

PDMAM is a water-soluble polymer with important proton acceptor properties, forming hydrogen-bonded IPCs with PAA,[13,33-35] precipitating out from water even at pH values as high as 3.7.[35] When the above described graft copolymers are mixed with PAA in aqueous solution in the low pH region, insoluble hydrogen-bonded IPCs between the

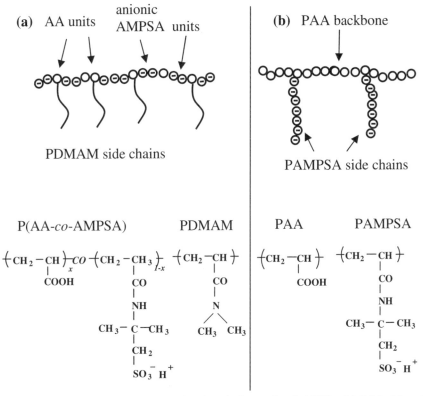

**Scheme 1.** Graft copolymers prepared to form hydrogen-bonded IPCs with PAA, (a), and PEG, (b), soluble at low pH. The chemical structure of the constituents, of which the graft copolymers are comprised, is also presented.

PDMAM side chains and PAA are formed. Nevertheless, the large fraction of the negatively charged AMPSA units contained in the graft copolymer backbone provides the IPCs formed with sufficient hydrophilicity, so that their solubility at low pH,[36] is finally assured.

In **Scheme 1(b)** an alternative is presented. Amine-terminated PAMPSA chains are grafted onto a PAA chain, through the same as above amide formation reaction.[32] The graft copolymer synthesized forms hydrogen-bonded IPCs with PEG, by means of its PAA backbone, which the solubility is ensured by its negatively charged PAMPSA side chains.

According to this approach, we need an anionic constituent, i.e., a strong acid, such as AMPSA, which could be used either for the synthesis of the backbone of a graft copolymer with a polybase as side chains, or for its side chains with a backbone of a weak polyacid, such as PAA. In both cases the hydrogen-bonded IPCs formed at low pH remain soluble due to the presence of these hydrophilic blocks. However, other anionic monomers, such as vinylsulfonic acid, instead of AMPSA, other weak polybases, such as polyacrylamide, poly(*N,N*-isopropylacrylamide), poly(ethylene glycol) or polyvinylpyrrolidone, instead of PDMAM, and other weak polyacids, such as PMAA, instead of PAA, could also be used.

## 3. Complexes formed between PAA and P(AA-*co*-AMPSA)-*g*-PDMAM graft copolymers

### 3.1. *Dilute solution study*

**Figure 1** presents the turbidity behavior of PDMAM/PAA90[i], P(AA-*co*-AMPSA36)-*g*-PDMAM50/PAA90[ii] and P(AA-*co*-AMPSA70)-*g*-PDMAM50/PAA90 mixtures in dilute aqueous solution as a function of

---

[i]The poly(acrylic acid) samples used in this work are shortly designated as PAAx, where x is the molecular weight of PAA in kDa.

[ii]The graft copolymers used throughout this work are designated as: P(AA-*co*-AMPSAx)-*g*-PDMAMy, where x is the mole percentage composition of the P(AA-*co*-AMPSAx) backbone in AMPSA units and y the weight percentage composition of the graft copolymer in PDMAM side chains.

pH.[31] The solution of the PDMAM/PAA mixture turns strongly cloudy as pH decreases at values lower than a critical value $pH_c$ = 3.75, while it remains transparent at higher pH. This behavior is due to the hydrogen-bonded IPCs formed between PDMAM ($M_w$ = 42 kDa), and PAA90 ($M_w$ = 90 kDa), which are insoluble at pH < 3.75, in good agreement with other results,[35] and reflecting the strong proton-acceptor ability of PDMAM.[34]

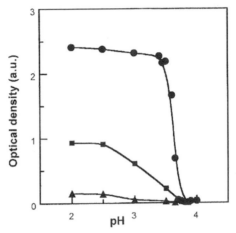

**Fig. 1**. Turbidimetric curves of the mixtures PDMAM/PAA90 (●), P(AA-*co*-AMPSA36)-*g*-PDMAM/PAA90 (■) and P(AA-*co*-AMPSA70)-*g*-PDMAM/PAA90 (▲) in buffer solutions versus pH, at 25 °C. The concentration of PAA90 and PDMAM is 2.0 x 10⁻³ g cm⁻³, while the concentration of the graft copolymers is 4.0 x 10⁻³ g cm⁻³. Reprinted with permission from Ref. 31. Copyright (2003) American Chemical Society.

On the contrary to the turbidimetric behavior of the PDMAM/PAA90 mixture, if the two graft copolymers, P(AA-*co*-AMPSA36)-*g*-PDMAM and P(AA-*co*-AMPSA70)-*g*-PDMAM, both containing 50 (w/w)% of PDMAM, but with a different backbone composition, that is containing 36 and 70 mol % respectively of anionic AMPSA units, are used in mixtures with PAA90, the turbidity, in the low pH region, decreases substantially. So, the mixture P(AA-*co*-AMPSA36)-*g*-PDMAM/PAA90 appears less turbid than PDMAM/PAA90 in the low pH region, while the solutions of the P(AA-*co*-AMPSA70)-*g*-PDMAM/PAA90 mixtures appear practically transparent in the whole pH region. This behavior has been attributed to the negative charge of the graft copolymers backbone, due to their AMPSA units. These results suggest that either interpolymer complexation is not favored when the PDMAM chains are covalently bound onto a negatively charged backbone or that the hydrogen-bonded IPCs formed are more hydrophilic and soluble in water.

The evidence, that such hydrogen-bonded IPCs are formed in these mixtures, is provided by their viscometric study in dilute solution. In **Fig. 2**, the dependence of the reduced viscosity ratio, $r_{\eta red}$, of the P(AA-*co*-AMPSA70)-*g*-PDMAM/PAA90 mixtures versus their PAA90 weight fraction, $W_{PAA90}$, in dilute aqueous solution, at several pH values, is shown. $r_{\eta red}$, defined by

$$r_{\eta_{red}} = \frac{\eta_{red}}{w_1(\eta_{red})_1 + w_2(\eta_{red})_2} \tag{1}$$

where $\eta_{red}$ is the reduced viscosity of the polymer mixture, $w_1$ and $w_2$ are the weight fractions of the two polymer components in the mixture and $(\eta_{red})_1$ and $(\eta_{red})_2$ are the corresponding reduced viscosities of the two pure components, is used to compare the reduced viscosity of the mixture with the additive value of the reduced viscosities of the two pure polymer components. Any deviation of $r_{\eta red}$ from unity offers a strong evidence of interpolymer interaction.

As we see in **Fig. 2**, the ratio $r_{\eta red}$ is around unity at pH = 3.75 and pH = 4.0, while it is lower than unity at lower pH values. Moreover, the curve for pH = 2.5 is systematically below the corresponding curve for pH = 3.5. These observations suggest that hydrogen-bonded IPCs of a compact structure are formed between P(AA-*co*-AMPSA70)-*g*-PDMAM and PAA90 as pH decreases to values lower than 3.75. Moreover, the pH decrease favors the compactness of these complexes, as it is usually observed with the hydrogen-bonded IPCs.[29,37,38]

The pH-controlled formation-dissociation of the hydrogen-bonded IPC is also revealed by the variation of the ratio $r_{\eta red}$ with pH. In **Fig. 3**, this variation is presented for the P(AA-*co*-AMPSA70)-*g*-PDMAM/PAA90 mixture with a composition $W_{PAA90} = 0.6$, i.e., the composition where the minimum was observed in **Fig. 2**. The ratio $r_{\eta red}$ decreases from around 1 at pH higher than 4 to about 0.4 as pH decreases

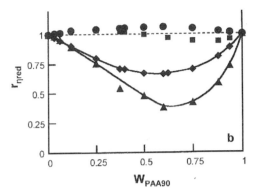

**Fig. 2.** Variation of the reduced viscosity ratio, $r_{\eta red}$, of P(AA-*co*-AMPSA70)-*g*-PDMAM/PAA90 polymer mixtures in buffer solutions as a function of their weight composition in PAA90, $W_{PAA90}$, at pH = 4.0 (●), pH = 3.75 (■), pH = 3.5 (♦) and pH = 2.5 (▲) at 25°C. The total polymer concentration is $2.5 \times 10^{-3}$ g cm$^{-3}$. Reprinted with permission from Ref. 31. Copyright (2003) American Chemical Society.

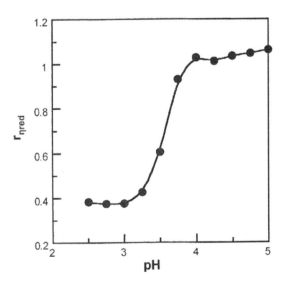

**Fig. 3.** Variation of the reduced viscosity ratio, $r_{\eta red}$, versus pH for an aqueous P(AA-*co*-AMPSA70)-*g*-PDMAM/PAA90 mixture with a PAA weight composition, $W_{PAA90} = 0.6$ at 25°C. Total polymer concentration c = $2.5 \times 10^{-3}$ g cm$^{-3}$. Reprinted with permission from Ref. 31. Copyright (2003) American Chemical Society.

to 3, and then it remains constant. This behavior induces that at pH > 4 interpolymer association does not occur and an almost ideal behavior is observed. As pH decreases at values lower than 4, a compact hydrogen-bonded IPC is formed between PAA and the PDMAM chains grafted onto the P(AA-*co*-AMPSA70) graft copolymer backbone.

### 3.2. *Small angle neutron scattering study*

The particles formed as soluble hydrogen-bonded IPCs at low pH, have been also studied with small angle neutron scattering (SANS). For this purpose, the P(AA-*co*-AMPSA82)-*g*-PDMAM graft copolymer, with a backbone consisting in 82% AMPSA and 18% AA units ($M_w$ = 270 kDa, measured in its sodium salt form), containing 48 wt % of PDMAM side chains ($M_n$ = 17 kDa), and shortly designated as G48, was used.[39]

**Figure 4** shows the variation of the SANS intensity, $I$, versus the scattering wave vector, $q$, for pure G48 and PAA90, and for G48/PAA90 mixtures, at four different unit mole polyacid / polybase ([PAA90] / [PDMAM]) ratios, r = 0.25, 0.5, 1, 1.5, in $D_2O$, at pH = 2. G48 and

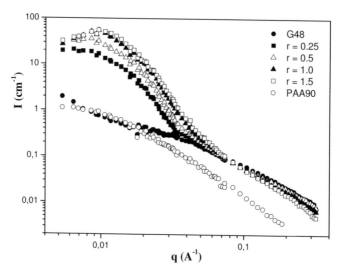

**Fig. 4**. I vs q for PAA90 (o), G48 (•) and G48/PAA90 mixtures at different unit mole ratios, r = [PAA]/[PDMAM], 0.25, (■); 0.5, (Δ); 1, (▲); 1.50, (□) in solutions in $D_2O$, at pH = 2. The concentration of G48 is 1.40 x $10^{-2}$ g/cm³ in all the solutions. The concentration of pure PAA90 solution is 1.00 x $10^{-2}$ g/cm³. Reprinted with permission from Ref. 39. Copyright (2006) American Chemical Society.

PAA90 pure solutions exhibit a typical (for polymers) neutron scattering behavior. Nevertheless, as some PAA90 is added in the G48 solution (for instance at r = 0.25), a noteworthy increase in the scattering intensity is observed in the low $q$ region, characteristic of a system structured at the scale $1/q$, i.e. approximately 100 Å. The only way to explain the peak in this rather low $q$ range is to assume the formation of bigger objects, formed by complexation. Anyway, in this low pH region, insoluble hydrogen–bonded IPCs are known to be formed between PDMAM and PAA.[31] $I$ increases further, as $r$ increases up to $r = 1$, while remains constant by further adding PAA90. The behavior observed is explained by the hydrogen-bonding complexation of PAA90 with the PDMAM side chains grafted onto the anionic P(AA-co-AMPSA) backbone of G48. As a result, compact colloidal complex particles, stabilized by the anionic chains of the graft copolymer backbone, should be formed, as presented in **Scheme 2**.

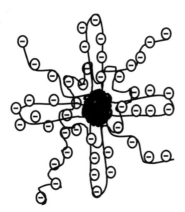

**Scheme 2.** Negatively charged colloidal particles formed through hydrogen-bonding interpolymer complexation of PAA with the PDMAM side chains of the graft copolymer P(AA-co-AMPSA)-g-PDMAM (G48), at low pH. Reprinted with permission from Ref. 40 Copyright (2007) American Chemical Society.

In **Fig. 5**, the maximum of the scattering intensity, $I_{max}$, related (among others) to the mass of the complexes, is plotted as a function of the unit mole ratio, $r$, for the G48/PAA90 polymer mixtures at pH = 2.0. $I_{max}$ is observed to increase with increasing $r$ until $r = 1.1$, which should

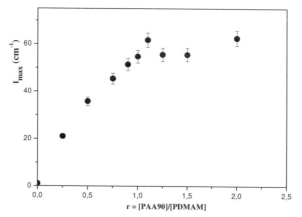

**Fig. 5.** Variation of the maximum in the scattering intensity, $I_{max}$, for the G48/PAA90 polymer mixtures in semidilute solution in $D_2O$, at pH = 2, vs. their unit mole ratio r = [PAA90] / [PDMAM]. Reprinted with permission from Ref. 39. Copyright (2006) American Chemical Society.

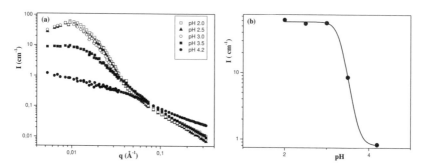

**Fig. 6.** (a) $I$ vs. $q$ for the G48/PAA90 mixture in semidilute solution in $D_2O$, for r = 1.10, at five different pH values: pH = 4.2 (●); pH = 3.5 (■); pH = 3.0 (○); pH = 2.5 (▲); pH = 2.0 (□);. (b) $I_{max}$ vs. pH for the same mixture. Reprinted with permission from Ref. 39. Copyright (2006) American Chemical Society.

correspond to the stoichiometry of the hydrogen-bonded IPC that is formed between PAA90 and PDMAM; then, a plateau is attained.

In **Fig. 6(a)**, the intensity $I$ vs. $q$ for the G48/PAA90 mixture in its stoichiometric composition (r = 1.10), at five different pH values, is presented. We observe that at high enough pH values (e.g., pH = 4.2) where the hydrogen-bonding complexation between G48 and PAA90 is prevented, due to the negative charge of the partially neutralized PAA90 chains, the scattering curve is representative of a macromolecular chain in solution, with a maximum in scattering intensity approaching 1 cm$^{-1}$. As pH decreases (e.g., down to pH = 3.5), the intensity at low $q$ increases by almost one order of magnitude, close to 10 cm$^{-1}$, indicating a chain organization in the system. At this pH (which is lower than pH$_c$ = 3.75), a hydrogen-bonded IPC between PAA90 and the PDMAM side chains of the G48 graft copolymer is formed, explaining the observed behavior. By further decreasing the pH value, the intensity at low q values increases further reaching a peak of 62 cm$^{-1}$ at $q$ = 0.0095 Å$^{-1}$ for pH = 2.0. It is also observed that the values of the scattering intensity for pH = 2.5 and pH = 3.0 are not much different, which indicates that at pH = 3.0, the complex formed has almost assumed its final form so that by further decreasing the pH value no considerable changes take place.

In **Fig. 6(b)** the maximum of intensity, $I_{max}$, is plotted as a function of pH. A sigmoidal curve was found to fit the experimental data quite well, showing that at around pH = 3.75, a transition occurs. Its origin should be related to the formation of the interpolymer complex due to hydrogen-bonding, and the structural organization of the system. In this pH region, the scattered intensity increases by two orders of magnitude as the value of the pH decreases from about 4 to about 3.

### 3.3. Characterization of the particles formed as soluble hydrogen-bonded IPCs at low pH

The colloidal particles formed in solutions of stoichiometric G48/PAA90 mixtures at pH 2.0 were recently characterized.[40] **Figure 7** shows the intensity time correlation functions $g^{(2)}(t)$ and the inverse Laplace transform (ILT) distributions for the G48/PAA mixture in solution at pH = 2.0 at six different concentrations. The time correlation functions curves obtained are generally indicative of a system comprised of colloidal particles. At low concentrations, lower than the critical concentration, c*, determined by viscometry and found equal to 7.0 x 10$^{-3}$ g/cm$^3$, **Fig. 7(a)**, **7(b)** and **7(c)**, single ILT distributions, around $t = 1 \times 10^{-3}$ s appear. At a higher concentration, **Fig. 7(d)**, close to c*, a broadening of the distribution to higher times appears, indicating a slowing of the diffusion times, explained by an increase in the

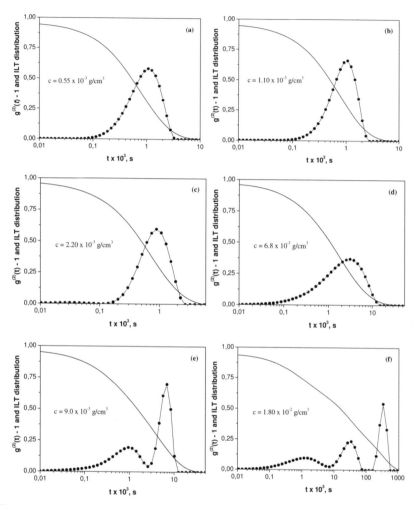

**Fig. 7**. Hydrodynamic radius, $R_H$, distribution for the particles formed by the mixture G48/PAA in aqueous solution at pH = 2.0, at different concentrations: (a), c = 1.1 x $10^{-3}$ g/cm$^3$; (b), c = 2.2 x $10^{-3}$ g/cm$^3$; (c), c = 4.5 x $10^{-3}$ g/cm$^3$; (d), c = 6.8 x $10^{-3}$ g/cm$^3$; (e), c = 9.0 x $10^{-3}$ g/cm$^3$; (f), c = 1.8 x $10^{-2}$ g/cm$^3$. Reprinted with prmission from Ref. 40. Copyright (2007) American Chemical Society.

interactions between the particles. This behavior is even more accentuated as concentration increases further, **Fig. 7(e)**, above c$^*$, where a second peak in the distribution curve appears at about one order of magnitude higher. Finally, at even higher concentration, **Fig. 7(f)**, a third peak appears at much higher time, while the correlation function curve is not anymore indicative of any independent particles in the system. The dramatic slow down in motion that occurs is explained by the formation of a transient network taking place in this high concentration region, **Scheme 3**.

From the relaxation time for each of the three dilute solutions, obtained by the peak of the corresponding ILT distribution curve, **Fig. 7(a)**, **7(b)** and **7(c)**, the translational diffusion coefficient, $D_T$, was calculated by means of the equation

$$D_T = (\tau q^2)^{-1} \qquad (2)$$

where $\tau$ is the relaxation time and $q$ the wave vector given by $q = 2\pi n \sin(\theta/2)/\lambda$, where n is the refractive index of the medium and $\lambda$ the wave length of the light beam. The translational diffusion coefficient value obtained by extrapolation to zero concentration, $D_0$, was related to the hydrodynamic radius, $R_H$, of the particles through the Stokes-Einstein equation,

$$R_H = K_B T/6\pi\eta_0 D_0 \qquad (3)$$

where $K_B$ is the Boltzmann constant, $T$ the absolute temperature, $\eta_0$ the viscosity of the solvent. A value equal to 105 nm was obtained for the hydrodynamic radius $R_H$ of the particles at infinite dilution.[40]

**Figure 8** shows the variation of the SANS intensity, $I$, versus the scattering wave vector, $q$, for G48/PAA90 at six different concentrations in $D_2O$, at pH = 2. At low q, roughly q < 0.01 $\text{Å}^{-1}$, typical Guinier scattering was found at low concentration, indicative of the finite size of aggregates. At intermediate q, 0.01 < q < 0.1 $\text{Å}^{-1}$, $I$ decreases following a scaling law of the form $I \sim q^{-d}$. As the values of the exponent d vary between 3.5 and 4.0, the presence of three-dimensional objects with smooth or fractal surfaces is indicated,[41] attributed to the insoluble hydrogen-bonded IPCs formed between the PDMAM side chains of the graft copolymer and the PAA chains.[31,42] At high q, q > 0.1 $\text{Å}^{-1}$, chain scattering is found attributed to the anionic backbone, consisting the hydrophilic shell of the colloidal particles formed (**Scheme 2**).

Furthermore, as the concentration becomes higher than $c^*$, the critical overlapping concentration, the form of the intensity curves changes, with

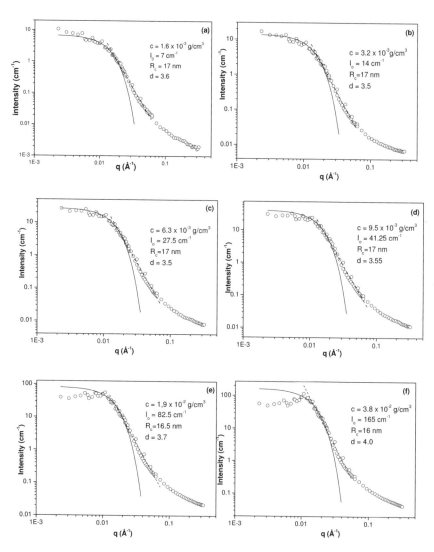

**Fig. 8.** SANS intensity variation vs. the wave vector q, for the G48/PAA polymer mixture in solution in $D_2O$, at pH=2, at different concentrations: (a), c = 1.6 x $10^{-3}$ g/cm³; (b), c = 3.2 x $10^{-3}$ g/cm³; (c), c = 6.3 x $10^{-3}$ g/cm³; (d), c = 9.5 x $10^{-3}$ g/cm³; (e), c = 1.9 x $10^{-2}$ g/cm³; (f), c = 3.8 x $10^{-2}$ g/cm³. Reprinted with permission from Ref. 40. Copyright (2007) American Chemical Society.

a tendency to shift to lower values at low $q$ and to exhibit a structural peak, at around 0.01 $\mathring{A}^{-1}$, which becomes clear only in the most concentrated solution, **Fig. 8(f)**, and reflects the interactions between the objects. By considering that it corresponds to the most probable distance between them, a cubic lattice model based on the mass conservation of the complex particles, with the distance between the particles given through $D = 2\pi/q_0$, can be applied. Since the volume, $V$, of each particle is estimated by $V = \varphi D^3$, where $\varphi$ is the volume fraction of the particles, their "dry" radius, $R_{dry}$, is calculated by the equation

$$R_{dry} = \sqrt[3]{\frac{6\pi^2\varphi}{q_0^3}} \tag{4}$$

In the case of the most concentrated solution, **Fig. 8(f)**, where $q_0 = 0.0108$ $\mathring{A}^{-1}$, and $\varphi = 1.85 \times 10^{-2}$, calculated by taking into account only the compact complex particles formed between the PDMAM side chains of G48 and the PAA chains and their mass density, $d_c = 1.28$ g/cm$^3$,[39] $R_{dry}$ comes out to be equal to 9.5 nm.

The low $q$ intensity region corresponds to a first approximation to the Guinier regime of the scattering of individual non-interacting, finite-sized objects;[43] their radius leads to a characteristic decrease in $I$, whose magnitude is related to their mass. If the objects are spheres of radius $R_c$:

$$I = I_0 \exp(-R_c^2 q^2/5) \tag{5a}$$

With

$$I_0 = \varphi \, \Delta\rho^2 \, V_0 \tag{5b}$$

where $V_0$ denotes the dry volume of an individual object, $\varphi$ the volume fraction of the objects and $\Delta\rho$ the scattering contrast between the solvent and the dry polymer.

Then, from Eq. 5(b), by taking $\Delta\rho = 5.0 \times 10^{10}$ cm$^{-2}$,[39] we obtain the value of 165 cm$^{-1}$ for the intensity at zero $q$, $I_0$. Using this value in the Guinier form expressed by Eq. 5(a), we obtain a relatively good fit for the data of **Fig. 8(f)**, if we use a value equal to 16 nm for the radius, $R_c$, of the compact particles We also observe that we have relatively good Guinier fitting for all the concentrations measured by using as $I_0$ the value occurring from the initially estimated quantity for the most concentrated solution, 165 cm$^{-1}$, adjusted each time proportionally to the concentration. The value for the radius of the particles obtained is practically stable at 16–17 nm. It should also be considered as a "wet" radius representing about 80% hydrated particles. Moreover, it should be compared to the hydrodynamic radius of the particles, $R_H$ = 105 nm, obtained from DLS measurements in dilute solution. It is noteworthy that this hydrodynamic radius includes not only the insoluble core of the compact hydrogen-bonding interpolymer complexes formed between PAA and the PDMAM side chains of the graft copolymer, but also a hydrophilic shell comprised by its anionic backbone, **Scheme 2**. The hydrophilic shell is comprised by loops and single strands of the anionic backbone, extended, due to their charge and the low ionic strength of the solution, while their length should be related with the distribution of the PDMAM side chains in the P(AA-*co*-AMPSA82) ionic backbone and its length, estimated to be over the 330 nm on the basis of its molecular weight.

From the volume of the particle, we can also obtain the molecular mass, $M_c$, of the dry complex particle through the equation

$$M_c = V \, d_c \, N_A \qquad (6)$$

where $N_A$ is Avogadro's number. Equation (5) gives $M_c$ = 2.8 x $10^6$ Da, corresponding to a value equal to 4.5 x $10^6$ Da for the whole particle. This core molecular weight value also implies that each particle contains about 90 PDMAM side chains involving more than six graft copolymer chains. This leads to the formation of a transient network explaining the increase in viscosity observed in semidilute solution,[31] and the gel formation also studied.[44]

From static light scattering (SLS) results, the weight average molecular weight, $M_w$, and the radius of gyration, $R_G$, of the colloidal nanoparticles were determined. A value of $M_w$ = 5.7 x $10^6$ Da was obtained for the molecular weight, comparable to the value M = 4.5 x $10^6$ Da, calculated after SANS measurements. An aggregation number equal to 6–8 can be calculated on the basis of these molecular weight values, showing that each colloidal nanoparticle should be comprised by 6–8 graft copolymer chains and 12–16 PAA chains. Regarding the radius of gyration, the value $R_G$ = 85 nm was obtained, which combined with the hydrodynamic radius, $R_H$ = 105 nm, found above, shows that the colloidal particles should be of a spherical form, as $R_G/R_H$ is close to the square root of 3/5.

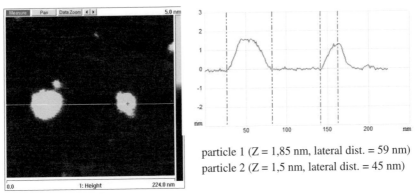

particle 1 (Z = 1,85 nm, lateral dist. = 59 nm)
particle 2 (Z = 1,5 nm, lateral dist. = 45 nm)

**Fig. 9.** Tapping mode AFM picture of the G48/PAA polymer mixture deposited on mica and observed under ambient conditions. Reprinted with permission from Ref. 40. Copyright (2007) American Chemical Society.

**Figure 9** represents an AFM image showing globular particles homogeneously distributed on the mica surface. The particles obtained are characterized by lateral dimensions of about 60 nm, with a height of about 1.5–2 nm. Assuming a tip radius of 10 nm (the actual size of commercial AFM tips is given between 5 nm and 15 nm), we can estimate a true lateral size around 22.5 nm. From these dimensions and assuming spherical cap geometry, the particles volume deposited on the substrate is about 3200 nm$^3$. It appears that this estimated volume is lower than the one of the insoluble core in solution, as determined by SANS (17150 nm$^3$), but it is in a fairly good agreement with the dry volume of the core, as it has been estimated to be 80% hydrated. The dimension and shape obtained of the nanoparticles observed by AFM should look like the one of the dry core. This finding emphasizes the fact that it is a multi-scale organized particle with a central hydrophobic core, hydrated up to 80%, comprised by the hydrogen-bonded IPC formed between PAA and the PDMAM side chains of G48, and an hydrophilic shell made of the P(AA-*co*-AMPSA82) anionic backbone of the G48.

Conclusively, the results obtained at pH = 2.0 revealed a structured system consisting of anionic colloidal nanoparticles. According to dynamic and static light scattering results spherical particles are formed with a hydrodynamic radius of about 105 nm. They are comprised by a compact core of PAA/PDMAM hydrogen-bonding interpolymer complexes, and a hydrophilic shell of anionic P(AA-*co*-AMPSA) chains. SANS measurements showed that the hydrophobic core presents a stable radius of 16 nm–17 nm. AFM revealed the formation of particles with a size approaching that of the hydrophobic core.

### 3.4. Semi-dilute solution study

In **Fig. 10**, the viscosity of the P(AA-*co*-AMPSA70)-*g*-PDMAM graft copolymer in mixtures with a high molecular weight polyacrylic acid

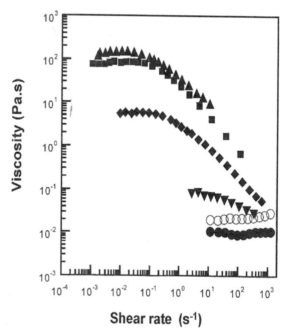

**Shear rate (s⁻¹)**

**Fig. 10.** Shear rate dependence of the viscosity of P(AA-*co*-AMPSA70)-*g*-PDMAM/PAA500 mixtures in semi-dilute aqueous solutions at pH = 3 and T = 25°C at different PAA500 concentrations, (▼): 5 x $10^{-3}$ g cm$^{-3}$, (♦): 1 x $10^{-2}$ g cm$^{-3}$, (■): 3 x $10^{-2}$ gcm$^{-3}$ and (▲): 6 x $10^{-2}$ g cm$^{-3}$. The curves for the two pure components are also given; (●): P(AA-*co*-AMPSA70)-*g*-PDMAM, 3 x $10^{-2}$ g cm$^{-3}$, the same in all mixtures and (○): PAA500, 6 x $10^{-2}$ g cm$^{-3}$. Reprinted with permission from Ref. 31. Copyright (2003) American Chemical Society.

sample, PAA500 ($M_w$ = 500 kDa), in semi-dilute aqueous solutions, at pH = 3 is plotted as a function of the shear rate.[31] For the sake of comparison, the corresponding curve of the pure PAA500 aqueous solution is also plotted. The viscosity of the two pure polymer solutions is rather low (of the order of $10^{-2}$ Pa.s) and its behavior versus shear rate is Newtonian In contrast, the viscosity of their mixtures increases significantly by increasing the PAA500 concentration, while its profile changes completely. At low shear rate the solution is still Newtonian, whereas for higher shear rates an important shear thinning behavior is observed. If shearing is stopped, the solutions recover their initial viscosity values and we obtain the same viscosity profiles when we repeat the experiments. Such a behavior is usual in associating polymer systems forming reversible networks through non-covalent cross-links, as it happens with the hydrophobically modified polymers.[45,46]

The above observed pH-controlled thickening behavior of the P(AA-co-AMPSA70)-g-PDMAM/PAA500 aqueous mixtures is nicely demonstrated in **Fig. 11**. In this figure, the pH dependence of the Newtonian plateau viscosity of a 1:1 (w/w) P(AA-co-AMPSA70)-g-PDMAM/PAA500 mixture in semi-dilute solution in water is presented and compared with the viscosity of its two pure components.

As expected, the viscosity of the two pure polymer components is relatively low, of the order of $10^{-2}$–$10^{-1}$ Pa·s. On the contrary, the mixture of the two polymers exhibits a characteristic pH-controlled thickening behavior. For pH values higher than 3.75, the viscosity is constant at about $10^{-1}$ Pa·s, that is, more or less, the additive value of the viscosities of its two pure polymer components in this pH range. As pH decreases from 3.75 to 2.5, a dramatic viscosity enhancement of about four orders of magnitude is observed, explained by the graft copolymer architecture which combines the proton acceptor ability of the PDMAM side chains with the strongly hydrophilic character of its negatively charged backbone.

An explanation of the above pH-responsive thickening behavior is schematically presented in **Scheme 3**. At pH > 3.75 no hydrogen-bonding interpolymer complexation takes place between PAA and the PDMAM side chains of the graft copolymer, due to the existence of a

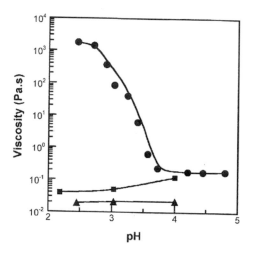

**Fig. 11.** pH variation of the Newtonean plateau viscosity, in semi-dilute aqueous solutions, of a 1:1 (w:w) P(AA-co-AMPSA70)-g-PDMAM/PAA500 polymer mixture, (●) and the two pure components, P(AA-co-AMPSA70)-g-PDMAM, (▲); PAA500, (■); at C = 6 x $10^{-2}$ g $cm^{-3}$ and T = 25°C. Reprinted with permission from Ref. 31. Copyright (2003) American Chemical Society.

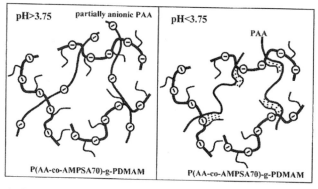

**Scheme 3.** A Schematic explanation of the pH-controlled thickening behavior of the P(AA-*co*-AMPSA70)-*g*-PDMAM/PAA500 mixtures in semi-dilute aqueous solution through reversible hydrogen-bonding association. Reprinted with permission from Ref. 31. Copyright (2003) American Chemical Society.

considerable fraction of anionic COO⁻ groups on the PAA chains. The viscosity is that of a polyelectrolyte semi-dilute solution. As pH decreases at values lower than 3.75, hydrogen-bonding association between PAA and the PDMAM side chains of the graft copolymer is taking place. This hydrogen-bonding interpolymer complexation should lead to phase separation, if the PDMAM homopolymer was used (**Fig. 1**). In the case above, a macroscopic phase separation is inhibited because the PDMAM chains are grafted onto P(AA-*co*-AMPSA70), a negatively charged, strongly hydrophilic, polymer backbone. As a consequence, a physical network is formed through the association of PAA with the PDMAM side chains of different graft copolymer chains.

Viscoelastic measurements, carried out for the P(AA-*co*-AMPSA80)-*g*-PDMAM60/PAA450 polymer mixture at pH values 2.0, 3.4 and 3.8, are shown in **Fig. 12**, presenting typical frequency sweeps and revealing the strongly pH-dependent thickening behavior of the polymer mixture studied.[44]

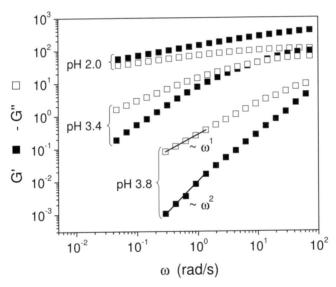

**Fig. 12.** Linear viscoelastic behavior of P(AA-*co*-AMPSA)-g-PDMAM60/PAA450 mixtures at pH 2.0, 3.4 and 3.8. The total polymer concentration is 6 % w/w and the weight ratio of the two polymers in the mixture is 1:1. Storage modulus, G', ■; loss modulus, G", □. Reproduced from Ref. 44.

At pH 3.8, the mixture behaves like a viscoelastic liquid, as the pulsation dependencies of G' and G" at low frequencies are proportional to $\omega^2$ and $\omega^1$ respectively, which characterize the terminal zone of the curve observed for viscoelastic liquids, where long-time dynamics are governed by reptation.[47] The reptation time, roughly corresponding to the G'-G" crossover, is lower than 0.06 s. At pH 3.4, the viscoelastic behavior of the mixture is similar to that at pH 3.8 but G' and G" moduli are greatly enhanced and the reptation time attains 0.6 s, showing a pronounced slowing down of the molecular dynamic. At pH 2.0, the mixture behaves like a gel, as G' modulus is higher than G" modulus and both are practically pulsation-independent. This sol/gel transition ensues from the graft copolymer architecture, combining the proton acceptor ability of the PDMAM side chains, which form strong hydrogen-bonded IPCs with and the highly hydrophilic character of the copolymer backbone, resulting from its high percentage composition (80%) in the strongly anionic AMPSA units. 3.75 is a critical pH value because hydrogen-bonding interpolymer complexation between PAA and PDMAM is prevented at pH > 3.75.[31] By decreasing pH, hydrogen-bonding complexation between PAA and the PDMAM side chains is progressively strengthened resulting in the formation of stickers along the graft copolymer anionic backbone, which hinder the molecular dynamic and enhance the viscoelastic properties.[48,49] At pH = 2.0, the PDMAM/PAA junctions are much more strengthened, due to the strong hydrogen-bonding complexation.

The influence of the graft copolymer weight composition in PDMAM is shown in **Fig. 13**. It shows the steady state viscosity versus shear stress of the P(AA-*co*-AMPSA)-*g*-PDMAM22/PAA450, P(AA-*co*-AMPSA)-*g*-PDMAM42/PAA450 and P(AA-*co*-AMPSA)-*g*-PDMAM60/ PAA450 polymer mixtures at pH = 2.0. For each solution, a cycle of increasing (full symbols) and decreasing shear stress (open symbols) has been applied. The viscous profiles observed for the P(AA-*co*-AMPSA)-*g*-PDMAM60 based mixture form a hysteresis loop, characteristic of thixotropic materials. When the shear stress increases, the polymer solution presents a high Newtonian viscosity, followed by a discontinuous shear thinning. In the decreasing stress mode, viscosity increases again, but with values considerably lower than those in the increasing stress run, a behavior representative of structured systems. On the contrary, the viscous behavior of the polymer mixtures of P(AA-*co*-

AMPSA)-*g*-PDMAM22 and P(AA-*co*-AMPSA)-*g*-PDMAM42 with PAA450 exhibits a well-defined Newtonian plateau at low shear stress, followed by a smooth shear thinning with no hysteresis loop. Such a rheological behavior is observed for polymer solutions with weak intermolecular interactions, as for neutral polymer solutions in the semi-dilute regime, where entanglements occur, or in weakly associative polymer solutions.[50] These results reveal that the mixture of the P(AA-*co*-AMPSA)-*g*-PDMAM60 graft copolymer with PAA450 at pH = 2.0 is structured, while the corresponding mixtures of P(AA-*co*-AMPSA)-*g*-PDMAM22 and P(AA-*co*-AMPSA)-*g*-PDMAM42 appear simply as dense macromolecular systems.

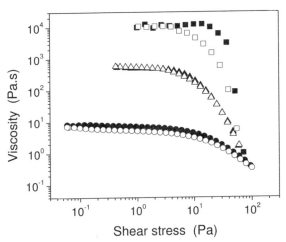

**Fig. 13**: Viscosity versus shear stress for the P(AA-*co*-AMPSA)-g-PDMAM22/PAA 450 (•, ○); P(AA-*co*-AMPSA)-g-PDMAM42/PAA 450 (▲, △) and P(AA-*co*-AMPSA)-g-PDMAM60/PAA 450 (■, □) mixtures at pH 2.0. Full symbols: data obtained by increasing stress; open symbols: data obtained by decreasing stress. Reproduced from Ref. 44. Copyright 2004 by The Society of Rheology. All rights reserved.

## 5. Complexes formed between PEG and PAA grafted with negatively charged PAMPSA chains

Except for grafting polybase chains onto an anionic backbone (e.g. PDMAM chains onto a PAMPSA backbone),[31] an alternative architecture has been also used for the preparation of soluble hydrogen-bonded IPCs by grafting PAMPSA chains onto a PAA backbone, **Scheme 1(b)**.[32]

These graft copolymers should be able to form hydrogen-bonded IPCs with PDMAM, but also with PEG or other polybases at low pH by means of their PAA backbone, while the negatively charged PAMPSA side chains should provide them with sufficient hydrophilicity, assuring their solubility.[36]

## 5.1. *Investigation of the complexes using turbidimetry and viscometry, Influence of pH and ionic strength*

**Figure 14** shows the variation of the optical density of mixtures of PEG, $M_w$ = 35 kDa, with PAA90, and the graft copolymers PAA-*g*-*x*PAMPSA4.5, with *x* = 14, 21 and 52 as a function of pH.[iii] The turbidimetric study was performed in pure 0.05 M citric acid-phosphate buffer solutions, **Fig. 14(a)**, and in the presence of 0.1 M NaCl, **Fig. 14(b)**. The concentration of each polymer was 2.5 mM in monomer interacting units, i.e. ether oxygens in the case of PEG and carboxylic groups in the case of PAA and the graft copolymers.

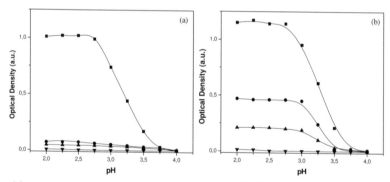

**Fig. 14**. (a). Variation of the optical density with pH for the mixtures of PEG with PAA90 (■), PAA90-*g*-14PAMPSA4.5 (●), PAA90-*g*-21PAMPSA4.5 (▲) and PAA90-*g*-59PAMPSA4.5 (▼) in 0.05 M CA/NaH₂PO₄ buffer solutions. The concentration of each polymer is 2.5 mM, on the basis of the monomer units participating in hydrogen-bonding interactions, i.e., ether oxygens and carboxylic units respectively. (b) The same as in (a), but in the presence of 0.1 M NaCl. Reprinted with permission from Ref. 32. Copyright (2006) American Chemical Society.

---

[iii]The graft copolymers used in this study are designated as PAA-*g*-*x*PAMPSA*y*, where *x* is the copolymer composition expressed as the weight percentage of the PAMPSA side chains, and *y* is the molecular weight of PAMPSA in kDa.

All mixtures were transparent at pH > 3.75, indicating that no insoluble hydrogen-hydrogen bonded IPCs are formed in this pH region. By decreasing pH, while the system PAA/PEG/H$_2$O, turns strongly turbid at pH < 3.75, when the graft copolymers PAA-*g*-*x*PAMPSA*y* are used instead of PAA, the turbidity of the solutions at pH < 3.75 is highly suppressed. It was attributed to the hydrophilic character of the anionic PAMPSA side chains of the graft copolymers. Moreover, while the mixtures PAA-*g*-*x*PAMPSA*y*/PEG in pure buffer solutions, **Fig. 14(a)**, appear practically transparent within the whole pH region studied, in 0.1 M NaCl, **Fig. 14(b)**, they show some turbidity. This behavior reflects a decrease in solubility of the possible hydrogen-bonded IPCs formed with PEG due to shielding of the negatively charged PAMPSA side chains of the graft copolymers.

However, as the considerable improvement in homogeneity of the mixtures of PEG with the graft copolymers above at low pH could be attributed also to a weakening in complexation, and not exclusively to the formation of soluble hydrogen-bonded IPCs, a viscometric study was undertaken to shed more light on the kind of the interaction taking place.[32]

The viscometry results obtained for the system PAA-*g*-21PAMPSA4.5/PEG/H$_2$O at different pH values were expressed by means of the reduced viscosity ratio, $(\eta_{sp}/c)_r \equiv r_{\eta red}$ (Eq. 1) and are presented in **Fig. 15(a)** as a function of the mixture unit molar composition, $N_{PEG}$, which represents the unit mole fraction of PEG in the mixture of the two polymers, given by the equation

$$N_{PEG} = n_{PEG} / (n_{PAA} + n_{PEG}) \tag{7}$$

where $n_{PEG}$ are the moles of the PEG monomer units and $n_{PAA}$ are the moles of the carboxylic groups in the graft copolymer.

For pH = 2.0, 2.75 and 3.5 $(\eta_{sp}/c)_r$ is lower than unity for all mixtures, indicating that a hydrogen bonded IPC of a compact structure is formed, getting stronger as pH decreases.

The effect of the ionic strength was investigated through a viscosity study in the presence of NaCl. As in **Fig. 15(b)** it is shown, the viscosity ratio $(\eta_{sp}/c)_r$ of the mixtures of PEG with the graft copolymer PAA-*g*-

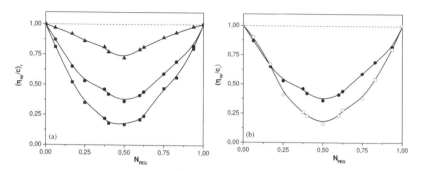

**Fig. 15.** (a). Variation of the reduced viscosity ratio, $(\eta_{sp}/c)_r$, with the mixture composition, $N_{PEG}$, for PAA90-g-21PAMPSA4.5/PEG mixtures in pure buffer solutions at pH = 2 (■), pH = 2.75 (●), and pH = 3.5 (▲). (b) The same as in (a) at pH = 2.75 (●), and in the presence of 0.1 M NaCl (○). Reprinted with permission from Ref. 32. Copyright (2006) American Chemical Society.

21PAMPSA4.5 at pH = 2.75 in 0.1 M NaCl (open circles) is lower enough than in pure buffer solution (solid circles). An increase in the ionic strength of the aqueous solution should have a double effect on the hydrogen-bonding interpolymer association, as it leads to an enhanced dissociation of the carboxylic groups of PAA, which is unfavorable to complexation, and it deteriorates the thermodynamic quality of the solvent with respect to polymers, leading to a strengthening of hydrophobic interactions and favoring, this way, the hydrogen-bonding interpolymer complexation.[51] In the present case, the deterioration of the thermodynamic quality of the solvent, due to the charge screening effect, mostly induced on the PAMPSA anionic side chains of the graft copolymer, should be the prevailing factor.

### 5.2. *Influence of the structural characteristics of the graft copolymers*

**Figure 16** presents the viscometry results for the mixtures of PEG with the graft copolymers PAA-g-xPAMPSA4.5 (x = 14, 21 and 52) and PAA-g-53PAMPSA20, at pH = 2.75. We observe that the $(\eta_{sp}/c)_r$ vs. $N_{PEG}$ curves shift to higher values as the weight percentage, x, of the graft copolymer in PAMPSA4.5 chains increases. In fact, we see that $(\eta_{sp}/c)_r$ takes values close to unity only for the mixture of PEG with the PAMPSA-richer copolymer, PAA-g-52PAMPSA4.5/PEG. This behavior suggests that, as it concerns the copolymer containing the lower molecular mass PAMPSA side chains (4.5 kDa), the formation of IPCs is inhibited when $x$ reaches 50 wt%. On the contrary, when the copolymer with the higher molecular mass PAMPSA side chains is used ($M_n$ = 20 kDa), hydrogen-bonding IPC formation with PEG is clearly observed at pH = 2.75, even for $x$ = 50 wt%.

The effect of the negatively charged PAMPSA side chains on the hydrogen-bonded IPCs formed is double. First, they form a protective hydrophilic corona, stabilizing the IPCs formed in the aqueous environment, obviously in the form of core-corona nanoparticles. Second, they act as structural defects, as it is the case of introduction of charged comonomers in the PAA chain.[28] However, due to the comb-type structure of these copolymers, the first effect is prevailing, while the second is weakened, so that soluble hydrogen-bonded IPCs are formed.

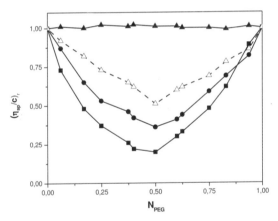

**Fig. 16.** Variation of $(\eta_{sp}/c)_r$ with $N_{PEG}$, for mixtures of PEG with PAA90-*g*-14PAMPSA4.5 (■), PAA90-*g*-21PAMPSA4.5 (●), PAA90-*g*-52PAMPSA4.5 (▲) and PAA90-*g*-53PAMPSA20 (Δ) at pH = 2.75 in pure buffer solutions. Reprinted with permission from Ref. 32. Copyright (2006) American Chemical Society.

### 5.3. *Size of the colloidal nanoparticles formed*

The hydrodynamic radius of the hydrogen-bonded IPCs formed between PEG and the graft copolymers PAA-*g*-xPAMPSAy at the stoichiometric composition, $N_{PEG}$ = 0.5, at pH = 2.0, in pure buffer solutions and in the presence of 0.1 M NaCl, was determined by DLS measurements. The results obtained are summarized in **Table 1**. In the absence of salt, the hydrodynamic radius of the aggregates is of the order of a few decades of nanometers (30 nm–50 nm), and it decreases as $x$ increases. This is explained by the increased hydrophilicity of the IPCs formed, due to the higher content of the graft copolymer in hydrophilic PAMPSA side chains. The addition of salt leads to a considerable increase of the size of the colloidal nanoparticles formed. This should be related with the charge screening effect on the anionic PAMPSA side chains and the change of the hydrophobic-hydrophilic balance in the IPC in favor of its hydrophobic character.

**Table 1**. Hydrodynamic radius, $R_H$, of the IPCs formed between the graft copolymers and PEG at their stoichiometric composition at pH = 2.0 in pure buffer solutions and in the presence of 0.1 M NaCl. Reprinted with permission from Ref. 32. Copyright (2006) American Chemical Society.

| Graft copolymer | $R_H$ (nm), no salt | $R_H$ (nm), in 0.1 M aCl |
|---|---|---|
| PAA-$g$-14PAMPSA4.5 | 48 | 87 |
| PAA-$g$-21PAMPSA4.5 | 33 | 83 |
| PAA-$g$-53PAMPSA20 | 30 | 82 |

These observations are schematically shown in **Scheme 4** where the colloidal nanoparticles formed are designed as spherical, for the sake of simplicity. Core-corona colloidal nanoparticles is supposed to be formed through hydrogen-bonding interpolymer interaction between PEG and PAA-$g$-xPAMPSAy. They should be comprised by an insoluble hydrophobic core of a PAA/PEG hydrogen-bonded IPC and a hydrophilic corona of anionic PAMPSA chains. They are smaller when the composition of the graft copolymer in PAMPSA side chains is higher, while, the addition of NaCl favors the formation of larger particles. This behavior is related with the role of the hydrophilic PAMPSA side chains. Their presence counterbalances the hydrophobic complexes formed by hydrogen-bonding interaction between PEG and the PAA backbone of the graft copolymers at low pH. However, this schematic description, based on the DLS results, must to be further confirmed by more studies, as for instance by SANS and AFM experiments.

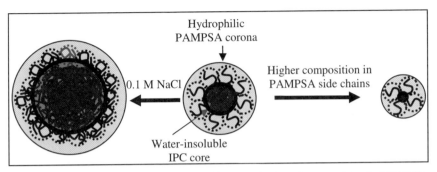

**Scheme 4**: Schematic illustration of the influence of salt and composition in PAMPSA side chains on the structure of the colloidal nanoparticles. Reprinted with permission from Ref. 32. Copyright (2006) American Chemical Society.

## 6. Conclusions

The comb-type copolymers P(AA-*co*-AMPSA)-*g*-PDMAM and PAA-g-xPAMPSAy, combine nicely two distinct characteristics, i.e., the ability of PDMAM or of PAA to form with weak polyacids, such as PAA, or nonionic polybases, such as PEG, respectively, at low pH, hydrophobic hydrogen-bonded IPCs, and the hydrophilic character of the negatively charged P(AA-*co*-AMPSA) backbone or PAMPSA side chains correspondingly. After these copolymers are mixed with PAA or PEG, respectively, in aqueous solution, at pH lower than 3.75, soluble hydrogen-bonded IPCs are formed. These complexes appear as core-corona colloidal nanoparticles, comprised by a hydrophobic core of an insoluble hydrogen-bonded IPC and a hydrophilic corona of anionic or neutral chains.

Finally, we have to point out that such complexes could be interesting candidates in drug delivery systems, when the oral route through the gastrointestinal tract should be followed, as they are stable in the pH range 2.0–3.5, and dissociate at pH higher than 4.0. Moreover, the complexes based on the anionic backbone copolymers grafted with neutral polybase chains, in mixtures with weak polyacids, may serve as pH-responsive thickening agents giving reversible gels at low pH.

# References

1.  E.A. Bekturov and L.A. Bimendina, *Adv. Polym. Sci.* 43, 100 (1980).
2.  E. Tsuchida and K. Abe, *Adv. Polym. Sci.* 45, 1 (1982).
3.  M. Jiang, M. Li, M. Xiang, and H. Zhou, *Adv. Polym. Sci.* 146, 121(1999).
4.  G. Bokias and G. Staikos, Recent Res. Devel. Macromol. Res. 4, 247 (1999).
5.  F. E. Baily Jr., R. D. Lundberg, and R. W. Callard, *J. Polym. Sci.* A2, 845 (1964).
6.  T. Ikawa, K. Abe, K. Honda, and E. Tsuchida, *J. Polym. Sci.Polym. Chem. Ed.* 13, 1505 (1975).
7.  I. Iliopoulos and R. Audebert, *Polym. Bull. (Berlin)* 13, 171 (1985).
8.  D. J. Hemker, V. Garza, and C. W. Frank, *Macromolecules* 23, 4411 (1990).
9.  V. V. Khutoryanskiy, A. V. Dubolazov, Z. S. Nurkeeva, and G. A. Mun, *Langmuir* 20, 3785 (2004).
10. O.V. Klenina and E.G. Fain, *Polym. Sci. U.S.S.R.* 23, 1439 (1981).
11. Eustace, D.J., Siano, D.B., and Drake, E.N., *J. Appl. Polym. Sci.* 35, 707, (1988).
12. G. Staikos, K. Karayanni, and Y. Mylonas, *Macromol. Chem. Phys.* 198, 205 (1977).
13. T. Aoki, M. Kawashima, H. Katono, K. Sanui, N. Ogata, T. Okano, and Y. Sakurai, *Macromolecules* 27, 947 (1994).
14. G. A. Mun, Z. S. Nurkeeva, V. V. Khutoryanskiy, G. S. Sarybayeva, and A. V. Dubolazov, *Eur. Polym. J.* 39, 1687(2003).
15. K. Karayanni and G. Staikos, *Europ. Polym. J.* 36, 2645 (2000).
16. G. A. Mun, Z. S. Nurkeeva, V. V. Khutoryanskiy, and A. D. Sergaziyev, *Colloid Polym. Sci.*, 280, 282 (2002).
17. Nurkeeva, Z. S., Khutoryanskiy, V. V., Mun, G. A., and Bitekenova, A. B., *J. Polym. Sci., Ser. B*, 45, 365, (2003).
18. T. Ozeki, H. Yuasa, and Y. Kanaya, *J. Controlled Release* 63, 287 (2000).
19. B. S. Lele and A. S. Hoffman, *J. Controlled Release* 69, 237 (2000).
20. M.-K. Chun, C.-S. Cho, and H.-K. Choi, *J. Controlled Release* 81, 327 (2002).
21. A. M. Mathur, B. Drescher, A. B. Scranton, and J. Klier, *Nature* 392, 367 (1998).
22. E. Umana, T. Ougizawa, and T. Inoue, *J. Membr. Sci.* 157, 85 (1999).
23. C. L. Bell and N. A. Peppas, *Adv. Polym. Sci.* 122, 125 (1995).
24. A. D. Antipina, V. Yu. Baranovskii, I. M. Papisov, and V. A. Kabanov, *Vysokomol. Soyedin.* A14, 941 (1972).
25. K. Sivadasan, P. Somasundaran, and N.J. Turo, *Coll. Polym. Sci.* 269, 131 (1991).
26. I. Iliopoulos and R. Audebert, *Eur. Polym. J.* 24, 171 (1988).
27. H. T. Oyama, D. J. Hemker, and C. W. Frank, *Macromolecules* 22, 1255 (1989).
28. I. Iliopoulos and R. Audebert, *Macromolecules* 24, 2566 (1991).
29. G. Bokias, G. Staikos, I. Iliopoulos, and R. Audebert, *Macromolecules* 27, 427 (1994).
30. A. Usaitis, S. L. Maunu, and H. Tenhu, *Eur. Polym. J.* 33, 219 (1997).
31. M. Sotiropoulou, G. Bokias, and G. Staikos, *Macromolecules* 36, 1349 (2003).
32. P. Ivopoulos, M. Sotiropoulou, G. Bokias, and G. Staikos, *Langmuir* 22, 9181 (2006).
33. T. P. Yang, E. M. Pearce, T. K. Kwei, and N. L. Yang, *Macromolecules* 22, 1813 (1989).
34. Y. Wang and H. Morawetz, *Macromolecules* 22, 164 (1989).
35. T. Shibanuma, T. Aoki, K. Sanui, N. Ogata, A. Kikuchi, Y. Sakurai, and T. Okano, *Macromolecules* 33, 444 (2000).

36. H. Dautzenbrg, W. Jaeger, J. Kotz, B. Philipp, C. Sheidel, and D. Stcherbina, *Polyelectrolytes: Formation, Characterization and Application* (Hanser, 1994).
37. H. Ohno, K. Abe, and E. Tsuchida, *Makrokol. Chem.* 179, 755 (1978).
38. V. Baranovsky, S. Shenkov, I. Rashkov, and G. Borisov, *Eur. Polym. J.* 28, 475 (1992).
39. M. Sotiropoulou, J. Oberdisse, and G. Staikos, *Macromolecules* 39, 3065 (2006).
40. M. Sotiropoulou, F. Bossard, E. Balnois, J. Oberdisse, and G. Staikos, *Langmuir* 23, 11252 (2007).
41. J. S. Higgins and H. C. Benoît, *Polymers and Neutron Scattering* (Oxford Science Publications, Clarenton Press, Oxford, 1994).
42. M. Zeghal and L. Auvray, *Europhys. Lett.* 45, 482 (1999).
43. P. Lindner and Th. Zemb, Eds., *Neutrons, X-rays and Light: Scattering Methods Applied to Soft Matter* (North-Holland, Delta Series, Elsevier, Amsterdam, 2002).
44. F. Bossard, M. Sotiropoulou, and G. Staikos, *J. Rheol.* 48, 927 (2004).
45. D. N. Schulz and J. E. Glass, *Polymers as Rheology Modifiers* (ACS Sump. Ser. 462, Am. Chem. Soc., Washington DC, 1991).
46. I. I. Potemkin and A. R. Khokhlov, *Polymer Gels and Networks*, Y. Osada and A. R. Khokhlov, Eds., (Marcel Dekker, New York, 2001), p. 47.
47. P.G. De Gennes, *Scaling concepts in polymer physics* (Ithaca, NY, Cornell University Press, 1979).
48. M. Rubinstein and A. Semenov, *Macromolecules* 34, 1058 (2001).
49. A. N. Semenov and M. Rubinstein, *Macromolecules* 35, 4821 (2002).
50. T. Aubry and M. Moan, *J. Rheol.* 40, 441 (1996).
51. V. Khutoryanskiy, G. Mun, Z. Nurkeeva, and A. Dubolazov, *Polym. Int.* 53, 1382 (2004).

# CHAPTER 3

## POTENTIOMETRIC INVESTIGATION OF HYDROGEN-BONDED INTERPOLYMER COMPLEXATION

Georgios Staikos[1*], Gina-Gabriela Bumbu[2], Georgios Bokias[3]

[1]*Department of Chemical Engineering, University of Patras,
GR-26504 Patras, Greece*
*[*]E-mail: staikos@chemeng.upatras.gr*
[2]*"P.Poni" Institute of Macromolecular Chemistry, 41A Gr. Ghica Voda Alley,
Ro.700487 Iasi, Romania*
[3]*Department of Chemistry, University of Patras, GR-26504 Patras, Greece*

## 1. Introduction

The intermacromolecular complexation through hydrogen bonding takes place mostly in aqueous solutions between complementary polymers, i.e., between weak polyacids (PA), such as poly(acrylic acid) (PAA) or poly(methacrylic acid) (PMAA), and Lewis polybases (PB), such as polyethyleneoxide (PEO), polyacrylamide (PAM), poly(N-vinyl-pyrrolidone) (PVPo), etc.[1-5] The structure of the polycomplex formed could be generally described by the scheme[6]:

$$- A - A - A -$$
$$\vdots \quad \vdots \quad \vdots$$
$$- B - B - B -$$

where A and B are the polyacid and the polybase monomer units, respectively.

One of the very first observations on the hydrogen bonding interpolymer complexation was the increased pH of the aqueous PA/PB solution, compared to the pH of the pure PA solution.[7-9] The pH increase observed is typically attributed to the displacement of the dissociation equilibrium of the PA to the undissociated form, due to the interpolymer complexation via hydrogen-bonding. In fact, when the polyacid is dissolved in aqueous solution, the carboxylic groups COOH are partially dissociated:

$$COOH \rightleftharpoons COO^- + H^+ \tag{1}$$

The dissociation equilibrium is characterized by the apparent dissociation constant, $K_d$, of the polyacid, according to the following expression:

$$K_d = \frac{\left[COO^-\right]\left[H^+\right]}{\left[COOH\right]} \tag{2}$$

where the $[COO^-]$ is the concentration of carboxylate groups, $[H^+]$ is the hydrogen ion concentration and $[COOH]$ is the concentration of the undissociated polyacid monomer units.

The hydrogen-bonded IPC is formed only with the participation of undissociated carboxylic groups of the polyacid, as described by the following complexation process (using the abbreviation B for the structural unit of the Lewis polybase PB) assuming a 1:1 stoichiometry:

$$COOH + B \rightleftharpoons COOH \cdots B \tag{3}$$

The association described by eq. (3) is characterized by an apparent complexation constant

$$K_c = \frac{[C]}{[COOH][B]} \tag{4}$$

where [C] is the concentration of the complexed COOH or B units, while [COOH] and [B] are the concentrations of free carboxylic acid and Lewis base units at equilibrium. Thus, as a consequence of the interpolymer association via hydrogen bonding, the concentration of free carboxylic groups in the solution decreases and the dissociation of the polyacid, Eq. (1), is shifted to the left-hand side, leading to the observed pH increase.

Although the determination of $K_c$ is always aimed in such potentiometric studies, we should mention that the quantity usually determined in most cases is the degree of complexation (linkage), $\theta$, defined as the ratio of the bound groups to the total of the potentially interacting groups, i.e.,

$$\theta = \frac{[PA]_0 - [PA]}{[PA]_0} = 1 - \frac{[PA]}{[PA]_0} \qquad (5)$$

where $[PA]_0$ and $[PA]$ are the molar concentrations of free polymeric acid groups before and after addition of the polybase, respectively.

## 2. Evaluation of potentiometric data

Based on the aforementioned simple considerations, potentiometry has been widely used for the study of the IPCs formed through hydrogen-bonding. The quantitative analysis of the pH changes by mixing a polyacid with a Lewis polybase in aqueous solution allows the determination of quantitative parameters of the complexation process, such as the degree of complexation (linkage), $\theta$, the apparent complexation constant, $K_c$, as well as the thermodynamic parameters of the complexation reaction, $\Delta H°$ and $\Delta S°$. However, due to the polymeric nature of the interacting species, the behaviour is not as simple as presented above. As a result, several approaches have been attempted for the potentiometric investigation of the hydrogen-bonded IPCs.

## 2.1. Potentiometric investigation of stoichiometric mixtures

The most simple and direct potentiometric approach of interpolymer complexation via hydrogen-bonding is the comparison of the pH of the pure polyacid with the pH of the PA/PB mixture at the same PA concentration. In the simplest case, namely in the absence of any other weak acid or base, the degree of linkage, $\theta$, can be determined as[9]

$$\theta = 1 - \left( \frac{\left[ H^+ \right]}{\left[ H^+ \right]_0} \right)^2 \qquad (6)$$

where $[H^+]$ and $[H^+]_0$ are the concentrations of hydrogen ions in the presence and in the absence of PB, respectively. This approach is based on the assumption that the dissociation constant of PA does not change significantly with pH, remaining practically the same in the pure PA aqueous solution and in the PA/PB aqueous mixture.

When the PA/PB mixture is equimolar and assuming a 1:1 stoichiometry, the complexation constant, $K_c$, is determined by the equation

$$K_c = \frac{\theta}{C_0 (1 - \theta)^2} \qquad (7)$$

where $C_0$ is the unit mole concentration of the polyacid or the polybase. This approach has been extensively developed and applied by Osada in the mid-seventies.[9-12] Thus, $K_c$ and $\theta$ were determined for the hydrogen bonding interpolymer complexation of PMMA or PAA with PEO or PVPo of various chain lengths, both in aqueous and aqueous – alcoholic media as a function of temperature. As seen from the results summarized in **Table 1**, $K_c$ depends strongly on chain length, temperature and alcohol content. From the temperature dependence of $K_c$, the change of the enthalpy, $\Delta H^0$, and of the entropy, $\Delta S^0$, of complexation can be

**Table 1.** Stability constants ($K_c$) of IPCs at various temperatures. Reproduced with permission from Ref. 12. Copyright 1980 by Taylor & Francis Ltd. http://www.informaworld.com

| Complex system | Molecular weight of PEO or PVP x $10^{-3}$ g/mol | Stability constant $K_c$ (L/mol) | | | | | |
|---|---|---|---|---|---|---|---|
| | | 10 °C | 20 °C | 30 °C | 40 °C | 50 °C | 60 °C |
| PMAA-PEO | 0.2 | 2 | 0 | 0 | 4 | 10 | 5 |
| | 1.0 | 3 | 0 | 0 | 5 | 14 | 16 |
| | 2.0 | 15 | 16 | 62 | 280 | 520 | 44 |
| | 3.0 | 10 | 68 | 250 | 470 | 790 | 1130 |
| | 7.5 | 760 | 2600 | 4700 | 6200 | 6800 | 7200 |
| | 20.0 | 1450 | 2800 | 4200 | 5000 | 5000 | 4580 |
| PAA-PEO | 2.0 | 1 | 10 | 8 | 3 | 0 | 0 |
| | 3.0 | 2 | 7 | 9 | 8 | 1 | 0 |
| | 7.5 | 16 | 18 | 24 | 20 | 13 | 3 |
| | 20.0 | 64 | 85 | 150 | 180 | 190 | 160 |
| PMAA-PVP | 10.0 | 950 | 920 | 1260 | 1850 | 3920 | 7560 |
| | 40.0 | 1460 | 1700 | 1780 | 2200 | 5010 | 11900 |
| | 160.0 | 14500 | 12200 | 14300 | 1800 | 22200 | 27400 |
| | 360.0 | 15500 | 18700 | 24700 | 33500 | 43500 | 48600 |
| PMAA-PEO in ethanol/water (% ethanol) | | | | | | | |
| 10% | 2.0 | 22 | 55 | 162 | 198 | 171 | 131 |
| 17% | | 328 | 268 | 207 | 118 | 61 | 49 |
| 23% | | - | 208 | 77 | 39 | 33 | 20 |
| 37% | | 0 | - | 0 | 0 | 0 | 0 |
| 50% | | 0 | - | 0 | 0 | 0 | 0 |

derived using the equations

$$\frac{d(\ln K_c)}{d\left(\frac{1}{T}\right)} = \frac{-\Delta H^0}{R} \tag{8}$$

$$\Delta H^o - T\Delta S^o = -RT \ln K_c \tag{9}$$

where R is the universal constant of the gases and T is the absolute temperature. For PMAA-PEO complexes, both $\Delta H^0$ and $\Delta S^0$ were found positive, indicating that hydrophobic interaction was a significant factor for the stabilization of the complexes. On the contrary, upon addition of ethanol, negative values are found (indicating solely hydrogen-bonding interactions), before the final destruction of the IPC at higher alcohol contents.

In the case of the PAA-PEO complexes in water, $\Delta H^0$ and $\Delta S^0$ are positive at low temperatures but change signs at higher temperatures,[13] indicating a less significant (compared to PMAA-PEO) hydrophobic contribution. Moreover, the addition of methanol[13] or dioxane[14] was found to destabilize the PAA-PEO complexes.

## 2.2. Potentiometric titrations

### 2.2.1. Potentiometric titration of a polyacid with a Lewis polybase

This method consists in titrating a PA solution with a concentrated PB solution or measuring the pH of solutions containing a constant PA concentration and increasing the PB concentration. The results are usually presented in terms of the PB/PA unit molar ratio, r.

In many cases, the potentiometric study is limited just in the presentation of the pH variation with r.[8,15-19] In this case, the stoichiometry of hydrogen-bonding complexation may be estimated, because the pH increase observed (as a result of complexation) levels off when stoichiometry is approached, as shown in **Fig. 1**.

It is also possible to estimate the variation of θ with r.[20-23] In fact, when no other charged species are present, θ may be estimated from Eq. (6).

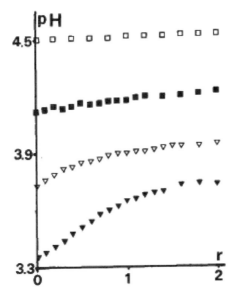

**Fig. 1.** pH variation with r for the complexation of PAA with PEG, at different neutralization degrees of the polyacid. Reproduced from Ref. 20. Copyright 1985 Springer-Verlang.

However, when the polyacid is partially neutralized or it is a copolymer of a weak acid and a strong electrolyte, the mass conservation laws and electroneutrality should be applied for the determination of $\theta$. Usually, $\theta$ is derived on the assumption that the dissociation constant of the polyacid, $K_d$, is not influenced by the pH change resulting from the IPC formation. As seen in **Fig. 2**, $\theta$ increases with r and finally levels off, that is the expected consequence of the mass action law. From such results, Iliopoulos *et al.*[20-22] have demonstrated that hydrogen-bonding complexation is much influenced by the existence of charged groups in the polyacid (originating either from neutralization or copolymerisation). It was shown that a charge in the content of about 10 mol % is preventing the complex formation.

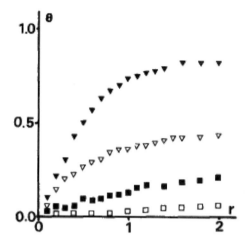

**Fig. 2.** Variation of θ with r for the experimental results of Fig. 1. Reproduced from Ref. 20. Copyright 1985 Springer-Verlang. With kind permission of Springer Science and Business Media.

## 2.2.2. Potentiometric titration of a Lewis polybase with a polyacid

This method includes a titration of the PB solution with a PA solution.[24-28] Typically, the results follow a behaviour similar to that presented in **Fig. 3.**[27] Both in the absence and in the presence of a hydrogen-bonding complexing polybase, pH decreases substantially with increasing the polyacid concentration. Obviously, this is more pronounced at the first stages of titration, where the concentration of the polyacid changes significantly. When the titrated solution contains a strongly complexing Lewis polybase, like hydroxyethylcellulose in **Fig. 3**, pH deviates to significantly higher values to those of the blank (pure water), the deviation being more pronounced at low polyacid concentrations.

For the estimation of θ, Eq. (6) can be applied. To improve the precision of the results, the concentration of carboxylate anions should be taken into account in the respective mass conservation law, as proposed by Hemker *et al.*[24] A major assumption in these calculations

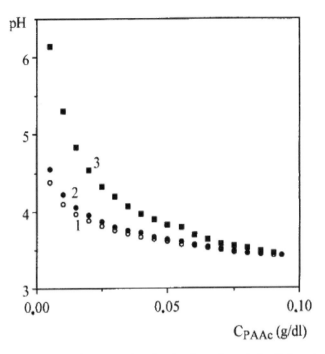

**Fig. 3.** Potentiometric results concerning the titration of   methyl cellulose (curve 2) or hydroxyethylcellulose (curve 3) with PAA. The blank experiment is shown with open symbols (curve 1). Reprinted with permission from Ref. 27. Copyright 2000 John Wiley and Sons, Inc.

is that the dissociation constant of the polyacid remains the same in the absence and in the presence of PB.

This is not always evident, as the pH deviation observed in some cases is very important. For this reason, Staikos and Bokias have proposed to use the blank experiment in order to determine the variation of $K_d$ with pH,[26] following the Kern semi-empirical equation[29]:

$$pK_d = a + b\, pH \qquad (10)$$

where a and b are empirical constants. The determination of the concentration of uncomplexed carboxylic acid groups (and $\theta$) can, thus, be derived from Eq. (2), using the value of $K_d$ at the actual pH of each PA/PB mixture each time.

The potentiometric titration of mixtures of two Lewis polybases with a weak polyacid was investigated by Chatterjee *et al.*[30-33] They found that in such ternary systems several breakpoints may be identified, related to the stoichiometry and relative strength of the hydrogen-bonding association of PA with the two PBs.

## 2.3. Potentiometric investigation under "iso-pH" conditions

The potentiometric titration under "iso-pH" conditions could be considered, in general terms, as the adaptation of the isoionic dilution method under low pH conditions.[34] In fact, isoionic dilution consists of diluting an aqueous polyelectrolyte solution with a salt solution of the same "effective" ionic strength as that arising from the polyelectrolyte.[35,36] Under such conditions, the effective ionic strength, I, upon dilution remains constant. When dealing with weak polyelectrolytes, like PAA in its acid form, isoionic conditions maybe achieved by diluting PAA with an HCl solution of the same pH as the pH of the polyelectrolyte ("iso-pH" conditions). Most important, apart from the stabilization of I, under "iso-pH" conditions, the increase of the dissociation degree of the polyelectrolyte due to dilution, is suppressed. Thus, the potentiometric investigation of the hydrogen-bonded IPCs under "iso-pH" conditions consists of mixing an aqueous PA solution with a polybase solution of the same pH (adjusted with HCl), as that of the polyacid solution. Under these circumstances, the pH increase of the mixture solution is due only to the change in the dissociation of the polyacid caused by the H-bonding formation, and not by dilution.

The application of this method is illustrated in **Fig. 4**, where the interpolymer complexation of PAA with PAM via hydrogen-bonding is investigated at pH = 3.00, by mixing PAA and PAM solutions of the same molar concentration and pH.[37] As seen in **Fig. 4a**, the pH of the mixtures deviates to higher values, due to interpolymer complexation, while the maximum pH is found at mole fraction of AA groups ($N_{PAA}$) around 0.5, i.e., under equimolar conditions. This, in fact, is a verification of the 1:1 stoichiometry of the complex formed.

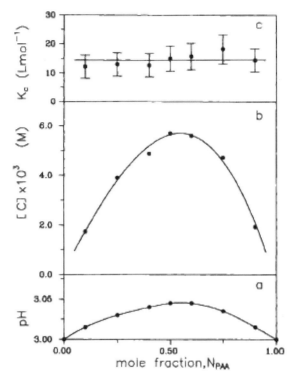

**Fig. 4.** Potentiometric study of the complexation of PAA with PAM under "iso-pH" conditions. Reproduced with permission from Ref. 37. Copyright Society of Chemical Industry. Permission is granted by John Wiley & Sons Ltd on behalf of the SCI.

This is also verified in **Fig. 4b**, where the concentration of the complex formed is presented as a function of $N_{PAA}$. [C] is determined from the mass conservation law of the carboxylic units through the relation:

$$[C] = [PAA]_0 - [COOH] - [COO^-] \quad (11)$$

where $[PAA]_0$ is the initial PAA concentration, $[COOH]$ is calculated from the Kern semi-empirical equation, Eq. (10), while $[COO^-]$ is obtained from the equation expressing the electroneutrality of the solutions:

$$[COO^-] = [H^+] - [Cl^-] \quad (12)$$

Finally, the complexation constant $K_c$ is calculated from Eq. (4). As seen in **Fig. 4b**, $K_c$ is constant over a wide range of $N_{PAA}$, showing that the potentiometric titration under "iso-pH" conditions permits the determination of a $K_c$ independent of the mixing composition in weak polyacid-polybase solution mixtures. The position of the maximum of the complex concentration [C] versus $N_{PAA}$ gives information about the effective ratio of the monomer units of the two interacting species in the IPC, that is the polybase units corresponding to a polyacid unit, known also as the stoichiometry of the complex.[37-41] The composition at which the concentration of the complex presents a maximum is considered to be the stoichiometric composition of the IPC. At the same time, by applying these new effective concentrations for PA or PB, $K_c$ should be constant. For example, in the complexation of PAA with poly(N-isopropylacrylamide), PNIPAM, it is found that each PAA unit corresponds to 1.5 PNIPAM units.

Finally, from the temperature-dependence of $K_c$, we can determine the characteristic thermodynamic functions of the complexation, namely the enthalpy, $\Delta H^o$, and the entropy, $\Delta S^o$, of the IPC formation, applying Eqs. (8) and (9). The results for some selected systems, determined under "iso-pH" conditions are presented in **Table 2**.

**Table 2.** Thermodynamic data for various hydrogen-bonded IPCs studied under "iso-pH" conditions.

| IPC | Stoichiometry | $\Delta H^o$ (kJ mol$^{-1}$) | $\Delta S^o$ (J mol$^{-1}$K$^{-1}$) | Ref. |
|---|---|---|---|---|
| PAAM/PAA | 2:3 | -75 | -226 | 37, 39 |
| PNIPAM/PAA | 3:2 | 160 | 620 | 34, 37 |
| PEG/PAA | 1:1 | - | - | 34, 38 |
| Hydroxypropylcellulose/ Maleic acid-styrene copolymer | 2:5 | 32 | 225 | 41 |

## 3. Concluding remarks

Potentiometry has been proved a popular technique for the investigation of the interpolymer complexation via hydrogen-bonding throughout the last forty years for a variety of reasons. First, potentiometry is a simple but sensitive method. Second, under the correct considerations, it permits the determination of several important thermodynamic parameters characterizing the interpolymer association, like $\theta$, $K_c$, $\Delta H^0$ and $\Delta S^0$.

When dealing with polymers like weak polyacids and Lewis polybases, the application of potentiometry is complicated, due exactly to the polymeric nature of the associating species. For this reason, several approaches have been developed. The aim of the present work was not to give an exhaustive list of the relating publications in the field. Mostly, we aimed at a brief and concise presentation and classification of the most established practical and theoretical approaches. We consider that the evolution of the potentiometric studies of hydrogen-bonded IPCs, as presented here, shows that potentiometry remains a powerful tool for a better and deeper understanding of these systems.

## References

1.  E.A. Bekturov and L.A. Bimendina, *Adv. Polym. Sci.* 41, 99 (1981).
2.  E. Tsuchida and K. Abe, *Adv. Polym. Sci.* 45, 1 (1982).
3.  I.M. Papisov and A.A. Litmanovich, *Adv. Polym. Sci.* 90, 140 (1989).
4.  E.A. Bekturov and L.A. Bimendina, *J. Macromol. Sci.-Rev. Macromol. Chem. Phys.* 37, 501 (1997).
5.  M. Jiang, M. Li, M.L. Xiang, and H. Zhou, *Adv. Polym Sci.* 146, 121 (1999).
6.  V. Yu. Baranovsky, A.A. Litmanovich, I.M. Papisov, and V.A. Kabanov, *Eur. Polym. J.* 17, 969 (1981).
7.  A.D. Antipina, V.Y. Baranowskii, and V.A. Kabanov, *Polym. Sci. USSR* 14, 1047 (1972).
8.  T. Ikawa, K. Abe, K. Honda, and E. Tsuchida, *J. Polym. Sci., Polym. Chem. Ed.* 13, 1505 (1975).
9.  Y. Osada and M. Sato, *Polymer Letters Ed.* 14, 129 (1976).
10. Y. Osada, *J. Polym. Sci., Polym. Chem. Ed.* 15, 255 (1977).
11. Y. Osada, *J. Polym. Sci., Polym. Chem. Ed.* 17, 3485 (1979).
12. E. Tsuchida, Y. Osada, and H. Ohno, *J. Macromol. Sci.-Phys.* 17, 683 (1980).
13. H. Ahn, E. Kang, C. Jang, K. Song, and J. Lee, *J. Macromol. Sci. – Pure Appl. Chem.* 37, 573 (2000).
14. Y. Cohen and V. Prevysh, *Acta Polym.* 49, 539 (1998).
15. S.M. Park, S.H. Jeon, and T. Ree. *J. Polym. Sci.: Part A: Polym. Chem.* 27, 4109 (1989).
16. T. Petrova, I. Rashkov, V. Baranovsky, and G. Borisov, *Eur. Polym. J.* 27, 189 (1991).
17. V. Baranovsky , S. Shenkov, I. Rashkov, and G. Borisov, *Eur. Polym. J.* 28, 475 (1992).
18. S. E. Kudaibergenov, L.A. Bimendina, and G.T. Zhumadilova, *Polym. Adv. Technol.* 11, 506 (2000).

19. M. Pinteala, T. Budtova, V. Epure, N. Belnikevich, V. Harabagiu, and B.C. Simionescu, *Polymer* 46, 7047 (2005).
20. I. Iliopoulos and R. Audebert, *Polym. Bull.* 13, 171 (1985).
21. I. Iliopoulos and R. Audebert, *Eur. Polym. J.* 24, 171 (1988).
22. I. Iliopoulos and R. Audebert, *Macromolecules* 24, 2566 (1991).
23. V.A. Prevysh, B.-C. Wang, and R.J. Spontak, *Colloid Polym. Sci.* 274, 532 (1996).
24. H.T.Oyama, D.J. Hemker, and C.W. Frank, *Macromolecules* 22, 1255 (1989).
25. D.J. Hemker, V. Garza, and C.W. Frank, *Macromolecules* 23, 4411 (1990).
26. G. Staikos and G. Bokias, *Makromol. Chem.* 192, 2649 (1991).
27. O. Nikolaeva, T. Budtova, V. Alexeev, and S. Frenkel, *J. Poly. Sci.: Part B: Polym. Phys.* 38, 1323 (2000).
28. F. Bian and M. Liu, *Eur. Polym. J.* 39, 1867 (2003).
29. W. Kern, *Phys. Chem Abt A* 181, 249 (1938).
30. S.K. Chatterjee, A. Malhotra, and L.S. Pachauri, *Angew. Makromol. Chem.* 116, 99 (1983).
31. S.K. Chatterjee and A. Malhotra, *Angew. Makromol. Chem.* 126, 153 (1984).
32. S.K. Chatterjee, K.R. Sethi, and G. Riess, *J. Macromol. Sci. – Chem.* 24, 859 (1987).
33. S.K. Chatterjee, J. B. Yadav, K.R. Sethi, and A.M. Khan, *J. Macromol. Sci. – Chem.* 26, 1489 (1989).
34. G. Bokias and G. Staikos, *Recent Res. Devel. Macromol. Res.* 4, 247 (1999).
35. D.T.F. Pals and J.J. Hermans, *J. Polym. Sci.* 3, 897 (1948).
36. D.T.F. Pals and J.J. Hermans, *J. Polym. Sci.* 5, 733 (1950).
37. G. Staikos, G. Bokias, and K. Karayanni. *Polym. Int.* 41, 345 (1996).
38. G. Bokias, G. Staikos, I. Iliopoulos, and R. Audebert, *Macromolecules* 27, 427 (1994).
39. G. Staikos, K. Karayanni, and Y. Mylonas, *Macromol. Chem. Phys.* 198, 2905 (1997).
40. C. Vasile, G.G. Bumbu, Y. Mylonas, I. Cojocaru, and G. Staikos, *Polym. Int.* 52, 1887 (2003).
41. G.G. Bumbu, C. Vasile, J. Eckelt, and B.A. Wolf, *Macromol. Chem. Phys.* 205, 1869 (2004).

# CHAPTER 4

## POLARIZED LUMINESCENCE AND NANOSECOND DYNAMICS IN THE STUDIES OF INTERPOLYMER COMPLEXES

Elizaveta V. Anufrieva, Mark G. Krakovyak, Tatiana N. Nekrasova, Rouslan Yu. Smyslov

*Institute of Macromolecular compounds, Russian Academy of Sciences, 31, Bolshoi pr., Saint-Petersburg, 199004, Russia*
*E-mail: polar@mail.macro.ru*

A development of method allowing to study interpolymer complexes (IPC) in diluted solutions ($<10^{-3}$ mol/l), so that IPC formation from polymer chains of complementary chemical structure is not accompanied by interchain contacts depending on the concentration of interacting polymer chains, is highly desired. The other important requirement to this method is the absence of its destructive effect on IPC. The third peculiarity of the method which permits to study the formation of IPC is its high sensitivity to the formation of IPC and its structural changes. The method of polarized luminescence (PL) satisfies all these requirements. This method allows to study the formation and structure of IPC by measuring nanosecond relaxation times $\tau$, and evaluating the changes in polarization of polymer luminescence with covalently attached luminescent labels (LL) having anthracene structure.

The possibility of measuring nanosecond relaxation times is related to nanosecond duration of the anthracene nucleus luminescence. The method is described in detail by Anufrieva and Gotlib.[1] Only nanosecond relaxation times allow determining the formation of interchain contacts during the formation of IPC and their changes during structural transformations of IPC. The high sensitivity of these evaluations can be explained by the comparability of nanosecond relaxation times with the duration of intermolecular interactions, responsible for the formation of IPC, hydrogen bonds and hydrophobic contacts.

Nanosecond relaxation times were measured by PL method for many polymers of variable chemical structure and for various structural transformations of macromolecules in the solution: coil-globule, coil-compact structure, coil-α-helix, and also for IPC of different types.[2] We have established the relationships relating intramolecular mobility of macromolecules (IMM), nanosecond relaxation times, $\tau$, with the changes in the nature of the solvent, its viscosity ($\tau \sim \eta_{solvent}$) and thermodynamic quality ($\tau \sim 1/[\eta]$, $[\eta]$ – intrinsic viscosity of the polymer solution). All the regularities of $\tau$ changes are strict and have good reproducibility.

We have also been able to answer the question why the massive anthracene nucleus does not break the dependences of the nanosecond mobility of the polymer chains in solution on the various changes in its environment. In the relaxation spectra of the luminescently-labeled polymer in solution only the vibrational motion of the main polymer chain, which accompanies the transferable movement of the anthracene nucleus in the side chain round the bonds of the main chain, is observed in the nanosecond interval. The rotational movement of the anthracene nucleus in the side chain occurs at time $< 10^{-10}$ seconds.

It is the oscillating adjustment of the main polymer chain to the movement of the anthracene nucleus in the side chain due to the great amount of monomer units in the polymer chain that reflects perceptibly the transformation of all interchain interactions inside the macromolecule and between macromolecules.

The conclusions concerning the IPC formation have been drawn on the basis of $\tau$ changes of each interacting molecules while labeling alternatively the component under research. This approach to the study of IPC formation and its structure turns out to be especially informative.[3] The data concerning the analysis of IMM changes during IPC formation were compared to the data of viscometric measurements to confirm the compaction of supramolecular formations which appear as a result of the intermolecular interactions.

For the assessment of IPC compaction, besides the values of $\tau$, the data based on the analysis of their interaction with low molecular weight organic ions: acridine orange (AO), auramine (AU), 8-aniline-naphtalene-1-sulfonate (ANS), also turned out to be informative. The portion of bound AO ions was evaluated by the change in AO mobility ($\tau$ changes from 1 ns for AO in solution to 100 ns for AO, bound with polymers in IPC). The changes in luminescence intensity of AU or ANS ions in IPC also provide valuable information about IPC structural features.

In the present report, we will discuss the following problems regarding the IPC formed mainly via H-bonding:

1. The structure and IMM of interacting molecules of homopolymers and their influence on the formation and structure of IPC;

2. IPC formed by homopolymers and copolymers;

3. Factors affecting IPC stability.

## 1. Structure and designation of polymers

We used polymers of different chemical structure to study the processes of formation, structuring and features of IPC, formed as a result of non-covalent interactions between macromolecules of complementary chemical structure in solutions.

I. Proton-donating structures: polyacrylic acid (PAA) and polymethacrylic acid (PMAA).

II. Proton-accepting structures: polyethylene glycol (PEG), Pluronic, poly-N-isopropylacrylamide (PNIPAM), poly-N-*iso*-propylmetha-crylamide (PNIPMAM), poly-N-*n*-propylmethacrylamide (PNPMAM), poly-N-vinylmethylacetamide (PVMAA), poly-N-vinylpyrrolidone (PVP), poly-N-vinylcaprolactam (PVC), polyacrylpyrrolidine (PAP) and copolymers VP-methacrylic acid (MAA), VP-acrylic acid (AA), and VP-crotonic acid (CA).

9-alkylanthracene was used as a luminescent label, which was covalently attached to macromolecules of the studied polymers. The ratio of one LL per thousand of monomer units was used in all studies. 9-anthrylmethylene fragment was either directly joined to C-atom of a polymer chain or covalently attached through an intermediate fragment R of the variable chemical structure (**Scheme 1**).

This structure of LL was chosen because anthracene and its 9-alkyl derivatives possess a set of chemical and photophysical characteristics, which are quite convenient for experimental research. In particular, they have high and differently directed reaction capacity and intensive luminescence characterized by quite high polarization and life time of the excited state in a nanosecond interval.

Scheme 1

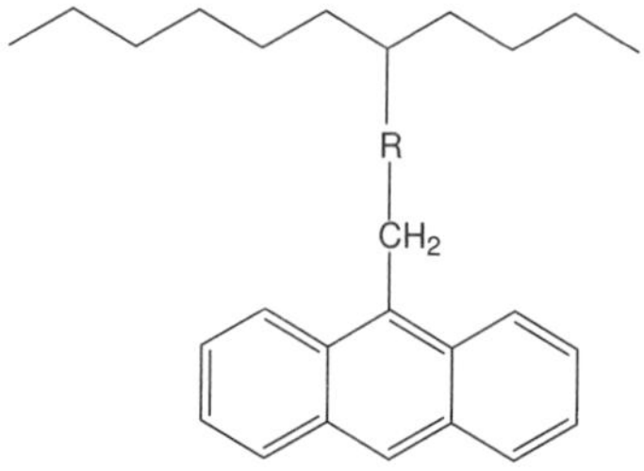

where R: -CO-O; -CO-NH-; -O-CO-NH-; -NH-CO-NH-, etc.

To obtain luminescently-labeled polymers of various classes with LL groups of the above-mentioned type, we used copolymerization with various monomers, containing anthracene, or reactions of functional groups of macromolecules with the reagents having the corresponding chemical structure, containing anthracene.[1]

## 2. Influence of chemical nature and IMM of individual polymer chains on their structure in IPC in aqueous solutions or in methanol

Formation of IPC by polymers, containing proton-accepting groups, and polycarboxylic acids takes place through H-bonding, however, hydrophobic contacts of non-polar groups of interacting molecules also play an important role.

From **Fig. 1**, it is clear that an increase in the number of non-polar groups in the macromolecules of interacting polymers leads to their intramolecular hindrance in IPC, which also grows together with increase in the number of these groups in a polymer. It can also be seen that for macromolecules with practically the same amount of non-polar groups the influence of H-bonds between the side chains on the structure of IPC is either strong (PNIPAM, **Fig. 1, 5c**) or weak (PNIPMAM, **Fig. 1, 6c**). The role of H-bonds between the side chains of these polymers is also revealed upon heating of their aqueous solutions, which leads to the formation of monomacromolecular globules.[4]

**Table 1** shows the data on the mobility of PMAA chains and poly-N-vinylamides in the IPC and in solution before the formation of the IPC. It can be concluded that the polymer chains hindered in the solution (PVC) form the IPC of more porous structure, compared to the polymer with

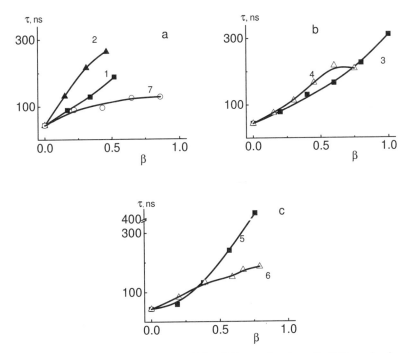

**Fig. 1.** Change in τ of PMAA* in water (pH = 3.5) as a function of addition of polymer with different proton-acceptor groups: PVP, β = [PVP]:[PMAA*] (1a), PVC, β = [PVC]:[PMAA*] (2a), PEG, β = [PEG]:[PMAA*] (3b), Pluronic, β = [Pluronic]: [PMAA*] (4b), PNIPAM, β = [PNIPAM]:[PMAA*] (5c), PNIPMAM, β = [PNIPMAM]: [PMAA*] (6c), and DNA, β = [DNA]:[PMAA*] (7a).

higher IMM.[5] Hence, the compaction of IPC depends on IMM of interacting polymer chains.

High IMM of PEG in solutions (τ ~ 1 ns) leads to the formation of more tightly packed IPC compared to the IPC formed by more hindered PNPMAM or PVP chains (τ ~ 20 ns).[3] Hindered DNA chain characterizing by low IMM, while interacting with PMAA forms IPC in which the structured PMAA molecules are not disturbed and the groups not included into this structure can interact with DNA.

**Table 1.** Relaxation times of PMAA and poly-N-vinylamides in the IPC ($\tau_{IPC}$) and in solution before the formation of IPC ($\tau_0$). The polymer ratio [PVP]:[PMAA*] = 1, [PVC]:[PMAA*] = 1.

| | | $\tau^{red}$, ns* | | | | | | |
|---|---|---|---|---|---|---|---|---|
| | PMAA | poly-N-vinylamide | | | | | | |
| | $H_2O$ pH = 4.5 | $H_2O$ | | Methanol | | Dimethyl-formamide | | |
| | $\tau_{IPC}$, ns | $\tau_{IPC}$, ns | $\tau_0$, ns | $\tau_{IPC}$, ns | $\tau_0$, ns | $\tau_{IPC}$, ns | $\tau_0$, ns | |
| PVP | 430 | 360 | 11 | 270 | 3 | 20 | 2.4 | |
| PVC | 430 | 120 | 20 | 160 | 6 | 26 | 5.5 | |

* The values of $\tau$ for their comparison in solvents of different viscosity $\eta$ are reduced to the same viscosity value $\eta_{red}$ using the equation $\tau^{red} = \tau \times \eta_{red} / \eta$, $\eta_{red} = 0.38$ sp[1].

**Figure 2** shows that formation of IPC between PMAA and poly-N-vinylamide) in methanol can be described in terms of the same dependencies irrespective of the content of non-polar groups in the macromolecules of poly-N-vinylamide (PVP, PVC), which sets these dependencies $\tau(\beta)$ apart from $\tau(\beta)$ for the same systems in water (**Figs. 2, 1**). Hence, the hydrophobic contacts of non-polar groups in the interacting macromolecules together with H-bonds govern the formation of IPC in aqueous solutions and affect their structure.

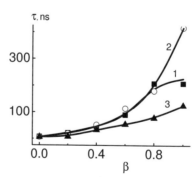

**Fig. 2**. Changes in $\tau$ of PMAA* in methanol upon addition of polymers with proton-accepting groups: PVP (1), PVC (2), and PAP (3).

Besides, from **Fig. 2** it follows that proton-accepting groups have different ability to form H-bonds, from high (PVP, PVC) to lower (PAP). The difference in the data for PVP and PAP may be related to the proximity of C=O groups in PAP to the main polymer chain, which hinders the participation of interacting groups in IPC formation.

## 3. Influence of IPC tertiary structure in water on the efficiency of addition reactions

From the analysis of IPC data in methanol (see above) it follows that IPC structure in aqueous solutions is determined not only by the formation of H-bonds between interacting polymer chains (secondary structure), but also by subsequent compaction with the formation of the "tertiary" structure. This is due to the inclusion of polar groups into H-bonds and decrease of their importance in the contacts with the solvent (water). Under such conditions non-polar groups find each other and promote IPC compaction. Local parts with high density of non-polar groups are formed in IPC. Luminescent low molecular weight organic ions with aromatic groups: AO, AU, ANS (**Fig. 3a** – $\theta_{AO}$, **b** – AU, **c** – ANS) can be used for their detection.

AO                    AU

ANS

In the absence of polycarboxylic acids, the polymers with proton-acceptor groups do not bind organic ions in aqueous solutions due to the lack of hydrophobic domains in macromolecules (**Fig. 3a, b**). When the compact IPC are formed upon addition of polycarboxylic acids, the number of polymer chains, incorporated into the complexes $\theta_{IPC}$ and the number of bound low molecular weight organic ions $\theta_{AO}$ increases. Their number is calculated by the change in nanosecond mobility of AO ions. The data for ions AU and ANS, not luminescent in water, but luminescent only in macromolecular environment have been obtained by the changes in luminescence intensity (**Fig. 3, (1-3)b, 1c**). The contribution of individual polycarboxylic acids into the changes of these characteristics is not very significant (**Fig. 3a, 2c**). The correlation between IPC formation and the changes in the parameters of the luminescence of organic ions, added to IPC, is so obvious that it can be used for quantitative characterization of complexation.[6]

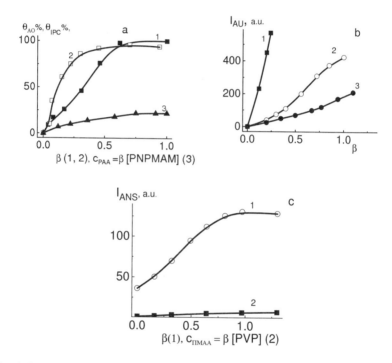

**Fig. 3.** $\theta_{AO}$ (1a), $\theta_{IPC}$ (2a), $I_{AU}$ (b), $I_{ANS}$ (c) as a function of $\beta$ = [polyacid]:[polymer] for IPC[(1, 2)a, (1, 2, 3)b and (1)c] and as a function of polyacid content $C_{POL} = \beta$ [polymer]. $\beta$ = [PAA]:[PNPMAM] (1, 2), (a), $\beta$ = [PAA]:[PVC] (1b), [PAA]:[PVP] (2b), [PAA]: [PVMAA] (3b), and [PMAA]:[PVP] (1 c). $C_{PAA} = \beta$[PNIPMAM] (3a), $C_{PMAA} = \beta$[PVP] (2c) at the constant content of polymer with proton-accepting groups.

## 4. Formation of IPC consisting of homo- and copolymers

The analysis of the data concerning the interaction between the homopolymer PVC and the copolymers VP with low content of proton-donating groups (20 mol. %) shows that IPC is formed at 20 mol. % content of methacrylic acid units as effectively as with homo PMAA. In the case of VP copolymers with 20 mol. % units of AA or CA, IPC is formed less efficiently or is not formed at all (**Table 2**). This difference is related to the decrease of succession (microblocks) length during the transition from MAA to AA units and statistical distribution of crotonic acid units at equal content of proton-donor carboxylic groups.

**Table 2**. PVC interaction with copolymers.

| IPC | $H_2O$, pH=3,5 | | Methanol | |
|---|---|---|---|---|
| | $\tau_{IPC}$ | $\tau_{PVC}$ | $\tau_{IPC}$ | $\tau_{PVC}$ |
| PVC* – PMAA | 880 | 40 | 1100 | 21 |
| PVC * – [MAA:VP (20:80)] | 800 | 40 | 21 | 21 |
| PVC * – [AA:VP (20:80)] | 120 | 40 | 21 | 21 |
| PVC *- [CA:VP (20:80)] | 40 | 40 | 21 | 21 |

It is seen from **Table 2** that IPC homopolymer – copolymer is formed only in water and is not formed in methanol. It means that the interaction of non-polar groups of homo- and copolymer as well as H-bonds proves to be the necessary condition of homo- and copolymer inclusion into IPC.[7]

## 5. IPC stability

The methods of determining the stability of IPC including poly-carboxylic acids and poly-N-vinylamides or other polymers with proton-acceptor groups are related to the ionization of polyacid in water or an amide solvent (N-methylpyrrolidone) addition.[8] The factors, stabilizing IPC in aqueous solutions – are "tertiary" structure elements, including non-polar groups of interacting polymers. A decrease in the number of non-polar groups in the IPC reduces stability of complexes. **Figure 4** presents the data, showing that IPC stability decreases, i.e., IPC is destroyed when the content of the destroying agent increases and ionization degree of polyacid increases: when PMAA is substituted by PAA during their interaction with PVC (**Fig. 4**), when PVC is substituted by PVP molecules (**Fig. 4**), or water – by dimethylformamide (**Fig. 5**).

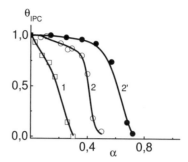

**Fig. 4.** Change in $\theta_{IPC}$ – portion of PAA* (1) or of PMAA*(2,2') in IPC with PVP (1, 2) or PVC (2') as a function of ionization degree of carboxylic groups, $\alpha$.

The $\theta_{IPC}$ value is determined with the help of the equation[1]:

$$\Theta^{IPC} / \tau_{IPC} + (1 - \theta^{IPC}) / \tau_0 = 1 / \tau_{obs} \qquad (2)$$

in which $\tau_{IPC}$, $\tau_0$, $\tau_{obs}$ are relaxation times of luminescently-labeled component in IPC, in water before IPC formation and observed during the PMAA ionization $\alpha$.

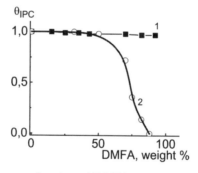

**Fig. 5.** Change in $\theta_{IPC}$ as a function of DMFA content in aqueous solution of IPC (PMAA-PVC*) (1) or (PAA-PVC*) (2).

## 6. Supramolecular chemical reactions with interpolymer complexes

IPC, formed as a result of non-covalent interactions between complementary molecules in solutions can also take part in such chemical processes as supramolecular exchange, substitution or addition reactions with low molecular weight or polymeric compounds.[9]

The study of molecular mechanisms of such reactions is necessary to get more detailed information about many natural processes with biopolymers in order to obtain IPC with certain features and use them in various areas of technology, biotechnology, medicine.[10]

## 6.1. *Exchange reactions*

During exchange reactions with IPC one of the macromolecules, constituting a complex, is exchanged by the other one of the same chemical structure **Scheme 2**.

$$IPC'(P_1\text{-}P_2) + P_2* \rightarrow IPC''(P_1\text{-}P_2*) + IPC'(P_1\text{-}P_2) + P_2$$

Studying supramolecular reactions at the molecular level is appropriate in highly diluted solutions using extremely sensitive methods which allow obtaining quantitative information. If we use the luminescently-labeled polymer as $P_2*$, while studying exchange reactions, then such reactions may be successfully studied by PL methods (having checked beforehand whether attaching LL affects the affinity between $P_2$ and $P_1$.) The high sensitivity of PL method in this case is due to the fact that relaxation times $\tau$, determined by PL method, characterizing intermolecular mobility of polymer chains, grow 10-100 times after their inclusion in IPC.

Our experiments have shown that interaction of $P_1$ with the mixture of $P_2$ и $P_2*$ under the conditions when ($[P_2] + [P_2*]$) in the reaction solution exceeds $[P_1]$, (here and further on in square brackets we indicate the mole concentration of polymer units in the solution, * stands for polymers containing covalently attached LL) leads to the formation of such IPC, in which the ratio $[P_2] / [P_2*]$ corresponds to the initial one. Hence, the addition of anthracene-containing LL to polymers (0.1–0.2 mol. %) in the studied cases does not affect their complex-forming properties.

After adding $P_2*$ polymer to the formed IPC' ($P_1$-$P_2$), an exchange reaction begins (**Scheme 2**), consisting of two stages: fast (several minutes) and slow (several hours, a day or more) (**Figs. 6, 7**). One can suggest that the presence of two stages in the process can be explained by the complexity of IPC structure.

Non-covalent inter-units interactions of complex-forming macro-molecules in solutions lead to the formation of two-traction sections – elements of IPC secondary structure. As a result of further interactions of ordered sections with each other a tertiary structure of supramolecular units is formed which leads to their further compaction and to a decrease in IMM of polymer chains, making up IPC, and, eventually to the deceleration of exchange reactions. Thus, all the factors, increasing the number and duration of contacts between $P_1$ and $P_2$ slow down exchange reactions.

IPC structural heterogeneity predetermines not only the presence of two stages in the exchange reaction, but also its intensity (**Figs. 6, 7**).

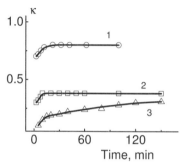

**Fig. 6.** Conversion of IPC'(PMAA-PVP) + PMAA* → IPC''(PMAA*-PVP)+PMAA exchange reactions κ versus time in aqueous solution at pH 3.3 (1) and 4 (2), and in ethanol (3). [PMAA]:[PMAA*]:[PVP] = 2:2:1.

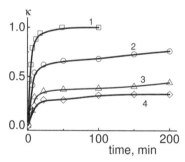

**Fig. 7.** Conversion of IPC'(PMAA- PEG) + PMAA* → IPC''(PMAA*-PEG)+PMAA exchange reactions κ versus time in aqueous solution at pH 4 for PEG of various $M \times 10^{-3}$ : 4(1), 6(2), 20(3), 40(4).

Therefore, the study of such reactions allows us to get important information about the structure and features of IPC, formed by complementary macromolecules of varied chemical structure under different conditions.

## 6.2. *Substitution reactions*

Substitution reactions with IPC may be represented by **Scheme 3**:

$$IPC'(P_1\text{-}P_2) + P_3 \rightarrow IPC''(P_1\text{-}P_3) + P_2$$

The study of these reactions is of special interest, as they play an important role in biological processes. Thus, the hierarchy of complex-forming capacity of biopolymers may underlie several consequently running reactions of supramolecular substitution, providing certain transport processes in biological systems. On the other hand, the study of substitution activities of certain polymers $P_3$ of varied chemical structure in relation to one of the IPC' components (i.e. $P_1$ or $P_2$) permits us to compare the complex-forming activity of different polymers and its relation to the peculiarities of chemical structure and molecular weight $P_3$.

If one of the polymers, constituting IPC' ($P_1$ or $P_2$), or the substituting macromolecular reagent $P_3$ is selected by attaching LL, the reaction may be quantitatively studied by PL method even when polymer content in solution is very low (**Fig. 8**).

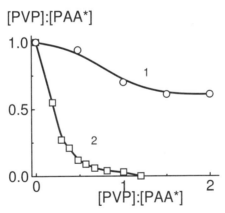

**Fig. 8.** Reaction of substitution IPC' (PVP-PAA*) + PMAA → PAA* + IPC''(PVP-PMAA) in water at pH 4 (1) and pH 5.4 (2). Initial relation [PVP]:[PAA*] = 1.

The results of the study of macromolecular substitution reactions, carried out by PL method, showed that the rate and the intensity of the reactions of this kind depend on the type, number and duration of new intermolecular contacts which appear during the reaction. In their turn, intermolecular interactions between the components of the reaction system depend on the structure of polymers in the reaction and the features of its environment (see **Fig. 8**). Preferred complexation of PVP with PMAA as compared with PAA is due to the stronger hydrophobic interactions. The influence of pH is due to the ionization of PMAA and to the changes in H-bonds.

## 6.3. *Addition reactions*

IPC are able to take part in supramolecular addition reactions, in particular – addition of organic ions (see section "Influence of IPC tertiary structure in water on the efficiency of addition reactions"). It is worth mentioning that these reactions can be carried out when single polymers, not bound in IPC, are much less capable of it or not capable at all. The given data show that the polymer system ability to add organic ions grows dramatically together with IPC formation (**Figs. 3 a, b, c**).

To study addition reactions and as well as other supramolecular reactions with IPC in diluted solutions, we used highly sensitive luminescent methods. However, the luminescence source was not the LL, attached to macromolecules covalently, but organic ions themselves. We used acridine orange (AO) and auramine (AU) cations, and 8-aniline-naphtalene-1-sulfonate (ANS) anions in their capacity.

## 7. Conclusion

Some of many examples of IPC formed by H-bonds are discussed here. It is clear that nanosecond IMM studies permit us to obtain information on many properties of IPC in solution. The most interesting of them are, of course, the relationship between IPC stability and reactions of IPC with high and low molecular weight compounds. The high sensitivity of luminescent methods used here points to its considerable promise.

## References

1.  E.V. Anufrieva, Yu.Ya. Gotlib, *Adv. Polym. Sci.* 40, 1 (1981).
2.  E.V. Anufrieva, *Pure and Appl. Chem.* 54, 533 (1982)
3.  V.D. Pautov, E.V. Anufrieva, M.G. Krakovyak, *Chinese J. of Polym. Sci.* 16, 268 (1998).
4.  E.V. Anufrieva, , T.D Ananieva, M.G. Krakovyak, V.B. Lushchik, T.N. Nekrasova., R.Yu. Smyslov, T.V. Sheveleva, *Polym. Sci. Ser. A*, 47, 189 (2005).
5.  E.V. Anufrieva, M.R. Ramazanova, M.G. Krakovyak, V.B. Lushchik, T.N. Nekrasova, T.V. Sheveleva, *Vysokomolec.soed. Ser. A*, 33, 1186 (1991).

6. E.V. Anufrieva, R.A. Gromova,. Yu.E. Kirsh, N.A. Yanul., M.G. Krakovyak, V.B. Lushchik, V.D. Pautov, T.V. Sheveleva, *Eur. Polym. J.* 37, 323 (2001).
7. E.V. Anufrieva, V.D. Pautov, M.G. Krakovyak, V.B. Lushchik, *Vysokomolec.soed. Ser. A*, 33, 1609 (1991).
8. E.V. Anufrieva, M.R. Ramazanova, M.G. Krakovyak, V.B. Lushchik, T.N. Nekrasova, T.V. Sheveleva, *Vysokomolec.soed. Ser. A*, 33, 256 (1991).
9. V.D. Pautov, E.V. Anufrieva, *Polym. Sci. Ser. A*, 34, 487 (1992).
10. J.-M. Lehn, *Supramolecular Chemistry. Concepts and Perspectives.* Weinheim; New York; Basel; Cambridge; Tokyo; VCH Verlagsgesellshaft mbH (1995).

# CHAPTER 5

# INTRAMOLECULAR POLYCOMPLEXES IN BLOCK AND GRAFT COPOLYMERS

Tatyana Zheltonozhskaya, Nataliya Permyakova and Lesya Momot

*Department of Macromolecular Chemistry, Faculty of Chemistry, Kiev National Taras Shevchenko University, Vladimirskaya St. 60, 01033, Kiev, Ukraine*
*E-mail: zheltonozhskaya@ukr.net*

A special class of intramolecular polycomplexes (IntraPCs), which are formed by block-type copolymers with chemically complementary polymer components, in particular, in linear block and graft copolymers, is considered in this review. Two different groups of IntraPCs stabilized either by electrostatic interactions or hydrogen bonds are distinguished within this class. At the same time, the main attention is paid to the peculiarities of synthesis, bulk structure, and stimuli-responsive properties of hydrogen-bonded IntraPCs. Their structure and properties are compared with those of the hydrogen-bonded intermolecular polycomplexes and block and graft copolymers with immiscible polymer components. New phenomena of the matrix block and graft copolymerization, caused by complex formation between the growing chain in one of the components and the initial covalently bound block of another one, are discussed. It is shown that IntraPC-forming block and graft copolymers have more homogeneous bulk structure compared to the corresponding InterPCs. A new approach to interpret the complicated stimuli-responsive behavior of IntraPCs in solutions is proposed. Finally, the applications of IntraPCs are highlighted.

## 1. Introduction

It is well known that intermolecular polycomplexes (InterPC) formed between polymer partners by cooperative non-covalent bonds of different nature belong to the special kind of polymer mixtures.[1-5] Analogously, the block-type copolymers (mostly linear block copolymers and graft copolymers) with chemically complementary blocks constitute a separate class of "intramolecular polycomplexes" (IntraPC)[6-7] or "intrapolymer complexes"[8] or alternatively "self-complexes".[9] Their properties differ essentially from those ones of block and graft copolymers with thermodynamically incompatible or slightly compatible polymer components, but resemble many properties of InterPCs.

## 2. Intramolecular polycomplexes

Spontaneous formation of InterPCs through a cooperative bond system requires achieving a particular critical chain length from both polymer partners, which depends on the energy of a single bond between them.[10-12] Additionally, the interacting polymers must be structurally and conformationally complementary.[10,13] The solvent nature is also of a considerable importance because its molecules can compete with active groups of polymers and can hinder the appearance of intermolecular bonds. From this point of view, block and graft copolymers containing chemically and structurally complementary blocks of sufficient length can be involved in intramolecular cooperative interactions, i.e., can form IntraPCs in suitable solvents. However, in grafted systems, an important difference is observed, for example, a covalent linking between the chemically complementary polymers such as poly(ethylene oxide) and poly(methacrylic acid) in the graft copolymers PMAAc-*g*-PEO[14] removes the critical chain length effect. Indeed, the corresponding IntraPCs were formed for all relative molecular masses of PEO and PMAAc. A comparative statistical mechanical analysis of the complex formation equilibrium between free and covalently bound (by one bond only) complementary oligomers and polymers[15] provided a theoretical explanation for experimental facts of strengthening in the complexation ability of polymer chains, when they are covalently attached one to another. It was shown that the loss in translational degrees of freedom is lesser in the case of the block copolymers than in the mixtures of interacting oligomers and polymers. For this reason, the complex formation in the block copolymers took place in a wider range of molecular weights, concentrations and ionization degrees (of PMAAc). It should be noted that in graft copolymers the average number of units between adjacent grafts could be considered as the average length of a reacting block in the main chain.

At the moment, two different groups of IntraPC are known. There are IntraPCs stabilized by electrostatic interactions similar to interpolyelectrolyte complexes. They arise in the so-called "double hydrophilic"[16] block copolymers and graft copolymers containing oppositely charged blocks.[17] These copolymers are called "ampholytic"[18] or "zwitterionic"[19] block and graft copolymers. The second group includes IntraPCs formed by hydrogen bonds[7] in the double hydrophilic copolymers, if their blocks are capable to hydrogen-bonding.

The first investigations of the zwitterionic block copolymers based on poly-2-vinylpyridine and poly(methacrylic acid) (P2VP-*b*-PMAAc) were published by Kamachi et al. in the early 1970s. The evolution of this direction was considered by Lowe and McCormick in details.[19] In recent years, the number of experimental and theoretical studies in this field increases progressively.[20-24] It is stipulated by unique micelle-forming and binding properties of zwitterionic block-type copolymers in aqueous medium,[17,25,26] which can be easily regulated by variation in the chemical nature and relative length of oppositely charged blocks as well as by adjusting pH values and salt concentration in solutions. Due to these properties zwitterionic block-type copolymers have many potential applications as pigment dispersants,[27] membranes in protein isolation and purification procedures,[28] and also as nanocontainers, nanoreactors and drug delivery systems, in the context of encapsulation of different kinds of molecules.[20] The novel perspective field is a development of "smart" zwitterionic stimuli-responsive films onto either flat or spherical surfaces by adsorption[29] or covalent grafting techniques[30-33] followed by their application as sensors and controlled-release forms.[34,35]

Among the first block-type copolymers with intramolecular hydrogen bonds the graft copolymers of polyacrylamide (PAAm) or poly(N-vinyl-2-pyrrolidone) (PVPD) onto poly(vinyl alcohol) (PVA) and polysaccharides were synthesized since 1950s.[36-39] However, the key idea about possible existence of these bonds between the main and grafted chains and also their influence on the properties of graft copolymers based on PVA-*g*-poly(N-vinyl-2-pyrrolidone) appeared only in 1976.[40] Historical aspects of the studies on graft copolymers with intramolecular hydrogen bonds were considered earlier in more details.[7] Development of analogous linear diblock and triblock copolymers with interacting components, such as PEO and PMAAc, began much later in the early 1990s.[41] Now this field develops very actively.

Our interest in the following parts of this chapter will be focused mainly on the second group of block and graft copolymers, forming IntraPCs stabilized by hydrogen bonds. Matrix aspects of synthesis and also their bulk structure, their behavior in solutions and some practically important properties of these copolymers are also highlighted.

### 3. Matrix phenomena in synthesis of block and graft copolymers

The problems related to the synthesis of block and graft copolymers have been widely discussed in the literature in the last decade.[7,16,42-46] It is due not only to a current interest in these polymers and their applications in nanotechnologies but also because of the recent development of novel polymerization mechanisms. These include the processes of regulated or "pseudoliving" radical synthesis such as "Stable Free Radical Polymerization" (SFRP),[47] "Atom Transfer Radical Polymerization" (ATRP)[48,49] and also "Reversible Addition-Fragmentation Transfer" (RAFT)[50] and "Molecular Design via Interchange to Xanthates" (MADIX).[51] At present all these mechanisms together with well-known processes of "living" ionic and ionic-coordination polymerization[52] are widely used for synthesis of block and graft copolymers of regulated molecular structure and narrow molecular-weight distribution (including the copolymers with chemically complementary components). The majority of these processes are carried out in non-aqueous solutions and often using protecting group techniques.[41] That is why it is not surprising that so-called "matrix effects",[53,54] well-known in synthesis of many InterPCs (Chapter 6), were ignored for a long time in the case of synthesis of block and graft copolymers with cooperatively interacting (in a given solvent) polymer components. In order to detect the existence of matrix phenomena in the processes of block and graft copolymerization, special studies were necessary.

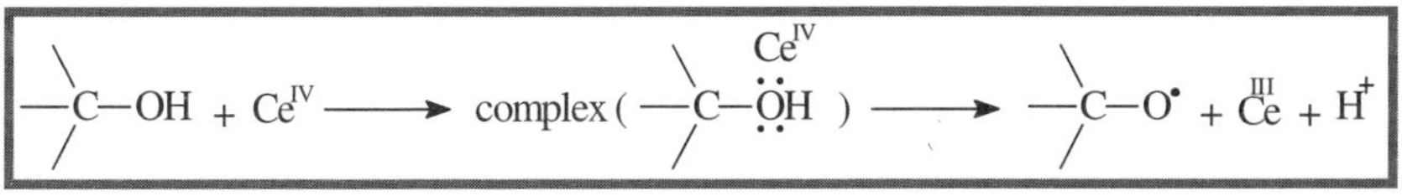

$$\overset{\backslash}{\underset{/}{-}}C-OH + Ce^{IV} \longrightarrow complex( \overset{\backslash}{\underset{/}{-}}C-\overset{..}{\underset{..}{O}}H ) \xrightarrow{Ce^{IV}} \overset{\backslash}{\underset{/}{-}}C-O^{\bullet} + \overset{III}{Ce} + H^{+}$$

**Fig. 1.** "Activation" of PVA hydroxyl groups by $Ce^{IV}$ ions in the graft copolymerization.

The first publication demonstrating the kinetic matrix effect in the course of PMAAc grafting onto starch in aqueous medium by the "grafting from" free radical mechanism refers to 1997.[55] The known redox reaction of $Ce^{IV}$ ions with hydroxyl groups of starch was applied for initiation of grafting polymerization. The rate of the graft propagation was found to be 10 times higher than that of MAAc homopolymerization. Later special kinetic investigations of analogous grafting of PAAm onto PVA initiated by $Ce^{IV}$ ions **(Fig. 1)** were carried out in order to confirm a hypothesis about the matrix character of the graft copolymerization and also to ascertain the influence of the density and length of growing grafts on the corresponding kinetic parameters.[7,56]

Spontaneous complex formation between free PVA and PAAm in aqueous solutions, which was preliminary demonstrated,[57] was used as a criterion of chemical complementarity of these polymers. The positive kinetic matrix effect (2–4 times increase in the rate of graft propagation compared to that of PAAm homopolymerization) and its dependence mainly on the density of the growing grafts were established. When the density of growing daughter chains became too high (~3 PAAm grafts per 100 units of the PVA main chain), the matrix effect disappeared. This important result was attributed to: (i) a decrease in the graft density growth in the average length of the reacting block of the matrix (i.e., the average distance between the adjacent grafts), which is accessible for complex formation with grafted chains, and (ii) an increase in steric hindrance to binding of the grafts with the main chain.

Unfortunately, in many other even relatively recent studies devoted to the grafting of PAAm or PAAc onto PVA or polysaccharides in aqueous medium,[58-61] the possible existence of the matrix effects was not considered.

An assumption about the possible matrix effect in the block copolymerization of PAAc with Pluronic macromolecules has also been

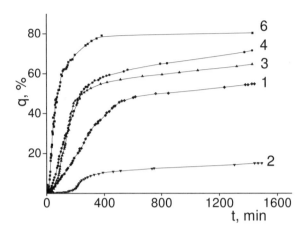

**Fig. 2.** The acrylamide conversion *vs.* time in PAAm homopolymerization *–1* and block copolymerization with PEG of molecular weight $1 \cdot 10^3$ *–2*, $6 \cdot 10^3$ *–3*, $1.5 \cdot 10^4$ *–4* and $4 \cdot 10^4$ *–5*. *T*=20°C.

proposed[62] to explain a very high speed of the process, but this hypothesis was not confirmed experimentally.

Recently, a special investigation of possible matrix effects in the radical block copolymerization of PAAm with PEG of different lengths ($M_v=1 \cdot 10^3 \div 4 \cdot 10^4$), also initiated by $Ce^{IV}$ ions (**Fig. 1**), was carried out.[63] The interaction between free polymeric partners in aqueous medium via hydrogen-bonding was confirmed previously.[64] The results of dilatometry measurements are shown in **Fig. 2** and **Table 1**.

**Table 1.** Kinetic parameters of synthesis of PAAm-*b*-PEO-*b*-PAAm triblock copolymers (TBC) and PAAm homopolymer.

| Polymer | $M_{vPEG} \cdot 10^{-4}$ | $\tau_0$, min | $V_p^{20} \cdot 10^5$, mol·dm⁻³·s⁻¹ | $V_p^{40} \cdot 10^5$, mol·dm⁻³·s⁻¹ | $q$, % |
|---------|------|-------|------|------|------|
| PAAm | - | 15.5 | 1.6 | 1.02 | 53.5 |
| TBC1 | 0.1 | 168.0 | 0.6 ($V_p^{10}$) | - | 14.2 |
| TBC2 | 0.6 | 54.9 | 5.76 | 5.62 | 63.4 |
| TBC3 | 1.5 | 31.4 | 4.62 | 4.19 | 69.5 |
| TBC4 | 4.0 | 25.0 | 15.5 | 7.92 | 80.3 |

$\tau_0$: time of the initial slow step.
$V_p^{20}$ and $V_p^{40}$ : the rates of the block copolymerization and homopolymerization at conversion $q$=20% and 40%.
$q$ : acrylamide conversion after 20 hours.

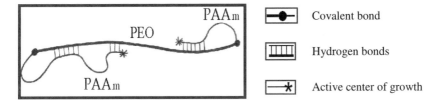

**Fig. 3.** Schematic representation of the matrix block copolymerization of PAAm with PEO at initial stage.

At $M_{vPEG} \geq 6 \cdot 10^3$, the parameters $V_p^{20}$, $V_p^{40}$ and $q$ in the block copolymerization essentially increased as compared with those ones in AAm homopolymerization. Taking into account the same conditions and molar concentrations of the reagents in all syntheses, it was concluded that positive kinetic matrix effect plays an important role in block copolymerization due to the interaction between growing PAAm blocks and PEO chains (matrices) at least on initial stage (**Fig. 3**). A more pronounced effect was observed when longer PEO chains were used. In particular, at $M_{vPEG} = 4 \cdot 10^4$ the rate $V_p^{20}$ of the block copolymerization was already nearly 10 times higher than the analogous value observed for synthesis of PAAm. This result is in good agreement with the well-known fact of increase in the complexation energy and InterPC stability upon lengthening of polymer chains.[12]

Another matrix effect found in this system is an increase in PAAm molecular weight (from $4.5 \cdot 10^4$ to $9.07 \cdot 10^5$), which was almost proportional to $M_{vPEG}$ within $6 \cdot 10^3 - 4 \cdot 10^4$. Thus, the influence of the intramolecular complex formation not only on the copolymerization kinetics but also on TBC molecular structure has been established. The majority of TBC samples synthesized were soluble in water in spite of the formation of IntraPC. It is not surprising, taking into account the availability of the two PAAm blocks in TBC and also their higher molecular weight unlike the PEO block. Small opalescence was observed only in solutions of TBC4 containing the longest PEO block (**Table 1**).

It should be noted that there is a principal possibility for the matrix process development not only along the same PEO chain (**Fig. 3**) but also along the neighboring PEO macromolecules. However, taking into account the above-mentioned theoretical calculations,[15] it can be assumed that the first situation should be accompanied by more gain in free energy of the system than the last one.

## 4. Hydrogen-bonding system in IntraPC stabilization

Among the numerous publications devoted to the block and graft copolymers, which form hydrogen-bonded IntraPCs, there are only few studies, where the structure of the bonds was elucidated and quantification of them at different conditions was carried out. At the same time such studies are of fundamental importance because they allow showing a relationship between a system of bonds and the properties of an IntraPC. Let's consider some specific examples.

**Block copolymers.** Hydrogen-bonding interactions in a series of the diblock copolymers PHEMA-*b*-PVPD (based on poly(2-hydroxyethyl methacrylate) and poly(N-vinyl-2-pyrrolidone)) and corresponding PHEMA/PVPD blends were elucidated by FTIR spectroscopy by Huang and co-workers.[65] In the copolymers, the polymerization degree of PHEMA block was constant ($P_n$=34) but $P_n$ of the second PVPD block changed from 30 to 80. FTIR spectra of thin films of the diblock copolymers and polymer blends and also individual polymers were recorded and analyzed in the regions of $v_{C=O}$ and $v_{OH}$ vibrations. The frequency difference between the free hydroxyl group absorption (3525 cm$^{-1}$) and hydrogen-bonded -OH group absorption of PHEMA was found to be noticeably greater in diblock copolymers ($\Delta v_{OH}$= 175 cm$^{-1}$) than in the polymer mixtures (160 cm$^{-1}$) and individual polymer (145 cm$^{-1}$). Moreover, the fraction of the hydrogen-bonded carbonyl groups of PVPD also turned out to be the highest in the case of diblock copolymers. In such a way, a very important conclusion about stronger hydrogen-bonding interactions in the copolymers, compared to analogous polymer blends, has been drawn. It allowed explaining the specific thermal behavior of PHEMA-*b*-PVPD diblock copolymers as compared to the blends that is discussed in the next section.

Interesting information about the hydrogen-bonding in PVPh-*b*-P4VP diblock copolymers, containing poly(vinylphenol) and poly(4-vinyl-pyridine) of a different relative length, as compared with that one in PVPh homopolymer was received by Kuo *et al.*[66] using FTIR

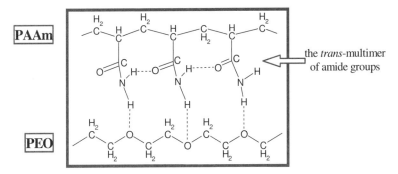

**Fig. 4.** The hydrogen bond system with participation of oxygen atoms of PEO and *trans*-multimers of PAAm amide groups.

spectroscopy. The ratio between molecular molar fractions of PVPh and P4PV corresponded to 29/71, 34/66, 50/50, 57/43 and 90/10. The authors correctly assigned the band of $v_{O-H}$ vibrations at 3125 cm$^{-1}$ to the hydrogen-bonding interactions between PVPh hydroxyl and P4VP pyridine blocks. According to the spectral data, the relative intensity of this band with respect to the bands at 3525 cm$^{-1}$ and 3350 cm$^{-1}$ (corresponding to $v_{O-H}$ vibrations of free hydroxyl groups and self-associated hydroxyl groups of PVPh) is the highest in the PVPh$_{35}$-b-P4VP$_{65}$ sample. It means that the number of H-bonds between both blocks is maximal exactly in the sample pointed. However, the choice of a band at 993 cm$^{-1}$, responsible for vibrations of P4VP, was quite unusual for quantification of H-bonds. As a result, the problem of estimation of the quantity of H-bonds between the blocks upon changing the copolymer content was not correctly resolved.

The temperature resistance of the intramolecular H-bond system in the PAAm-b-PEO-b-PAAm triblock copolymer ($M_{vPEO}=4\cdot10^4$, $M_{vPAAm}= 3{,}18\cdot10^5$) was examined by FTIR spectroscopy in another study.[67] Preliminary investigations of thin films of PEO, PAAm, and TBC, which were performed by FTIR at T=20°C using the computer separation of strongly overlapping bands in the Amide I and Amide II region by the spline method,[68] have revealed the following H-bond sequences between the components in TBC (**Fig. 4**).

This structure was confirmed by the low-frequency shift (in 23 cm$^{-1}$) of the $v_{C-O-C}$ vibration band of PEO in the TBC spectrum as compared

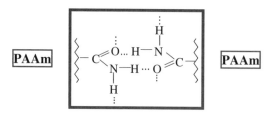

**Fig. 5.** The *cis-trans*-multimers of amide groups arisen between two drawn together segments of the same or different PAAm chains.

with the spectrum of individual PEO (1113 cm$^{-1}$) and also by the essential increase in the effective length of the *trans*-multimers of the amide groups in TBC unlike the individual PAAm. Additionally, a certain reduction in the number of the hydrogen-bonded *cis-trans*-multimers of amide groups (**Fig. 5**) at the transition from PAAm to TBC was shown.

With a temperature increase from 20 to 200°C only a small reduction in the intensities of all the characteristic $v_{C-O-C}$, $v_{C=O}$ (Amide I), $\delta_{N-H}$ (Amide II) and $v_{N-H}$ vibration bands in the TBC spectrum without visible alterations in the band positions took place. Based on the computer processing of spectra recorded at low and highest temperatures (T=35°C and 200°C), insignificant reduction in the number of *cis-trans*- and *trans*-multimers of amide groups (on 9.5% and 14.7% correspondingly) simultaneously with lengthening of the *trans*-associates in 1.35 times was established. These facts pointed a high thermostability of all the types of hydrogen bonds in TBC and led to alternative interpretation of hydrodynamic behavior of TBC solutions under the temperature influence.[67]

The above-mentioned examples dealt with the block copolymers containing only nonionic hydrogen-bonding blocks. In these systems, the quantification of hydrogen bonds by spectroscopy methods only such as FTIR, NMR is possible. But, in the IntraPCs formed by block copolymers composed of nonionic (PEO, PVA, PVPD, polysaccharides) and polyacid (PAAc, PMAAc and others) blocks, this task can be solved simpler, by the potentiometric titration method, similarly to H-bond quantification in the corresponding InterPCs.[10,69] For this purpose, a sample of an homopolyacid, whose the chain length is comparable to the

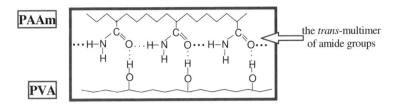

**Fig. 6.** H-bond sequence between the main chain and the grafts in PVA-*g*-PAAm.

analogous polyacid block in the copolymer, is only necessary. Unfortunately, such a simple approach has not yet been applied.

**Graft copolymers.** The system of hydrogen bonds in graft copolymers is of a special interest from the viewpoint of the influence of the graft number (graft density) and length on the process of the intramolecular complex formation and structure of IntraPCs. On the other hand, the hydrogen-bonding interaction between neighboring grafts can be an additional factor of the IntraPC stabilization.

The role of the density and chain length of grafts in the formation of intramolecular hydrogen bonds in PVA-*g*-PAAm graft copolymers was shown in a number of studies.[70-72] Two series of the copolymers: (i) with commensurable graft length ($M_{vPAAm}$~$1\cdot10^5$) but variable number $N$ of the grafts (from 25 to 49), and (ii) with constant $N=9$ but different graft length ($M_{vPAAm}$ rose from $3.72\cdot10^5$ to $5.1\cdot10^5$) were investigated by IR spectroscopy. Special normalization of spectra and their computer processing in the Amide I and Amide II regions were carried out. Note that the molecular weight of the main chain in all PVA-*g*-PAAm samples was invariable ($M_{vPVA}=8\cdot10^4$). It was shown that the structures of the graft copolymer macromolecules are stabilized by an extensive hydrogen bond network including H-bonds between the main chain and the grafts (**Fig. 6**) and also H-bonds between segments of the grafted chains such as *cis-trans*-multimers of amide groups (**Fig. 5**).

An increase in $N$ led to substantial changes in the hydrogen bond network (**Fig. 7**). Indeed, the number of H-bonds such as *the main chain–the grafts* was relatively greater at small $N=25$ value (case **a**) and then diminished with $N$ growth up to 49. Simultaneously, the quantity of H-bonds of *the graft–the graft* type increased and achieved maximum

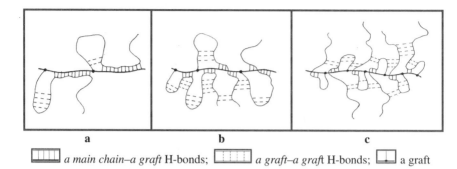

a *main chain–a graft* H-bonds; a *graft–a graft* H-bonds; a graft

**Fig. 7.** Schemes illustrating the hydrogen bond system in PVA-*g*-PAAm in the cases of relatively small (**a**), high (**c**) and intermediate (**b**) graft density.

value at $N$=49 (case **c**). At the same time the case **b,** reflecting a large number of both H-bond types in PVA-g-PAAm was also identified at an intermediate $N$=31 value. This situation was attributed to an enhancement of the steric hindrances with $N$ growth for realization of thermodynamic affinity between PVA and PAAm.

Interesting effects of partial "detachment" and stretching of grafted chains from the main chain (**Fig. 8**), caused by an increase in the molecular weight of the grafts until reaching a certain critical value ($M^* \approx 4.3 \cdot 10^5$ at $N$=9) has been revealed by IR spectroscopy studies of the second series of PVA-*g*-PAAm samples. In fact, the phenomenon of destruction of the IntraPC structure in the copolymers at $M_{PAAm} > M^*$ was observed. The reason for this is probably the overlapping and interaction of sufficiently long grafted chains at large distances from the main chain. These results created a basis for correct interpretation of all the peculiarities in the PVA-*g*-PAAm behavior in bulk and in solution.

A more complicated system of H-bonds was established by FTIR spectroscopy in PVA-*g*-P(AAm-*co*-AAc) graft copolymers with different content of AAc units ($A$=18 and 32 mol %).[73] The copolymers were synthesized by alkaline hydrolysis of the parent PVA-*g*-PAAm sample having the following characteristics: $M_{vPVA}$=9.4$\cdot$10$^4$, $M_{vPAAm}$= 1.68$\cdot$10$^5$, $N$=9, and $A$=1 mol %. Normalization and processing of FTIR spectra of the copolymers was carried out in accordance with the methodology reported earlier.[72] The main purpose of these studies was to explain the unusual phenomenon of reduction in the copolymer solubility in water

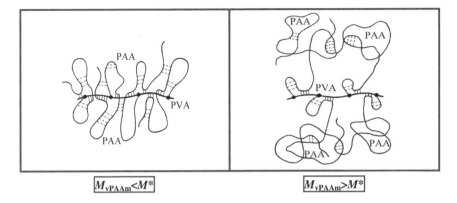

$$M_{vPAAm} < M^*$$ $$M_{vPAAm} > M^*$$

**Fig. 8.** Schematic representation of the effect of "detachment" and stretching of grafts from the main chain since some critical molecular weight of them.

until its complete loss when the part of the ionic AAc units in the grafted chains increased from $A=1$ to 66 mol %. Besides, the characteristic bands, well known and described in the spectrum of the initial graft copolymer, the appearance of two novel $v_{C=O}$ and $v_{O-H}$ vibration bands at 1711 cm$^{-1}$ and 2200 cm$^{-1} < v < 2800$ cm$^{-1}$ in the spectra of modified graft copolymers was identified. Their intensities increased with $A$ growth. These facts and other alterations in PVA-$g$-P(AAm-$co$-AAc) spectra, unlike the spectrum of the unmodified graft copolymer, pointed out the formation of mixed cyclic dimmers between –COOH and –CONH$_2$ groups of the grafts (**Fig. 9**).

These H-bond structures were shown to be energetically stronger than the *cis-trans*-multimers of the amide groups (**Fig. 5**) and also the H-bonds between PVA and PAAm chains (**Fig. 6**). Moreover, the higher the $A$, the greater the number of these H-bond structures. In this case, the hydrogen-bonding interactions between grafted chains in the macromolecules of the modified graft copolymers (in fact additional IntraPC formation by means of the grafts) should be intensified with $A$ growth. It may probably lead even to their separation from the main chain, due to the competition. Thus, strengthening in the hydrogen-bonding interactions between modified grafts was considered to be the main reason for the decrease in the PVA-$g$-P(AAm-$co$-AAc) solubility in water, when the fraction of AAc units grew.

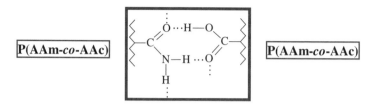

**Fig. 9.** The structure of mixed cyclic dimmer between carboxylic and amide groups.

The existence of strong hydrogen-bonding interactions in the grafted chains was also assumed in the PVA-*g*-poly(N-isopropylacrylamide-*co*-AAc) and PVA-*g*-P(NIPAAm-*co*-MAAc) graft copolymers, on the basis of their swelling behavior upon increase in temperature.[74] Unfortunately, this assumption was not confirmed by any method. Additionally, possible hydrogen-bonding interactions such as *the main chain–the grafts* was not taken into consideration.

## 5. Peculiarities of the bulk structure of IntraPC

The structure of the block and graft copolymers with cooperatively interacting polymer components in the bulk state has attracted less attention than their behavior in solutions. At the same time, such investigations are of great importance from several viewpoints. On the one hand, these studies allow confirming the fact of intramolecular complex formation and relating the structural features observed with the molecular architecture of the corresponding block and graft copolymers. On the other hand, these studies can demonstrate some differences between bulk structures of the copolymers (IntraPC) and polymer blends (InterPC) formed by the same polymer components at the same ratios. Finally, the studies of morphology are very important for practical application of IntraPC as separating membranes,[74,75] sorbents,[76] solid polymer electrolytes,[77,78] and also as hydrogels, nanospheres and films in drug-delivery systems.[8,14,79-83]

At present, numerous studies, concerning the bulk structure of the hydrogen-bonded InterPC forming polymer blends, and showing a difference from the structure of the ordinary miscible polymer blends, are published.[3,84,85] According to them, some cases in the blend morphology (at a given temperature) depending on properties (amorphous or crystalline) of individual polymers and their ratios should be distinguished. Let's consider them from a general view, in order to show further peculiarities of the bulk structure of IntraPCs. Note that our consideration does not concern the blends including block copolymers.

**Both polymers forming InterPCs are amorphous.** In this case their blends in a bulk state were found to be amorphous and homogeneous on molecular level irrespective of the polymers ratio. The amorphous structure and homogeneity of these materials were confirmed by different methods. In particular, DSC thermograms showed an appearance of a single glass transition temperature $(T_g)$ at all polymer compositions. The glass transition temperature changed between $T_{g1}$ and $T_{g2}$ of the pure polymers as in the following systems: phenoxy polymer/poly(methyl acrylate-*co*-methyl methacrylate) and PAAc/poly-(2-hydroxyethyl vinyl ether) (PHEVE),[86,87] or turned out to be even higher than $T_g$ of the most rigid-chain component as in poly(vinyl-phenol)/poly(4-vinylpyridine) (PVPh/P4VP) system.[88] In any case, the dependences of $T_g$ on the composition for the blends of cooperatively interacting polymers significantly deviates from the weight-average law proposed by Fox[89] for the ordinary compatible polymer mixture under the assumption that the heat capacity jumps in the glass transitions for both components are close ($\Delta C_{p1} \approx \Delta C_{p2}$):

$$\frac{1}{T_g} = \frac{w_1}{T_{g1}} + \frac{w_2}{T_{g2}} \qquad (1)$$

where $w_1$ and $w_2$ are the weight fractions of one and another polymers in a blend.

Among the others relationships suggested by Gordon-Taylor, Couchman-Karasz, Kowacs, Brown-Kowacs etc.[90] to describe the $T_g$-composition dependence in the miscible polymer blends, the Kwei equation is used often for the blends with specific interchain interactions such as hydrogen bonds[3]:

$$T_g = \frac{w_1 \cdot T_{g1} + k \cdot w_2 \cdot T_{g2}}{w_1 + k \cdot w_2} + q \cdot w_1 \cdot w_2 \qquad (2)$$

where $k$ and $q$ are the fitting parameters. It is assumed that $q$ reflects the strength of hydrogen-bonding in a system. In addition, the Couchman modified equation was also applied in structural studies of some analogous blends.[84] It should be noted that a single $T_g$ based on DSC analysis signifies the polymer mixing in a blend on a scale of about 20–40 nm.[91]

Appearance of a single glass transition in the InterPC structure, which $T_g$ value was intermediate between those of the pure components, has been confirmed also by the dynamic mechanical thermal analysis (DMTA) in PAAc/methylcellulose blends.[92]

Polymer miscibility in the blends of InterPC forming amorphous polymers on a scale of less than 20 nm was also demonstrated by solid-state $^{13}$C NMR investigation using cross-polarization and a magic angle sample spinning (CP/MAS) technique.[66,93] Two polymer pairs: PVPh/P4VP and poly[styrene-*co*-*p*-(1,1,1,3,3,3-hexafluoro-2-hydroxy-propyl)-α-methyl styrene [PS(OH)]/PMMA were used in these experiments. Important parameter such as the spin-lattice relaxation time in the rotating frame ($T_{1\rho}^{H}$) was determined through the delayed-contact $^{13}$C CP/MAS experiments according to the following equation:

$$M_\tau = M_0 \cdot \exp[\, -\tau / T_{1\rho}^{H} \,] \quad , \tag{3}$$

where $\tau$ is the delay time applied in the experiment and $M_\tau$ is the corresponding resonance intensity. The $T_{1\rho}^{H}$ value was found from the slope of the experimental linear dependence of $\ln(M_\tau/M_0)$ *vs* $\tau$. A single composition dependent $T_{1\rho}^{H}$ was obtained for InterPCs suggesting that they are homogeneous to a scale where spin diffusion occurs within the time $T_{1\rho}^{H}$. The values for InterPCs were noticeably lower than for the corresponding individual polymer components.[66]

The compatibility on molecular level of the InterPC forming amorphous polymers in the blends was also revealed by other methods such as SEM (PAAc/PHEVE blend),[87,92] TEM, and non-radiative energy transfer (NRET) fluorospectroscopy (PS(OH)/PMMA blend).[3,94]

Note that similar behavior was found for the blends of the amorphous oppositely charged polyelectrolytes (PAAc/poly(ethylene imine), PMAAc/chitosan, PAAc/poly(ethylenepiperasin) and others), which formed InterPC by electrostatic interactions.[4]

**One of the InterPC components is crystallizable.** In this case, a bulk structure of the blends essentially depended on polymer composition and InterPC architecture. Actually, when InterPC formation occurred between separate macromolecules and the blend composition was close to the stochiometric or so-called the "characteristic" content ($\varphi_{char}$, base-mol$_1$/base-mol$_2$) of InterPC (at which both polymers were quantitatively bound with each other),[10] the blend structure was amorphous and homogeneous on molecular level. The crystallizable component lost its ability to crystallize due to strong complex formation. Such a situation was observed in the following systems: PVA/PAAc and PVA/PMAAc ($\varphi_{char}=1$) by DSC, $^{13}$C CP/MAS NMR and SEM methods,[95-97] PEO/PAAc ($\varphi_{char}=1$) by DSC,[98] PEO/PMAAc ($\varphi_{char}=1$) by WAXS and SAXS methods,[99] PVA/PVPD by DSC, DMTA and $^{13}$C CP/MAS NMR,[100-103] and some others.

The bulk structure of these blends remained amorphous and homogeneous (with a single $T_g$ and without crystalline regions) also in the range $\varphi < \varphi_{char}$, when there was an excess of the amorphous component. Thus, complete miscibility between the excessive amorphous polymer and the InterPC in the blend was revealed. Another situation was observed in the range $\varphi > \varphi_{char}$, when the crystalline component was in excess. In this case, since certain $\varphi$ value in addition to amorphous regions of complete polymer miscibility (which ones showed a single $T_g$) also the crystalline regions of the excessive polymer arose in the blend structure.[96,100,101,103,104] Their melting temperature ($T_m$) was found to be lower than that of the pure crystalline polymer. Such effect, known as the melting temperature depression, is caused by strong polymer interactions in the amorphous regions and is described by Nishi and Wang equation (4)[105]:

$$\Delta T_m = T_m^0 - T_m = -T_m^0 \cdot (V_{u1} / \Delta H_{u1}) \cdot B \cdot \varphi_2^2 \qquad (4)$$

where 
$$B = R \cdot T_m^0 \cdot (\chi_{12} V_{u2}) \qquad (5)$$

Here, the subscripts 1 and 2 designate the crystalline and amorphous polymers, $T_m^0$ is the melting temperature of the pure crystalline polymer, $T_m$ is the melting temperature of the blend, $V_u$ is the molar volume of the repeating units in polymers, $\Delta H_u$ is the enthalpy of fusion per mole of

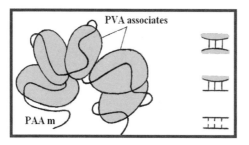

**Fig. 10.** Structure of InterPC formed by PVA and PAAm.

repeating units, $B$ is the interaction energy density of two polymers, $R$ is the gas constant and $\chi_{12}$ is the thermodynamic interaction parameter. Using this approach, based on DSC measurements, the thermodynamic affinity between polymers in amorphous regions in the structure of many InterPCs and ordinary polymer mixtures has been evaluated.[91,103]

At the same time, the cases of InterPC formation between macromolecules of the amorphous polymer and self-associates of the crystallizable polymer in solution, in particular, in the PVA/PAAm system[57] (**Fig. 10**), are also known. The InterPC formed by PVA and PAAm had a constant composition ($\varphi_{char}=5$ base-mol$_{PVA}$/base-mol$_{PAAm}$ at $M_{vPVA}=4\cdot10^4$ and $M_{vPAAm}=4.4\cdot10^6$) irrespective of the polymer ratio.[64] It resulted in appearance of both completely miscible amorphous regions and also small crystalline domains in the bulk structure of PVA/PAAm blend even at $\varphi=\varphi_{char}$. The DSC analysis of these blends showed the presence of a single $T_g$, and also $T_m$, whose value was lower as compared to the melting point of pure PVA. Such morphology can represent a totality of micro- or nano-size crystalline particles of PVA distributed in homogeneous matrix formed by compatible PVA and PAAm segments. However, the most interesting effect was observed when $\varphi<\varphi_{char}$ (i.e., in excess of PAAm). Just as the state of crystalline PVA domains did not change (their parameters $T_m$ and $\Delta H_m$ remained the same), in the amorphous regions a phase separation was also displayed. Besides the regions of complete compatibility of PVA and PAAm, other regions composed of PAAm chains only were appeared. In fact, an excess of PAAm was immiscible with the InterPC. Such a conclusion was drawn after the analysis of the two glass transitions in the corresponding DSC thermograms.

Unlike InterPCs, formed by covalently unbound polymer components, block and graft copolymers with chemically coupled polymer partners form the IntraPC. The covalent bonds intensify the intramolecular interactions and result in a stronger dependence of the IntraPC bulk structure on the copolymer architecture. Unfortunately, only a few studies were reported in this area. There is limited information about the influence of the number and the relative length of the reacting blocks and also of the grafted chain density (in graft copolymers) on the bulk morphology of these polymers. At the same time, most of these questions were elucidated theoretically[106-110] and experimentally[110-115] in the case of block and graft copolymers with thermodynamically incompatible components. Special types of the microphase-separated or super-crystalline structures formed by block and graft copolymers with immiscible polymer components, even in amorphous state, were described. Below, we will consider the structure of the IntraPCs formed by block and graft copolymers.

**Block copolymers.** An interesting recent study[66] was focused on the evaluation of the bulk structure of a series of PVPh-*b*-P4VP diblock copolymers, containing only amorphous components, in comparison with the corresponding polymeric blends. Five copolymer samples mentioned above (see Sec. 4) with variable relative length of the blocks and also PVPh and P4PV homopolymers were synthesized and characterized. They had $M_n=(1\div2.2)\cdot10^4$ and narrow polydispersity indexes, PDI=$1.05\div1.17$. The existence of strong H-bonds between both polymers in the blends (InterPC) and in the copolymers (IntraPC) was proved by FTIR spectroscopy. DSC and $^{13}$C CP/MAS NMR analyses of the copolymers and polymer mixtures were carried out in order to examine the miscibility between the polymers on a few dozens nanometer scale.

Both the polymer blends and the diblock copolymers showed a homogeneous amorphous structure in all composition regions but the levels of polymer mixing in the presence or absence of covalent bonds between them were different. A single glass transition was observed in all DSC thermograms and its $T_g$ values were practically in all cases higher than those of the most "hard" component ($T_{gPVPh}$=180°C). Nevertheless, the positive $T_g$ deviations were essentially greater for the diblock copolymer samples but not for the blends. Processing the DSC data according to the Kwei equation (2) led to the values $q$=100 and 185 for the blends and copolymers, respectively. These results pointed to stronger interaction and mixing in covalently bound polymeric partners. An analogous conclusion was drawn on the basis of $^{13}$C solid-state NMR data. In all experiments the $T_{1\rho}{}^H$ values found with the help of equation (3) were noticeably lower for the block copolymers compared to the blends. Using $T_{1\rho}{}^H$ values, the upper spatial scale of the spin-diffusion path length $L$ was estimated:

$$L = (6D \cdot T_{1\rho}^{H})^{1/2} \qquad (6)$$

where $D$, typically about $10^{-16}$ m$^2 \cdot$s$^{-1}$, is the effective spin-diffusion coefficient, which depends on the average proton to proton distance and the dipolar interaction. In such a way, the upper limit of the domain size was calculated to be 2 nm for the blends and 1.5 nm for the copolymers. Therefore, the degrees of homogeneity of the amorphous structures in the diblock copolymers were relatively higher than those in the mixtures.

The study dealt with poly(hydroxyethyl methacrylate) (PHEMA)-*b*-PVPD diblock copolymers and blends of PHEMA and PVPD homopolymers led to similar conclusions. [65] Two amorphous, chemically complementary, polymer components were used and analogous experimental approaches and techniques were chosen. However, in this case the length of one of the blocks (PHEMA) was constant while the length of the other block was varied (see Sec. 4). Molecular weights of the copolymers changed in the range of $M_n$=(0.82÷1.4)·$10^4$ and PDI was relatively small ($M_w/M_n$=1.25-1.32). The DSC analysis established the composition-dependent single $T_g$'s for both the copolymers and the blends, implying that polymer segments formed an homogeneous amorphous phase at all composition regions, due to hydrogen-bonding interactions. Single $T_g$'s detected were significantly higher than those predicted by the Fox equation (1). In addition, the positive deviations were found to be much higher for the block copolymers than for the polymer mixtures. By fitting DSC data to the Kwei equation (2), two

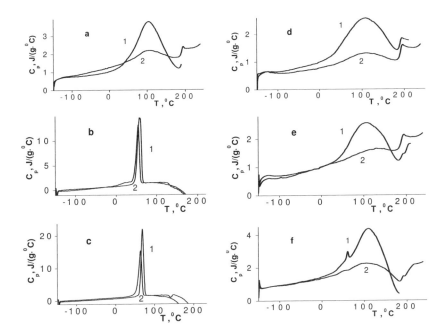

**Fig. 11.** The dependences of specific heat capacity on temperature for PAAm with $M_v = 6.3 \cdot 10^5$ (**a**), PEG with $M_v = 3 \cdot 10^3$ (**b**) and $4 \cdot 10^4$ (**c**), and TBC with $M_{vPEO} = 3 \cdot 10^3$ (**d**), $4 \cdot 10^4$ (**e**) and $1 \cdot 10^5$ (**f**). The 1-st *–1* and 2-nd *–2* scans.

values of $q=80$ and 185 were obtained, which reflected the strength of hydrogen-bonding consistently in the mixtures and block copolymers. These facts led to the conclusion about a higher level of polymers' mixing in the amorphous structure of the copolymers in comparison with the blends. Further $^{13}$C CP/MAS NMR studies have demonstrated full correlation with DSC analysis. Indeed, the upper limit of the domain size $L$ (established by determining $T_{1\rho}^{H}$ and using the relationship (6)) was equal to 1.2–1.6 nm for the blends and 0.9–1.5 nm for the copolymers that fully confirmed the above conclusion.

The morphology of the IntraPCs formed by PAAm-*b*-PEO-*b*-PAAm triblock copolymers, with a crystallizable central block, was tested in the studies.[68,116,117] Seven TBC samples with growing $M_{vPEO}$ in the range of $3 \cdot 10^3 \div 1 \cdot 10^5$ and consistently increasing $M_{vPAAm}$ from $4.5 \cdot 10^4$ to $9.07 \cdot 10^5$, pure PEG and PAAm samples, and also two PEG/PAAm blends, which compositions corresponded to those of TBCs with $M_{vPEO} = 3 \cdot 10^3$ and

$4 \cdot 10^4$, were studied by DSC and WAXS. All the polymer samples were freeze-dried from aqueous solutions. Some DSC thermograms are shown in **Fig. 11**.

These results demonstrate an intense endothermic peak of water evaporation and a single glass transition ($T_g$=190.9°C) for pure PAAm (**a**), an intense endothermic melting peak ($T_m$=58÷70°C, increasing with $M_{vPEG}$ growth) for pure PEG (**b,c**), a weak endothermic melting peak of PEO in TBC, appearing only at $M_{vPEO}$=1·10$^5$ (**f**), and finally, two glass transitions in TBC structure, which are displayed since $M_{vPEO}$=1.5·10$^4$ at first in the 1st runs only (**e**) and then in the 2nd runs also (**f**). In all cases, a separate glass transition for PEO segments ($T_g$=-57 °C according to literature data) was not displayed. Using the DSC data, the main parameters of glass transitions ($T_g$, $\Delta T_g$, $\Delta C_p$) in amorphous regions and melting process ($T_m$, $\Delta T_m$, $\Delta H_m$) in PEO crystalline regions and also the crystallinity degree, $X_{cr}$, of PEO chains in homopolymers, blends and TBC were determined. For the copolymers $X_{cr}=\Delta H_m/(\Delta H_m^0 \cdot w_1)$, where $\Delta H_m$ and $\Delta H_m^0$ are the melting enthalpies for TBC and 100% crystalline PEO homopolymer consequently but $w_1$ is the weight fraction of the PEO blocks.

The DSC analysis has showed high values of $X_{cr}$=80÷88% for the initial PEG samples but in TBCs either they completely lost their ability to crystallize (at $M_{vPEO}$<1·10$^5$) or sharply reduced their crystallinity degree (in 2.8 or 7.9 times at $M_{vPEO}$=1·10$^5$ according to the data of 1-st or 2-nd runs in **Fig. 11**, **f**). In the last case, a noticeable $T_m$ depression ($\Delta T_m$=10°C) was also observed. Two possible reasons for these effects were analyzed: (i) IntraPC formation by H-bonding interaction of PEO and PAAm blocks (**Fig. 4**) and (ii) a special location of the crystallizable PEO blocks between two long amorphous PAAm blocks in TBC. The second factor was related to the known phenomena of crystallinity reduction or even its loss by a crystallizable polymer, covalently-bound with one and especially two blocks of another thermodynamically immiscible amorphous polymer, having a molecular weight higher than some critical value.[110,111,118] However, taking into account the analogous effects in both PEO/PAAm blends, namely, $X_{cr}$ reduction to 8%–20% and $T_{mPEG}$ depression to 5°C–8°C in the initial blends (data of the 1st scans) and practical disappearance of the crystalline phase in the blends passed through a molten state (data of the 2nd scans), the conclusion about a predominant contribution of the IntraPC formation in a sharp change of the PEO crystalline properties in TBC was drawn. A stronger reduction in PEO crystallinity in TBC in contrast to the blends was also pointed out.

The situation in the amorphous regions of the copolymer structure was also discussed. The appearance of two glass transitions in the 1st runs of TBC thermograms at $M_{vPEO} \geq 1.5 \cdot 10^4$ pointed on microphase separation in the amorphous regions of the initial structure. Their glass transition temperatures, $T_g^1 = 187.0 \div 188.5$°C and $T_g^2 = 202.0 \div 204.6$°C, were found to be lower and higher than $T_{gPAAm}$, respectively. At the same time both experimental $T_g^1$ and $T_g^2$ exceeded $T_g$ values predicted by both Fox equation (1) and Couchman-Karasz equation (7), which is used for $T_g$-composition analysis of compatible polymer blends with specific interactions:[90]

$$\ln\left(\frac{T_g}{T_{g1}}\right) = \frac{w_2 \cdot \Delta C_{p2} \cdot \ln\left(T_{g2}/T_{g1}\right)}{w_1 \cdot \Delta C_{p1} + w_2 \cdot \Delta C_{p2}} \tag{7}$$

Here, $\Delta C_{p1} = 0.25$ J·(g·degree)$^{-1}$ and $\Delta C_{p2} = 0.55$ J·(g·degree)$^{-1}$ are the heat capacity jumps for one (PEO) and another (PAAm) pure component in a blend. These calculations also suggested that strong interactions between the blocks in TBC existed. The $T_g^1$ and $T_g^2$ parameters in the copolymers were attributed to the regions of miscibility of polymer components and domains formed by excessive PAAm segments, not interacting with PEO. The number of such excessive segments in TBCs increased with increase in $M_{vPEO}$, that is why the value of the second heat capacity jump ($\Delta C_p^2$) in the copolymer series also grew. It is important that the bulk structure of all the copolymers, passed through a molten state (excluding only the sample with $M_{vPEO} = 1 \cdot 10^5$), became fully homogeneous, which is clearly seen from DSC thermograms (**Fig. 11**, the 2nd scans) showing a single glass transition. Its $T_g = 188.0 \div 190.6$°C was close to the $T_g^1$ value in the corresponding microphase-separated TBC sample. Such behavior of the IntraPC forming TBC was directly opposite to that of block and graft copolymers with thermodynamically immiscible polymer components. For the later ones, a microphase separation in the bulk structure is only intensified after their melting.[119]

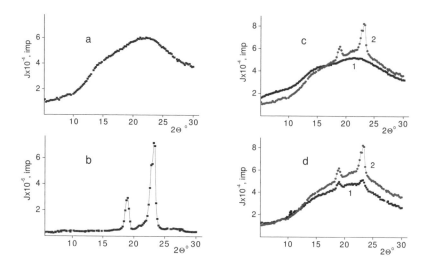

**Fig. 12.** WAXS intensities *vs* scattering angle for PAAm with $M_v$=6.3·10$^5$ (**a**), PEG with $M_v$=6·10$^3$ (**b**) and also TBC with $M_v$=6·10$^3$ (**c**) and 1·10$^5$ (**d**). Experimental data $-1$ and additive curves calculated for polymer blends of the same compositions as in corresponding copolymers $-2$. $T$=25°C.

Inhomogeneity of amorphous regions characterized by two glass transitions was revealed also in the initial bulk structure of both PEG/PAAm blends but that one disappeared in the blends passed through a molten state. In contrast to the phase-separated structure of the first blend (containing PEG with $M_v$=3·10$^3$), the initial structure of the corresponding TBC sample was fully homogeneous (**Fig. 11, d**). This result and also the above-mentioned effect of a sharper reduction in the PEO crystallinity in TBC than in the blends confirmed stronger interaction and mixing of covalently bound PEO and PAAm chains. As to the difference between TBC bulk structures obtained from a solution and through a molten state, one can assume that this effect is related to a special aggregation (self-assembly) of the IntraPC forming copolymers in an aqueous medium (see Sec. 6).

The state of the TBC structure at the level of the nearest surrounding of separate polymer chains was examined by wide-angle X-ray scattering (WAXS). The examples of diffractograms are presented in **Fig. 12**. WAXS profile for PAAm (**a**) contained two diffuse overlapping maximums that indicated the existence of two systems of planes of a paracrystalline lattice in an amorphous polymer structure. The first maximum with lower intensity ($2\Theta \sim 15°$) reflected a side periodicity in the PAAm chain location but the second one having higher intensity ($2\Theta = 22.1°$) was related to the periodical disposition of the flat H-bonded *cis*-dimmers of PAAm amide groups in the *cis-trans*-multimers (**Fig. 5**).

WAXS profiles for all PEG were identical (**b**). They showed two known crystalline peaks at $2\Theta = 19.0°$ and $23.1°$, corresponding to an ordinary PEO helical (7/2) crystalline form. For all the TBC samples with $M_{vPEO} < 1 \cdot 10^5$ the crystalline peaks of PEO in the difractograms were absent (**c**, curve 1), whilst the appearance of these peaks was predicted by the additive curves (**c**, curve 2). Similar to the pure PAAm the TBC difractograms demonstrated two diffuse overlapping maximums with reduced relative intensity of the second one (at $2\Theta = 22.1°$) that pointed on the amorphous structure of the copolymers. A reduction in the relative intensity of the second maximum in the TBC profiles was in full agreement with the partial destruction of *cis-trans*-multimers of amide groups in TBC as compared with pure PAAm, which was established by FTIR (see Section 4). Weak crystalline peaks of PEO over amorphous maximums were displayed only in WAXS profile of TBC with a PEO block of maximum length ($M_v = 1 \cdot 10^5$). Due to strong overlapping of the amorphous maximums, the effect of PEO and PAAm interaction on the side periodicity in the macrochain disposition (as it was made in the study[99] for the InterPC PEO/PMAAc) could not be shown. But the correlation between WAXS and DSC data in the context of full loss of crystallinity by PEO chains with $M_v < 1 \cdot 10^5$ in TBC was obvious.

**Graft copolymers.** They have a more complicated architecture than linear block copolymers, that is why the influence of such factors as the number $N$ (density) and length of grafts, and also possible crystalline properties either of the main or grafted chains on the copolymer structure should be established. Unfortunately, in many studies concerned the IntraPC forming graft copolymers their molecular parameters were not determined. It does not allow identifying the role of every factor separately and comparing bulk structures of graft copolymers analogous in architecture, including immiscible and extremely miscible polymer components. The later comparison would be very desirable because of the well-known phenomenon of increase in the mutual mixing of even immiscible main and grafted chains due to the covalent bonds between them.[7]

The overwhelming majority of structural studies in this area is dedicated to the IntraPC forming graft copolymers with amorphous and crystalline components. The graft copolymers, such as poly(*n*-buthylacrylate-*co*-acrylic acid)-*g*-PEO,[120] with a variable fraction of AAc units in the amorphous main chain and uniform crystallizable PEO grafts, whose number *N* per macromolecule changed from 6 to 16, is one of the examples. They were synthesized by the "grafting through" technique[45] using *n*-buthylacrylate, acrylic acid, and the methacrylate terminated PEO macromonomer. The formation of H-bonds between PEO and poly(AAc) units and the crystalline properties of these copolymers were studied. The graft copolymers demonstrated a crystallinity diminishing with increase in AAc content, which was adequate to decrease in *N*.

Structural studies of P(MA-*co*-AAm)-*g*-MEPEO graft copolymers, synthesized by the "grafting through" method also, using AAm and the methacrylate terminated methyl ether of PEO as macromonomer were carried out in the study.[78] The copolymer structure did practically not contain MEPEO crystalline regions and exhibited a single glass transition with a relatively low $T_g$. Obviously, such a behavior was related to hydrogen-bonding interaction between the main and the grafted chains.

Graft copolymers of St-*g*-PAAm and Am-*g*-PAAm, with crystallizable main chains and amorphous grafts, were synthesized by $Ce^{IV}$ initiated grafting of PAAm "from" two starch and three amylose backbones of different length at the constant molar ratio $[Ce^{IV}]/[AAm]= 2.14 \cdot 10^{-4}$ and variable polymerization times.[121] In such a way, the authors changed the density and length of the grafts. However, both parameters in the final copolymers were not controlled. Their morphology was studied by WAXS and strong mutual mixing of the chemically complementary main and grafted chains was observed in all the copolymers. It resulted in a full loss of the St and Am chains ability to crystallize. Indeed, WAXS profiles of the copolymers demonstrated only one diffuse amorphous maximum at $2\Theta=17.0 \div 17.5°$ in contrast to weak crystalline peaks existing in the difractograms of pure St and Am samples.

The crystalline properties of the main chain in two series of PVA-*g*-PAAc and PVA-*g*-PMMA graft copolymers were probed by DSC.[122] Three PVA samples of unknown molecular weight were grafted with PAAc and PMMA chains using the "grafting from" method with $Ce^{IV}$ initiation. Unfortunately, the so-called "grafting degree" (the relative increase in PVA weight after grafting) was the only copolymer molecular parameter controlled in this study. Due to this reason, DSC data could not be well interpreted. However, the authors reported the following effects: (i) a strong $T_m$ depression in the copolymers, in contrast with the initial PVA samples, up to $\Delta T_m = 45°C$ for PVA-*g*-PAAc and 65°C for PVA-*g*-PMMA, and (ii) a lowering in $\Delta H_m^0$, especially sharp in the case of the second graft copolymer type. These effects confirm the decrease in the crystallinity of PVA in the copolymer structure due to the strong interaction between the main and the grafted chains in the amorphous regions.

The influence of both the number $N$ (density) of amorphous grafts and their molecular weight (chain length) on the bulk structure of the IntraPCs forming graft copolymers with a crystallizable main chain was established for the first time by detailed DSC studies of two series of PVA-*g*-PAAm graft copolymers.[7,123,124] The system of hydrogen bonds in these copolymers was considered in Section 4, but the corresponding DSC thermograms are shown in **Fig. 13**.

The first important result obtained from these experiments was the complete disappearance of the PVA crystalline properties in all the copolymers. However, the amorphous copolymer structure remained completely homogeneous (on a scale of some dozens of nm) with a single glass transition only in a definite region of $N$ and $M_{vPAAm}$ values, which were lower than some critical $N^*$ and $M_v^*{}_{PAAm}$ numbers, correlated with each other. Thus, the second significant result from DSC analysis has been achieved. In fact, in PVA-*g*-PAAm series with comparable graft length ($M_{vPAAM} \sim 1 \cdot 10^5$) and increasing graft number two glass transitions in the initial copolymer structure (in the 1st DSC scans) at $N > N^* \approx 25$ were identified (**Fig. 13**, **a**, curves *2-4*).

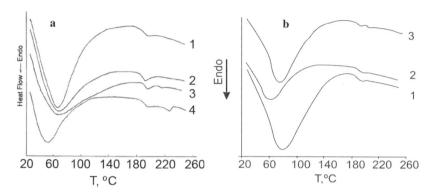

**Fig. 13**. DSC thermograms (the 1-st runs) for two PVA-*g*-PAAm series ($M_{vPVA}=8\cdot10^4$) with different number of grafts (**a**): $N=4$ ($M_{vPAAm}=1.12\cdot10^6$) –*1*, 25 ($M_{vPAAm}=1.21\cdot10^5$) –*2*, 31 ($M_{vPAAm}=0.63\cdot10^5$) –*3* and 49 ($M_{vPAAm}=0.78\cdot10^5$) –*4*, and also with variable graft length and $N=9$ (**b**): $M_{vPAAm}=3.7\cdot10^5$ -*1*, $4.3\cdot10^5$ –*2* and $5.1\cdot10^5$ –*3*.

Both glass transitions were displayed also in the 2nd scans of *3* and *4* curves showing a clear microphase separation in the corresponding copolymer structures passed through a molten state. The analogous picture was observed in the PVA-*g*-PAAm series with a constant, relatively small $N$ but with increasing $M_{vPAAm}$ (**b**, curves *1-3*). In this case, the appearance of two glass transitions, in both the 1st and 2nd DSC scans, took place at $M_{vPAAm} > M_v*_{PAAm} \approx 4,3\cdot10^5$ (curve *3*). At the same time, for the additional copolymer sample, having the smallest graft density ($N=4$) and the greatest graft length ($M_{vPAAm}=1.12\cdot10^6$), only one glass transition was identified in both the DSC scans, revealing a completely homogeneous amorphous structure (**a**, curve *1*). The later result meant that the lower $N$ allows higher $M_{vPAAm}$ to be achieved with the copolymer structure remaining amorphous and homogeneous. In contrast, the higher $N$ will allow the smaller $M_{vPAAm}$ to maintain a homogeneous copolymer structure.

The $T_g^1$ values (186.0÷190.5°C) for all the microphase-separated PVA-g-PAAm samples were essentially higher than those found by $T_g$-composition analysis according to the Fox equation (1), but approached to $T_g$=188.6÷189.3°C, predicted by the Couchman-Karasz equation (7).[71] They were ascribed to the glass transitions in the regions of full miscibility of the H-bonded PVA and PAAm segments. Unlike this, the $T_g^2$ and $\Delta C_p^2$ values, which noticeably increased with increasing $N$ (**Fig. 13, a**, curves 3,4), were attributed to the glass transitions in the domains formed by the very rigid PAAm segments, not interacting with the main chain.

The appearance of a separate PAAm microphase in the amorphous PVA-*g*-PAAm structure at $N>N^*$ or $M_{vPAAm}>M_v^*{}_{PAAm}$ was in full agreement with a redistribution of hydrogen bonds in the copolymer macromolecules upon increase in the number and length of the grafts (see Section 4). From this point of view, the abnormally high density and length of grafts in the copolymer macromolecules led to obstacles in the realization of thermodynamic affinity between PVA and PAAm. The steric hindrances were the main factor at $N>N^*$, while overlapping and interaction of the very long grafts at large distances from the main chain was the main one in the case of $M_{vPAAm}>M_v^*{}_{PAAm}$. It is clear that $N^*$ and $M_v^*{}_{PAAm}$ values cannot be constant and should depend on the molecular architecture of the graft copolymers.

It is known that an incomplete microphase separation of segments and appearance of the regions enriched either by one or another component in a copolymer structure is possible in block and graft copolymers.[125] In this situation, the Koberstein-Leung approach[126] allows evaluating the compositions and the homogeneity degree of the microphases. This approach was used in the case of the PVA-*g*-PAAm sample with the longest grafts (**Fig. 13, b**, curve 3) in order to determine the PVA content in the regions of polymer miscibility ("soft" microphase characterized by $T_g^1$) and to verify their homogeneity.[71] The PVA weight fraction ($\omega_1^1$) in the "soft" PVA-rich microphase was calculated by the following equation:

$$\omega_1^1 = T_{g1} \cdot (T_g^1 - T_{g2}) / [T_{g2} \cdot (T_{g1} - T_g^1) + T_{g1} \cdot (T_g^1 - T_{g2})] \qquad (8)$$

Here $T_{g1}$, $T_{g2}$ and $T_g^1$ are the glass transition temperatures for the pure PVA and PAAm, and also for the "soft" PVA-rich microphase in the PVA-*g*-PAAm structure, respectively. Thus, the value of $\omega_1^1 = 0.011$ was obtained. Furthermore, the overall weight fraction of the "soft" PVA-rich microphase ($\omega^j$) in the copolymer was estimated according to the following equation:

$$\omega^j = \Delta C_p^1 /[\Delta C_{p1} \cdot \omega_1^j + \Delta C_{p2} \cdot (1 - \omega_1^j)] \tag{9}$$

where $\Delta C_{p1}$, $\Delta C_{p2}$ and $\Delta C_p^1$ correspond to the heat capacity jumps for PVA, PAAm and the "soft" PVA-rich microphase of the copolymer. The value of $\omega^j$ was found to be 0.937. Taking into account the $\omega_1^1$ and $\omega^j$ numbers, and also the real weight fraction ($w_1$=0,018) of PVA in the given copolymer, established by elemental analysis, it could be concluded that the main part of PVA segments is actually located in the "soft" microphase, uniformly distributed among a great number of PAAm segments. Such homogeneous microregions formed the basis of the PVA-g-PAAm bulk structure. At the same time, the weight fraction of the "hard" microphase, containing practically solely PAAm segments with abnormally high rigidity, was relatively insignificant ($\omega^2$=0.063).

## 6. Stimuli-responsive behavior of IntraPC in solutions

Our attention in this section will be focused on the behavior of natural (non-cross-linked) macromolecules of IntraPC forming block and graft copolymers, under the influence of their molecular architecture and concentration, and also such stimuli as solution pH, temperature, salt additives, and hydrodynamic shear field. The situation in the IntraPC forming networks, composed of chemically complementary polymers, is described elsewhere.[8,14,75,127,128]

Due to the dual nature of the IntraPC forming copolymers, they combine properties of InterPCs and ordinary amphiphilic block or graft copolymers, and show a very complicated behavior in solutions. A general approach to understand the alterations in IntraPC state, in response to different stimuli, is the examination of some competitive equilibriums correlated with each other. In the simplest case of the IntraPC forming block copolymer containing A and B blocks with the (A)—X-H proton donor groups, capable to dissociate in solutions, and (B)—Y proton acceptor groups, these equilibriums can be written down as follows:

$$(A)\text{—}X\text{-}H \overset{K_a}{\Leftrightarrow} (A)\text{—}X^- + H^+ \tag{10}$$

$$(A)\text{—}X\text{-}H + Y\text{—}(B) \overset{K_{H-bond}}{\Leftrightarrow} (A)\text{—}X\text{-}H\cdots Y\text{—}(B) \tag{11}$$

$$A\text{—}B \overset{K_{IntraPC}}{\Leftrightarrow} IntraPC(A\cdot B) \tag{12}$$

$$Z_{A\cdot B} \cdot IntraPC(A\cdot B) \overset{K_{M(A\cdot B)}}{\Leftrightarrow} Micelle(A\cdot B\text{-core}) \text{ (or aggregate)} \tag{13}$$

$$Z \cdot A\text{—}B \overset{K_{InterPC}}{\Leftrightarrow} InterPC(A\text{—}B)_Z \tag{14}$$

$$Z_A \cdot A\text{—}B \overset{K_{M(A)}}{\Leftrightarrow} Micelle(A\text{-core}) \tag{15}$$

$$Z_B \cdot A\text{—}B \overset{K_{M(B)}}{\Leftrightarrow} Micelle(B\text{-core}) \tag{16}$$

Here, $K_a$ is the effective dissociation constant of the A polyacid block, which depends on the strength of the acid groups, their distribution along the polymeric chain, and the dissociation degree $\alpha$ (and on the solution pH):

$$pK_a = pH - \lg\frac{\alpha}{1-\alpha} = pK_a^0 + 0.434 \cdot RT \frac{\partial G_{el}}{\partial \alpha} \tag{17}$$

In the formula (17), $pK_a$ and $pK_a^0$ are negative logarithms of the effective and characteristic dissociation constants, and $(\partial G_{el}/\partial \alpha)$ is the change in the electrostatic contribution into the Gibbs's free energy during the process of the polyacid ionization. Further, $K_{H-bond}$ is the effective constant of the cooperative binding of (A)—X-H and (B)—Y groups. In fact, that is an analogue of the effective constant of inner equilibrium in the reactions of InterPC formation,[129] which depends on the strength of separate H-bonds, their cooperativity, and the degree of binding ($\theta$) of the groups:

$$pK_{H-bond} = \frac{\theta}{(1-\theta)\cdot C_0} = pK_{H-bond}^0 \cdot e^{[-\varphi(\theta)]} \tag{18}$$

Here, $C_0$ is the total concentration of one (or another) type of acidic groups in solution, $pK_{H-bond}^0$ is the negative logarithm of the constant of the first H-bond formation, and $\varphi(\theta)$ is the cooperativity function.[129] The relation (18) was developed under the assumption about equal length of two reacting polymer chains and can be used in the case of a symmetric IntraPC forming block copolymer. The following constant $K_{IntraPC}$, reflecting the intra-association of the whole A and B blocks within the same macromolecule, can be considered as an analogue of the equilibrium constant $K_v$ of the InterPC formation on macromolecular level (the case $\theta=1$), which was found to be a function of a chain length[12]:

$$K_v = K_1^v = e^{v \cdot (-\Delta G_1^0 / RT)} \qquad (19)$$

In the formula (19), $v$ is the number of reacting units in a shorter "oligomeric" component (or a shorter block in the diblock copolymer) but $K_1 = e^{-\Delta G_1^0 / RT}$ is the effective constant of binding of one unit of the shorter component with a unit of the longer one. It is clear that if the components are homopolymers, then $v \equiv P_n$.

The next equilibrium constant $K_{M(A \cdot B)}$ characterizes the process of the IntraPC hydrophobic micellization (if the A-*b*-B diblock copolymer is asymmetric, i.e., $P_{nA} > P_{nB}$ or vice versa) or simply hydrophobic aggregation (if the copolymer is symmetric, $P_{nA} \approx P_{nB}$). At the same time, $K_{M(A)}$ and $K_{M(B)}$ are the constants of the copolymer micellization due to hydrophobic interactions of unbound with each other A or B blocks in an aqueous medium. It is known that the micelle formation equilibriums, similar to (13), (15) and (16), sharply shift to the right, when the copolymer concentration and/or the solution temperature become higher than some critical values (*CMC* and $T_{CMT}$). In this case, the standard Gibbs free energy of micellization per mole of copolymer chains, transferred at some temperature from the ideal dilute solution to the micelle cores, is given by the following equation[130,131]:

$$\Delta G^0 \approx RT \cdot \ln(CMC) - RT \cdot Z^{-1} \cdot \ln([A_m]) \qquad (20)$$

where $Z$ is the aggregation number and $[A_m]$ is the concentration of the micelles. When $Z$ is high, the second term is very small, therefore:

$$\Delta G^0 \approx RT \cdot \ln(CMC) \qquad (21)$$

Simultaneously, the standard Gibbs free energy of micellization at some concentration $C$ and critical temperature is expressed by:

$$\Delta G^0 \approx RT_{CMT} \cdot \ln C \qquad (22)$$

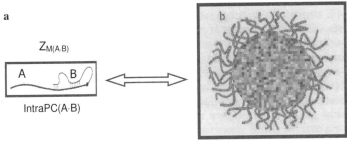

Fig. 14. Scheme of micellization of the asymmetric IntraPC forming A-*b*-B copolymer in initial state. IntraPC formed by a single macromolecule (**a**) and a "crew-cut" micelle with friable core of A and B segments formed by Z macromolecules (**b**) are shown.

It is well known that hydrophobic micellization of block copolymers in water is an entropy driven process.[44]

The last constant $K_{\text{InterPC}}$ of the equilibrium (14) characterizes the inter-association of A and B blocks in different macromolecules, which results in the formation of friable associates or even physical gels (if the copolymer concentration is high enough). The system of equilibriums (10)-(16) describes practically all situations developed in aqueous solutions of the IntraPC forming block and graft copolymers at any conditions. If both interacting A and B blocks contain only nonionic proton donor and proton acceptor groups, the first equilibrium must be excluded. Let's now consider the correlation between these equilibriums.

In the initial state (without external stimuli), the IntraPC forming copolymers take an active part mainly in the equilibriums (10)-(13). Indeed, the spontaneous formation of IntraPC(A·B) by macromolecules of a double hydrophilic block copolymer via cooperative hydrogen bonds leads to arising a new amphiphilic copolymer (**Fig. 14, a**) with hydrophobic blocks composed of bound A and B segments (the case of an asymmetric copolymer) or compact globules (the case of a symmetric one). Further, either the process of a micellar self-assembly (**Fig. 14, b**), depending on the relative length of hydrophobic (complex) and hydrophilic (surplus) blocks, or a hydrophobic aggregation of globules develops. The intensity of micellization or aggregation (the aggregation number $Z_{\text{M(A·B)}}$) and the final micellar morphology (spherical "hairy" or "crew-cut" micelles with hydrophobic "core" and hydrophilic "corona", worm-like or rod-like micelles, or vesicles[44,132,133]) are determined, in the absence of external stimuli, by the chemical nature, relative and whole length of the A and B blocks, and also by the copolymer concentration and its thermodynamic affinity to the solvent. Under the term "chemical nature", the presence of hydrophobic groups, the strength of (A)—X-H acid groups and (A)—X-H···Y—(B) hydrogen bonds, and also a structural complementarity of the A and B blocks, are implied. Note that the appearance of separate H-bonds in the complexed "core" between the unbound A and B units, belonging to different IntraPCs, can be an additional driving force for micellization.

An analogous situation remains also when external stimuli stabilize the system of hydrogen bonds. In particular, a decrease in the solution pH and temperature, and an increase in the concentration of a low-molecular-weight electrolyte enhance the system of H-bonds. When the H-bond system in the copolymer begins to be destroyed (for example, upon increasing the solution pH and temperature) but A and B blocks still interact with each other, a process of inter-association of the blocks according to the equilibrium (14) can take place. Inter-association of the A and B blocks, belonging to different copolymer chains, can be compared with the known reactions of macromolecular exchange, which develop in the InterPC solutions upon essential weakening in the cooperative bond system between the polymeric partners.[4]

Finally, when the system of cooperative bonds turns out to be practically destroyed by external stimuli, the copolymers chains begin actively to participate in the equilibriums (15) or (16). This kind of micellization with formation of dense A- or B-cores (**Fig. 15**) is realized if water becomes a poor solvent with respect to A or B blocks under the new external conditions, and the copolymer concentration is $C_{A\text{-}b\text{-}B} \geq CMC$. Unlike hydrophobic micellization, initiated by segregation of non-polar bound A·B segments in water (**Fig. 14**), the micellization process driven by hydrophobic interactions between separate poorly soluble A or B blocks of amphiphilic block copolymers is widely studied.[44,113,134,135] Different theoretical approaches based on the scaling and mean field conceptions allowed relating the main parameters of the "crew-cut" and "hairy" micelles, such as the "core" radius and the

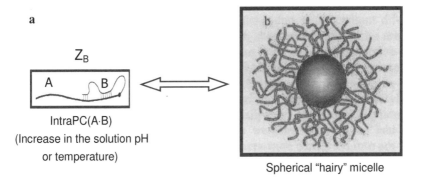

**Fig. 15.** Schematic representation of the A-*b*-B copolymer micellization driven by destruction of the H-bond system. Initial IntraPC (**a**) and a spherical "hairy"micelle with a dense core formed by poorly soluble B segments (**b**) are exhibited.

"corona" thickness, and also the aggregation number $Z$, with the characteristics of A and B blocks.[136-138]

The behavior of IntraPC forming graft copolymers, especially with a high graft density, can differ from the above general scheme proposed.

These peculiarities and their reasons will be discussed below. It should be noted that other stimuli, capable to destroy the system of hydrogen bonds between the blocks (in particular, the hydrodynamic shear field and additives of some organic substances or even a strong dilution of a solution), could also influence the state of IntraPC forming block and graft copolymers in solutions. The conception about the participation of the IntraPC forming copolymer macromolecules in a series of competitive and correlated equilibriums is very useful and significantly simplifies the discussion of the experimental results.

**Block copolymers.** The state of a well-characterized sample of the asymmetric diblock copolymer based on PEO and PMAAc (PEO$_{177}$-*b*-PMAAc$_{21}$, PDI=1,8) in dilute aqueous solution ($C$=1 kg·m$^{-3}$) under the effect of solution pH and temperature, and also addition of urea was investigated by Gohy *et al.*[139] The main purpose of this study was to follow the formation of an InterPC between two different block copolymers containing oppositely charged blocks of PMAAc and poly(2-vinylpyridinium) and also nonionic blocks of PEO. At the same time the structure of the above IntraPC forming block copolymers has been also probed by dynamic light scattering (DLS). It was shown that an asymmetric $PEO_{177}$-$b$-$PMAAc_{21}$ copolymer, in the region pH<6.1 and at $T$=20°C, formed spherical and practically monodispersed (PDI≈0.08) micelles with a "core" of hydrophobic PEO·PAAc H-bonded segments and a "corona" of the hydrophilic uncomplexed PEO segments (as in **Fig. 14**). The apparent hydrodynamic radius of the micelles was constant, $R_h$=55 nm. At pH≥6.1, a full disintegration of the IntraPC and micellar structures occurred, due to a significant ionization of PMAAc according to equilibrium (10) and essentially free PEO-$b$-PMA⁻Na⁺ chains existed in solution. Similar micellar (and IntraPC) destruction took place also upon addition of urea ($C$=1 kg·m⁻³), which is an excellent competitor in the processes of H-bond formation. In this case, the main role of the hydrogen-bonding in the PEO-$b$-PMAAc micellization under the given conditions was clearly demonstrated.

An interesting phenomenon was observed at low pH with the temperature increase up to 80°C. The values of $R_h$ and the scattered intensity grew at the temperature increase up to 55°C and then did not change. Simultaneously, the initially clear PEO-$b$-PMAAc solution became opalescent. The authors attributed such a behavior to reorganization of the micellar structures, namely, to the destruction of initial micelles and formation of new micelles (as in **Fig. 15**) due to the deterioration of the PMAAc block solubility near 50°C. Unfortunately, this hypothesis was not fully confirmed experimentally and the question about the possible micellar rearrangements remains open. However, an alternative explanation of the results is also possible from the viewpoints of strengthening in hydrophobic interactions, which stabilize the structure of the IntraPC upon increase in temperature/or changing in micellar morphology (for example, from small spherical to large-scale cylindrical micelles) without destruction of the IntraPC.

The situation with the same diblock copolymer, but having a longer polyacid block, has been also reported.[9,140] Two copolymers such as $PEO_{113}$-$b$-$PMAAc_{207}$ and $PEO_{122}$-$b$-$PMAAc_{294}$ were obtained in the sodium salt forms and their structures in dilute ($C$=0.1-1.0 kg·m⁻³) and for the first sample in semi-dilute ($C$=5 kg·m⁻³) aqueous solutions in a wide pH region were probed using DLS, potentiometric titration, and fluorescence spectroscopy. In order to conduct the later experiments, based on the nonradiative energy transfer (NRET) and other fluorescence techniques, pyrene or naphthalene labels as energy acceptor and donor, correspondingly, were linked to –COOH groups of the PMAAc blocks.

Based on these studies, four states of the block copolymers in dilute solutions at $T=25°C$ were identified. In the region of low pH<4.5, ($\alpha_{PMAAc}<0.07$) formation of an IntraPC and hydrophobic micellization occurred according to the equilibriums (11)-(13). The micellar "core" consisted of the complexed PEO and PMAAc parts (**Fig. 14**) and the "corona" included unbound polyacid segments. At pH=3.5, the apparent micellar radius $R_h$ was equal to 45 nm (for the second sample) and it sharply decreased with pH increase up to 4.5. The NRET efficiency, evaluated by the intensity ratio, $I_{Py}/I_{Np}$, in mixed solutions of the Py- and Np-labeled copolymers, also decreased sharply in this pH region for both samples. At the same time, the steady-state and time-resolved fluorescence experiments in solutions of Py-labeled copolymers have fixed constant values of the $I_1/I_3$ ratio (the intensity of the (0.0) band to that of the (0.3) band of pyrene), and the average lifetime $\tau_{av}$ of pyrene within this pH region. Hence, the micropolarity around pyrene probes did not change. These observations indicated a consequent destruction of the initial micellar structure due to ionization of carboxylic groups in the "corona" without noticeable destruction of the IntraPC. Such a behavior was different from that of micelles of the same copolymer, but with longer PEO blocks (see above). The stability of these micelles in a wider pH region was ensured by the nonionic character of the PEO chains in the "corona".

In the range $4.5\leq pH\leq5.5$ ($0.07\leq\alpha\leq0.3$), the $R_h$ value was found to be minimal (~8 nm) and practically unchanged. Additionally, the NRET efficiency weakly decreased with pH growth up to a minimum value, the $I_1/I_3$ values were slightly enhanced, but the $\tau_{av}$ parameters for both the copolymers first remained constant (before pH~5.2) and then began to decrease sharply. Therefore, within this pH region, due to disintegration of the initial micelles, essentially free copolymer macromolecules with still interacting PMAAc and PEO blocks (the IntraPC unimers), existed.

Full destruction of the IntraPC formed by the copolymers took place at further ionization of the polyacid blocks in a relatively narrow interval 5.5<pH<6.0 (0.3<$\alpha$<0.42). This conclusion was based on a sharp rise in the $I_1/I_3$ ratio up to a constant value achieved at pH=6.0 and a strong reduction of the $\tau_{av}$ parameter in the copolymer solutions. Interestingly, within this pH region, only a small increase in the $R_h$ number was observed. The authors attributed this phenomenon to a cooperative release of the hydrophobic interactions between the PMAAc methyl groups, but it could be interpreted more correctly by a small inter-association of the blocks according to equation (14). It should be noted that the potentiometric titration data, in particular, the $pK_a$ *vs* $\alpha$ curves[9] calculated according to equation (17) demonstrated an evident conformational transition, such as *globule-coil*, for both copolymers exactly at $\alpha$>0.3. This fact also confirms the above conclusion and indicates the cooperative character of the IntraPC destruction. A fully extended conformation of the PEO-*b*-PMAAc macromolecules (the forth state) was achieved only at pH≥6.0.

The stability of the copolymer micelles at pH=3.8, towards the chain exchange, and a mechanism of this process, were also important aspects of a study.[9] The NRET experiments were carried out upon mixing the low pH solutions of the Np- and Py-labeled copolymers. Both exchange mechanisms such as *insertion/expulsion* of single chains and *merging/splitting* of the micelles have been revealed and their kinetic parameters have been determined. The second slower mechanism was dominative, especially in the case of the longer PEO-*b*-PMAAc copolymer.

In semi-dilute aqueous solutions of PEO$_{113}$-*b*-PMAAc$_{207}$ in its Na-form (at high pH) the presence of large friable aggregates along with free copolymer macromolecules was suggested.[140] Addition of NaNO$_3$ caused chain compaction accompanied by destruction of the initial aggregates. At the same time, in a salt-free copolymer solution the macroscopic phase separation proceeded at pH<3.8. The authors proposed to consider this phenomenon as a result of intermicellar hydrogen-bonding of uncharged –COOH groups in the "corona". However, the possibility of simple changing in micellar morphology in semi-dilute solution should also not be excluded. Studying the temperature effect at pH=4.5 have shown that the hydrodynamic diameter of micellar aggregates grew from $D_h$=146 nm to the constant value of 158 nm, when the solution temperature increased from 20°C to 70°C. The proposed mechanism of the temperature influence (analogously to the conception by Gohy *et al.*[139]) included a transition from micelles with PEO·PMAAc complexed "cores" (**Fig. 14**) to micelles with hydrophobic PMAAc "cores" (**Fig. 15**) due to achievement of the PMAAc cloud point (T~50°C). Unfortunately, the possible alternative explanation mentioned earlier, was not taken into consideration.

The appearance of micelles was also observed in dilute aqueous solutions of the above-mentioned asymmetric triblock copolymers PAAm-*b*-PEO-*b*-PAAm with nonionic components (the hydrolysis degree of the amide groups was ~1%) and a variable block length.[63] Using static light scattering ($\Theta$=90°), the *CMC* values were determined (**Fig. 16**) and the standard free micellization energies were calculated using the formula (21) (**Table 2**). Also, two depolarization coefficients, $\Delta_v$ and $\Delta_u$, of the scattered light at vertically polarized and unpolarized falling light, and the optical anisotropy parameter $\delta^2$ for scattering particles were determined using the following relations[141]:

$$\Delta_v = \frac{H_v}{V_v} \qquad \Delta_u = \frac{H_u}{V_u} \qquad \delta^2 = \frac{10 \cdot \Delta_u^0}{6 - 7 \cdot \Delta_u^0} \qquad (23)$$

Here, $H_v$, $H_u$ and $V_v$, $V_h$ are the horizontal and vertical components of scattering at vertically polarized and unpolarized falling light, and $\Delta_u^0$ is the depolarization coefficient extrapolated to $C$=0. It is known that the $\Delta_v^0$ and $\Delta_u^0$ coefficients characterize the anisotropy of a matter and the dimension (form) of scattering particles (macromolecules or associates). The parameter $\delta^2$ reflects both these contributions in optical anisotropy.

The results have shown that micellization of TBC4 chains with longer PEO and PAAm blocks began at lower molar concentration (lower *CMC*) and their micelles were more stable (higher -$\Delta G^0$). It is not surprising taking into account an increase in Inter- and IntraPC stability upon lengthening the interacting polymer chains (see equation (19)). Moreover, the matter of TBC macrocoils and micelles was established to be optically isotropic ($\Delta_v^0$=0) but a size or asymmetry of micelles formed

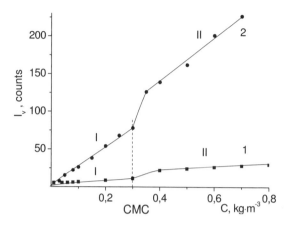

**Fig. 16.** Intensity of scattering of the vertically polarized falling light *vs* concentration of TBC1 in aqueous solutions ($M_{vPEO}=3\cdot10^3$ and $M_{vPAAm}=4.5\cdot10^4$) *–1*, and TBC4 (see **Table 1**) ($M_{vPEO}=4\cdot10^4$ and $M_{vPAAm}=1.21\cdot10^5$) *–2*. The I-st and II-nd regions correspond to TBC unimers and micelles. $T=25°C$, pH~7.

**Table 2.** Thermodynamical and optical parameters of TBC micellization.

| Copolymer | $CMC \cdot 10^6$ mol·dm$^{-3}$ | $-\Delta G^0$, kJ·mol$^{-1}$ | $\Delta_v^0$ | | $\Delta_u^0$ | | $\delta^2$ | |
|---|---|---|---|---|---|---|---|---|
| | | | I | II | I | II | I | II |
| TBC1 | 3,80 | 30,9 | 0 | 0 | 0,06 | 0,05 | 0,11 | 0,08 |
| TBC4 | 0,44 | 36,3 | 0 | 0 | 0,04 | 0,13 | 0,07 | 0,26 |

by TBC4 chains proved to be essentially greater (higher $\Delta_u^0$ and $\delta^2$) than those of TBC1. The last conclusion is correlated with a weak opalescence of the TBC4 solutions at $C>0.3$ kg·m$^{-3}$ (against clear TBC1 solutions at all concentrations) and their higher intrinsic viscosity.

The temperature effect on the micellar structures in the TBC solutions is shown in **Fig. 17**. In the entire temperature region used in this study, the PEO and PAAm homopolymers were well soluble in water and their reduced viscosity only slightly decreased upon heating. Unlike this, the viscosity change for TBCs had a more complicated character (**Fig. 17**). A minimum was observed in the interval $35°C<T<45°C$, reproduced upon cooling (curve 3). Based on the high thermostability of the H-bonds between PEO and PAAm (Sec. 4) and the well known fact of strengthening of the hydrophobic interactions up to a maximum in that temperature region, an increase in the $\eta_{red}$ values of TBC1 in aqueous

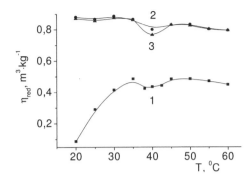

**Fig. 17.** Reduced viscosity *vs* concentration for aqueous solutions of TBC1 –*1*, TBC4 –*2* (heating) and –*3* (cooling). $C_{TBC}$=0.4 kg·m$^{-3}$; the heating (cooling) rate was ~5°C/20 min.

solution within $20°C \leq T \leq 35°C$ was interpreted by an extension of the separate micelles or by their aggregation with strengthening of the stabilization factor. But the appearance of viscosity minimums at $T$~40°C was attributed to the contraction of the micellar "cores" (without any destruction) resulting from the maximal effect of the same factor.

Now, we will discuss the behavior of IntraPC forming block copolymers composed of polyacids and polyacrylamide derivatives. They have attracted a great interest during the last years because PAAm derivatives such as poly(N-isopropylacrylamide) (PNIPAAm) and poly(N,N-diethylacrylamide) (PDEAAm) exhibit lower critical solution temperature (LCST) at around 32°C and possess an abnormally high thermo-sensitivity. The behavior of the pH- and thermo-sensitive highly asymmetric diblock copolymer PAAc$_{45}$-*b*-PDEAAm$_{360}$ (PDI=1.12) in aqueous solutions ($C$=0.6-5.2 kg·m$^{-3}$) was reported in the studies.[142,143] DLS and spectrophotometry (at $\lambda$=500 nm) were used to examine the copolymer solution properties below and above its LCST in alkaline (pH>9) and acidic (pH<4) regions. At $T$=20°C and pH=3.9 the dilute opalescent copolymer solution ($C$=0.9 kg·m$^{-3}$) contained the mixture of separate polydispersed micelles with $\langle R_h \rangle_z$=43.8 nm and large aggregates with $\langle R_h \rangle_z$ >300 nm. According to the authors, this strong aggregation in acidic solutions of the double hydrophilic copolymer was stipulated by formation of PAAc-"core" micelles due to the shift of the equilibrium (10) to the left and/or by penetration in micellar "cores" of the 1.1-

diphenilhexylic end copolymer groups remained free from the polymerization initiator. A more simple and correct explanation can be related to the formation of IntraPCs by the copolymer macromolecules, followed by their self-assembly into micelles with PAAc·PDEAAm- "cores", because InterPCs between polyacids and polyacrylamide derivatives such as PNIPAAm and poly(N,N-dimethylacrylamide) (PDMAAm) are well known.[80,144,145] When the temperature increased up to 44°C, a macroscopic precipitation occurred in time. It can be concluded from these experiments that the precipitation process is developed due to the insolubility of the PDEAAm segments in the "corona".

In the alkaline region (pH=12.8) at $T$=21°C, mainly the copolymer unimers ($\langle R_h \rangle_z$=4.7 nm) coexisting with small fraction (~0.05 wt.-%) of large aggregates ($\langle R_h \rangle_z$=101 nm) were revealed. The fact indicated a complete destruction of the initial micellar structures together with IntraPC upon increase in pH, that was caused by –COOH groups ionization. Above the LCST ($T$=45°C), only monodispersed spherical "crew-cut" micelles with PDEAAm-"cores" and $\langle R_h \rangle_z$=21.5 nm were observed.

The possibility to develop stable micelles, based on IntraPC forming diblock copolymers such as *core-shell-corona* with the same "core" and constant aggregation number, but stimuli-responsive "shell" and "corona", was demonstrated in a recent publication by Li *et al.*[146] First, the diblock copolymer precursor P$t$BA$_{30}$-$b$-PNIPAAm$_{68}$ (PDI=1.23) including poly(*tert*-butyl acrylate) was obtained and then its incomplete acidic hydrolysis (up to 70%) was carried out. The presence of about 30% of insoluble *tert*-butyl acrylate units, statistically distributed along the PAAc chains, in the final asymmetric P($t$BA-*co*-AAc)$_{30}$-$b$- PNIPAAm$_{68}$ copolymer was turned out to be enough to resolve this problem. Static and dynamic light scattering (SLS and DLS), and TEM confirmed the peculiarities of the copolymer behavior in dilute aqueous solution ($C$=0.2 kg·m$^{-3}$) under the influence of solution pH and temperature.

The block copolymer was directly dissolved in water at room temperature. Nearly monodispersed spherical micelles with $R_h$=39.8 nm were identified at pH=5.8 and $T$=25°C. In their center, a dense "core" consisted of water-insoluble *t*BA units was identified. The "core" was covered by "shell" of partially complexed PAAc·PNIPAAm units and further was surrounded by "corona" of excessive PNIPAAm segments. From DLS data the gyration radius of micelles, $R_g$, was also determined. The ratio $R_g/R_h=1.01$ was found to be much larger than that for a uniform sphere (~0.775) indicating a friable micellar structure. With increase in $T$ from 25 to 50°C a progressive collapse of PNIPAAm chains in the "shell" and "corona" was developed, which led to a significant decrease in $R_h$ up to the constant value of 25 nm achieved already at $T=35$°C. Simultaneously, $R_g$ and $R_g/R_h$ values lowered to 15.9 nm and 0.64, respectively. It is significant that the collapse of PNIPAAm segments was accompanied by destruction of the Inter- and IntraPCs in the "shell" and also by release of PAAc segments into the "corona" without ruining of the initial micelles. Indeed, no precipitation was observed with increase in temperature, moreover, the apparent molecular weight, $M_w$, of micelles determined by SLS proved to be unchanged in the entire temperature region. Thus, new micellar structures contained the same *t*BA "core", collapsed PNIPAAm "shell" with separate PAAc segments stretched outside, and also PAAc "corona" at elevated temperatures. These data have shown an existence of a competition in the "shell" between the complex formation of PAAc and PNIPAAm segments (according to equilibriums (12) and/or (14)) and self-association of PNIPAAm chains.

In the acidic region, the micelles obtained at pH=5.8 and $T=25$°C underwent a natural shrinking due to the protonation of -COO⁻ groups and strengthening of the hydrogen-bonding interactions in the PAAc·PNIPAAm complexed "corona". It resulted in reduction of the $R_h$, $R_g$ and $R_g/R_h$ values to 32.8 and 30.2 nm, and 0.92 nm, respectively (at pH=3). The sharpest shrinking or collapse of micelles ($\Delta R_h \approx 5.5$ nm) reminded that a phase transition took place in the narrow pH region from 4.8 to 4.4. At the same time, in the weakly basic region (pH=7.5) the micelle dimensions only slightly increased up to $R_h=40$ nm.

Interesting studies of other IntraPC forming diblock copolymers contained hydrophobic poly(*p*-hydroxystyrene) and hydrophilic PMAAc (PHOS-*b*-PMAAc) were published recently by Mountrichas & Pispas.[147] Five copolymer samples with variable molecular weights ($M_w$=(0.98-

2.56)·$10^4$; PDI=1.05-1.15) and different relative block length were synthesized and investigated in dilute aqueous solutions ($C$=1 kg·m$^{-3}$) in the pH range from 2.5 to 11.2 by means of DLS, fluorescence spectroscopy with pyrene probe and potentiometric titration. However, the main attention was paid to three copolymer samples such as PHOS$_{39}$-$b$-PMAAc$_{125}$, PHOS$_{91}$-$b$-PMAAc$_{53}$ and PHOS$_{123}$-$b$-PMAAc$_{28}$, having nearly constant $M_w$=(1.54-1.72)·$10^4$. Their behavior in solution was essentially different, due to variation of the relative lengths of hydrophilic and hydrophobic blocks. Because of the limited solubility of all the samples in water, their aqueous NaOH solutions with pH=12 or higher were obtained in the beginning. In these solutions fully ionized - COO$^-$Na$^+$ groups and only partially ionized -O$^-$Na$^+$ groups of the blocks were present.

The asymmetric PHOS$_{39}$-$b$-PMAAc$_{125}$ copolymer with the longest polyacid block demonstrated the highest solubility in an alkaline pH region, however even at pH=11.2 large friable micelles (or micellar aggregates) with $R_h$≈130 nm existed. Their "cores" were stabilized not only by hydrophobic interactions but also by hydrogen bonds between PHOS segments. It can be assumed that these micelles belonged to the "hairy"-type. Another picture was observed at the same pH in solutions of PHOS$_{91}$-$b$-PMAAc$_{53}$. Lengthening of the "core"-forming block and shortening of the "corona"-forming block led to formation of the denser and smaller in size micelles ($R_h$=22 nm). DLS data for the third copolymer with the longest PHOS block were absent but its lowest solubility in water among the samples was mentioned.

Upon decrease in pH, no monotonous alterations in $R_h$, the light scattering intensity $I_{90}$ at $\Theta$=90°, and the intensity ratio $I_1/I_3$ of pyrene probe were observed. An initial decrease in $R_h$ for the first sample was followed by a sharp increase up to a maximal value of 130 nm (at pH=8) and further fall before $R_h$=88 nm at pH~6. Simultaneously, in spite of a general decrease in $I_1/I_3$ and increase in $I_{90}$ values with reduction of pH to 3, a small minimum for $I_1/I_3$ and maximum for $I_{90}$ at pH~8 and vise versa, a small maximum for $I_1/I_3$ and minimum for $I_{90}$ at pH~6, were noticeable. The authors interpreted these results, and also a strong polydispersity of the scattering particles in the intermediate pH region, by existence of loose micelles with a PHOS-"core", a PMAAc-"corona", and secondary interactions, such as hydrogen-bonding between polymer blocks of the same or different macromolecules. This conception, assuming the existence of the PHOS-"core" in the micellar center can be accepted only in the range 6.5<pH<8.5. But near pH=6, in our opinion, the intensive rearrangement of the micelles, especially of their "cores", due to competition between both the IntraPC formation and self-assembly of the hydrophobic PHOS chains (according to (12) and (15) equilibriums), may take place. In this case, the lowest $R_h$ and $I_1/I_3$ numbers (59 nm and 1.2 at pH=3), and also the highest $I_{90}$ value, which were found in the study, should characterize new dense micelles formed by shortened or "folded" $PHOS_{39}$-$b$-$PMAAc_{125}$ chains as a result of IntraPC formation. It is interesting that macroscopic aggregation of the copolymer upon pH<4 developed very slowly (after standing for 24 hours).

The micellar reconstruction in the intermediate pH region was displayed more clearly in the solutions of $PHOS_{91}$-$b$-$PMAAc_{53}$ and other copolymer samples, where $P_{n\,PHOS}$ was higher than $P_{n\,PMAAc}$. The $R_h$ value for $PHOS_{91}$-$b$-$PMAAc_{53}$ first rapidly decreased from 22 nm at pH=11.2 to 13 nm at pH~6 and then began to increase slowly up to 17.5 nm at pH=4.6. But, the most important effect was found at pH<6.5 as the appearance of separate micelles along with micellar aggregates. Formation of these aggregates and even precipitation at pH<4 occurred very quickly and their dimension increased with a decrease in pH. These observations confirmed the transfer of PHOS segments from the "core", where they were placed in the alkaline pH region, into the micellar "corona" that promoted the micellar aggregation and was possibly resulted from a micellar reconstruction.

Using the fluorescence spectroscopy, the *CMC* values for all the copolymers were also determined at pH=7. The lowest *CMC*= $6.4 \cdot 10^{-2}$ kg·m$^{-3}$ was found for $PHOS_{166}$-$b$-$PMAAc_{66}$ with the longest hydrophobic block.

One more work in the field[148] was devoted to a pH-sensitive assembly of a linear diblock copolymer based on PEO and dextran, which was partially (in 60 mol %) converted into carboxymethyldextran ($60$-$CMD_{68}$-$b$-$PEO_{140}$) by carboxymethylation. This copolymer attracted special interest because of the presence of both carboxylic and hydroxyl groups, capable to hydrogen-bonding with oxygen atoms of PEO, in the polysaccharide chain. In order to obtain an absolutely linear block copolymer, the coupling reaction between ω-amino methoxy poly(ethylene glycol) and end-lactonized dextran was carried out. Self-assembly of the copolymer in dilute aqueous solution in the pH range from 2.0 to 12.0 was probed by SLS and DLS. At pH=2, only large micelles, composed of a CMD·PEO-"core" and a PEO-"corona" with $R_h$=100 nm, and the apparent aggregation number $Z$~40 chains, were revealed. An increase in the solution pH initiated a decrease in $R_h$ and $Z$ values, which led to ~50 nm and ~30 chains at pH~5. However, the size distribution in $2 \leq pH \leq 5$ range remained unimodal. Thus, the authors were able to show an interesting phenomenon of gradual micellar "melting", which was stipulated by ionization of CMD carboxylic groups in the "core" (resulted in the "core" swelling) and accompanied by $Z$ reduction.

Another situation was found at pH>5. According to the DLS data, the size distribution became bimodal presenting a contribution from small objects ($R_h$~3-5 nm), which grew with increase in pH. This effect meant that since pH~5 a significant destruction of micelles took place. In alkaline solutions, where –COOH groups of CMD were fully ionized, the micellar structures were absent. Instead of them, individual macromolecules and also very loose aggregates including approximately 3 copolymer chains have been identified by SLS.

The above discussion showed that a great number of "smart" stimuli-responsive functional materials can be developed based on IntraPC forming block copolymers.

**Graft copolymers.** Due to the special molecular architecture graft, copolymers cannot form such well-regulated IntraPCs and further micellar structures in solutions as linear block copolymers. Moreover, under certain conditions, namely, at high enough density and length of grafted chains, the appearance of IntraPC can be problematic at all because of the intramolecular steric hindrances and/or the effect of "detachment" of long grafts from a main chain (see Sec. 4). Additionally, the state of graft copolymers in solutions strongly depends on the position of the more insoluble polymer component in the macromolecules, that is on the selective action of a solvent with respect to the main or the grafted chains in that or other external conditions. When the main chain is the most insoluble component, the interesting phenomenon of the main chain collapse and formation of monomolecular micelles after destruction of the intramolecularly bonded system can be observed. Here, we will consider the existing experimental results concerning the IntraPC forming graft copolymers.

The problem to establish the peculiarities of the natural and stimuli-responsive behavior of the PVA-*g*-PAAm graft copolymers in aqueous solutions in dependence on the quantity $N$ and length of the grafts was addressed in a lot of studies.[6,7,71,72,149-151] Three series of the graft copolymers: (i) with comparable $M_v$ of the grafts but different $N$ (PVA-*g*-PAAm1-3), (ii) with the constant relatively small $N$ but variable $M_v$ (PVA-*g*-PAAm4-6), and (iii) with altering $N$ and $M_v$ (PVA-*g*-PAAm7-9), and also of an additional PVA-*g*-PAAm10 sample with the smallest $N$ and largest $M_v$ were used for this purpose. The molecular parameters and the system of hydrogen bonds, and additionally the bulk structure of the copolymers of the first two series and of the last copolymer sample were elucidated earlier in Sections 4 to 5. In the third series, $N$ increased from 25 to 42, whilst $M_v$ of the grafts decreased from $3.72 \cdot 10^5$ to $1.63 \cdot 10^5$. The properties of these polymers in dilute and semi-dilute solutions were probed by SLS, DLS, rheology, and interferometry.

All the copolymers except PVA-*g*-PAAm1 ($N=25$, $M_v=1.21 \cdot 10^5$), formed fully homogeneous aqueous solutions. The dilute solution ($C=1$ kg·m$^{-3}$) of the sample above was opalescent and not fully homogeneous at room temperature, but its homogeneity and transparency noticeably increased with temperature growth. One of the Zimm's plots, calculated from SLS data, is shown in **Fig. 18**, and the main parameters determined for the scattering particles are presented in **Table 3**. Note that PVA and PAAm dilute solutions demonstrated practically the same refractive index increments, $\partial n/\partial C=0.157$ and $0.163$ cm$^{-3} \cdot$g$^{-1}$ correspondingly, that is why the $\partial n/\partial C$ values found in PVA-*g*-PAAm aqueous solutions did not depend much on the copolymer architecture.

SLS data showed that almost all the copolymers were aggregated, even in dilute solutions. Only the 5th and 10th samples turned out to be in a molecularly dispersed state. In the first copolymer series, where the

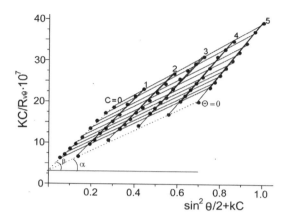

**Fig. 18.** Zimm plot based on the data of static light scattering of vertically polarized light in solutions of PVA-$g$-PAAm4 ($N$=9, $M_v$=3.75·10$^5$) at $\lambda$=436 nm, $\Theta \leq 65°$ and $T$=20°C. $C$=0.2 –$1$, 0.4 –$2$, 0.6 –$3$, 0.8 –$4$ and 1.0 kg·m$^3$ –$5$.

**Table 3.** Molecular and thermodynamical parameters of the graft copolymers.

| Copolymer | $M_w$ ·10$^{-6}$ | $M_{w\ \text{PVA-g-PAAm}}$ ·10$^{-6}$ | $N$ | $Z$ | $A_2 \cdot 10^9$ m$^3$·mol·kg$^{-2}$ | $\langle R^2 \rangle^{1/2}_z$ nm |
|---|---|---|---|---|---|---|
| PVA-$g$-PAAm2 | 29.07 | 2,41 | 37 | 12 | - 0.02 | 362 |
| PVA-$g$-PAAm3 | 7.84 | 3,90 | 49 | 2 | 0.03 | 193 |
| PVA-$g$-PAAm4 | 6,92 | 3,46 | 9 | 2 | 1,18 | 299 |
| PVA-$g$-PAAm5 | 3,81 | 3,81 | 9 | 1 | 1,55 | 253 |
| PVA-$g$-PAAm6 | 18,83 | 4,70 | 9 | 4 | 1,66 | 777 |

$M_w$ : the weight-average molecular weight of scattering particles.
$M_{w\ \text{PVA-g-PAAm}}$ : the weight-average molecular weight of separate macromolecule.
$Z$ : the aggregation number.
$A_2$ : the second virial coefficient.
$\langle R^2 \rangle^{1/2}_z$ : the z-average radius of inertia.

graft density was relatively high but the graft length was relatively small, the $A_2$ value remained close to zero in spite of its slight increase with $N$. Therefore, water could be assumed as a $\Theta$-solvent with respect to these copolymers. The decrease in the $Z$ and $\langle R^2 \rangle^{1/2}_z$ values observed simultaneously with the $A_2$ reduction led to an important conclusion about a noticeable improvement in the thermodynamic quality of water as a solvent with respect to the copolymers with $N$ growth. Such a conclusion, was correlated with a higher aqueous solubility of the grafts (PAAm), unlike the main chain (PVA), and also with the aforementioned alterations in the system of hydrogen bonds *vs N* (Sec. 4). In the second copolymer series, with the constant and relatively lower graft density, but essentially greater graft length, the $A_2$ values were found to be nearly two orders higher than those in the first series. Moreover, they rose with increasing $M_{vPAAm}$, thus indicating an important contribution of the graft length in the solubility of the PVA-*g*-PAAm macrocoils. Unlike $A_2$, the $Z$ and $\langle R^2 \rangle^{1/2}_z$ parameters changed not monotonously with lengthening of the grafts (**Table 3**).

The picture observed was in full agreement with the effect of partial "detachment" of the grafts from the main chain at $M_{v\ PAAm} > 4.3 \cdot 10^5$ (Sec. 4). Indeed, using $Z$ and $\langle R^2 \rangle^{1/2}_z$ values from **Table 3** and assuming a spherical shape for the copolymer aggregates, it was showed that the volume of a single macromolecule in the PVA-*g*-PAAm4 aggregates was much lower than that in the PVA-*g*-PAAm6 aggregates. Therefore, the graft "detachment" in PVA-*g*-PAAm6 resulted in a sharp swelling of its macrocoils. In this situation, inter-association between the main and grafted chains of different macromolecules could be realized (see the equilibrium (14)).

Hydrophobic interactions play an important role in InterPC stabilization and aggregation.[144,152,153] In order to verify whether hydrophobic regions were present in the macromolecules and aggregates of the IntraPC forming PVA-*g*-PAAm graft copolymers, the method of benzene solubilization was applied.[150] The results obtained, not only confirmed the existence of hydrophobic areas, but also showed the relationship between their volume $V$ and two declared copolymer parameters. The highest volume, $V=794$ nm$^3$, per macromolecule, was characteristic for the sample PVA-g-PAAm1, which was distinguished by its worst solubility in water. With $N$ increase in the first copolymer series, $V$ diminished to 98 nm$^3$, that was correlated with an improvement in the copolymer solubility in water. The $V$ values measured in solutions of the copolymers of the second series changed unexpectedly with $M_{vPAAm}$ growth. In particular, they turned out to be almost equal (382 and 410 nm$^3$) in the PVA-g-PAAm4,5 solutions, but increased sharply in the solutions of PVA-g-PAAm6 ($V=593$ nm$^3$), where the most loose

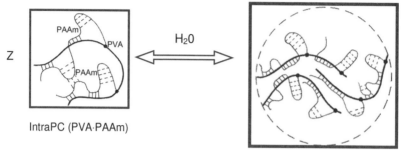

**Fig. 19.** Reversible hydrophobic self-assembly of the IntraPC forming PVA-*g*-PAAm macromolecules in dilute aqueous solutions without external stimuli.

copolymer aggregates existed. A special investigation of this surprising effect showed that the swollen copolymer macromolecules were able to bind benzene not only by dispersion interactions with hydrophobic areas, but also by hydrogen bonds with proton-donor groups.

The following scheme (**Fig. 19**), represents a possible structure of the separate PVA-*g*-PAAm macromolecules and their aggregates. It could be assumed also that individual (non aggregated) macromolecules of PVA-*g*-PAAm5,10 formed in dilute aqueous solutions the monomolecular micelles with complexed PVA·PAAm-"core" and PAAm-"corona".

It is well known that a significant dilution of a solution can shift the equilibrium of the InterPC complex formation on macromolecular level towards the individual polymer components and lead to the InterPC destruction.[10] Stability of IntraPC in PVA-*g*-PAAm macromolecules with respect to this stimulus was examined by viscometry and SLS. Some of the experiments for the first copolymer series are shown in **Fig. 20**. In the range $0.15 \leq C \leq 0.8$ kg·m$^{-3}$ the molecular and supramolecular structures of the declared copolymer samples were stable in solutions but at $C<0.15$ kg·m$^{-3}$ a sharp increase in the $\eta_{red}$, $\Delta_v$, and $\Delta_u$ values occurred, implying destruction of both aggregates and IntraPCs into separate macromolecules. Furthermore, it was established that such behavior was typical only for the graft copolymers with a high enough number of –COOH groups in the grafted chains (~5-8 mol %), which were formed because of partial hydrolysis of the amide groups.

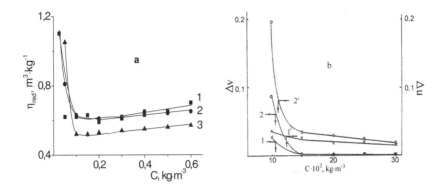

**Fig. 20**. Concentration dependences of the reduced viscosity (**a**) and the depolarization coefficients (**b**) ($\Theta=90°$, $\lambda=436$ nm) for PVA-$g$-PAAm1 $-1$(**a**), PVA-$g$-PAAm2 $-2$(**a**) and $1,1'$(**b**), and also PVA-$g$-PAAm3 $-3$(**a**) and $2,2'$(**b**). $T=25°C$.

The samples, which were examined immediately after their synthesis (the –COOH group content ~1 mol %), demonstrated either the linear dependence of $\eta_{red}$ *vs* concentration or a decrease in the $\eta_{red}$ values at $C<0.15$-$0.20$ kg·m$^{-3}$. The first case reflected full stability of aggregates and IntraPCs in the entire concentration region but the last one indicated the destruction of macromolecular aggregates only without destruction of IntraPCs. Thus, the instability of IntraPCs in some copolymer macromolecules, with respect to dilution of the solutions, was related only to the polyelectrolyte nature of the partially hydrolyzed PAAm grafted chains.

The temperature influence was studied by viscometry and DLS.[6,71,151] For all the copolymers very complicated but, at the same time, similar dependences of the intrinsic viscosity and the Huggins viscometric constant *vs* temperature were revealed. The most interesting results obtained are shown in **Fig. 21**. A sharp decrease in [$\eta$] to a minimum that was accompanied by a sharp "jump" of $K$ to a maximum in the narrow temperature region from 30 to 35°C, pointed the possible conformational transition that is likely related to the destruction of the IntraPC formed by the graft copolymers. Similar extreme dependences of [$\eta$] *vs* $T$ were observed in solutions of other IntraPC forming graft copolymers such as AC-$g$-PAAm (in acetic and formic acids) and DAC-$g$-PAAm (in formic

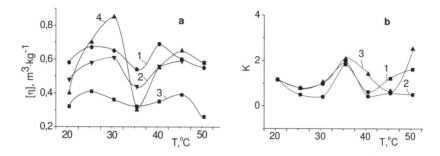

**Fig. 21**. Temperature dependence of intrinsic viscosity (**a**) and the Huggins viscometric constant (**b**) in aqueous solutions of PVA-*g*-PAAm7 –*1*, PVA-*g*-PAAm8 –*2*, PVA-*g*-PAAm9 –*3* and PVA-*g*-PAAm10 –*4*. *1,1′* (**b**), and also PVA-*g*-PAAm3 –*3*(a) and *2,2′* (**b**). *T*=25°C.

acid), which contained the acetylcellulose and diacetylcellulose main chains,[154] and also PVA-g-PEO (in water).[155] These results were explained by changes of the solvent quality with respect to the main and grafted chains caused by changes in temperature. In order to elucidate the nature of this conformational transition, further DLS studies of the thermo-sensitive behavior of PVA-*g*-PAAm were carried out (**Fig. 22**).

The correlation function of the scattered light by the PVA-*g*-PAAm10 solution, demonstrated two exponential contributions at all temperatures that resulted in bimodal distributions of the diffusion coefficients (**Fig. 22**). Taking into account: (i) the narrow molecular-weight distribution of the PVA-*g*-PAAm10 sample (GPC), (ii) the rigid structure of its coils in aqueous solution (data of the high-speed sedimentation), and (iii) the validity of the criterion $q \cdot R_g > 1$, where $q$ is the scattering vector and $R_g$ is the radius of gyration of macrocoils, in the experiment, the bimodal distributions were attributed to the translational and translational-rotational motions of the scattering particles. The effective translational and rotational diffusion coefficients ($D_t$ and $D_r$) were established from these data. Analysis of all information showed that as temperature increased from ~30°C to 35°C, a collapse of the main chain, due to dissociation of a large number of intramolecular H-bonds (in fact due to IntraPC destruction), and display of different solubility of PVA and PAAm in water, were developed. Such a rearrangement in the

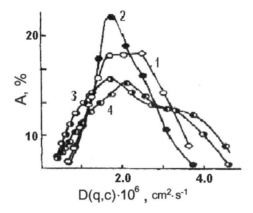

**Fig. 22**. Diffusion coefficient distributions in the PVA-*g*-PAAm10 solution at $T$=25°C –*1*, 30°C –*2*, 35°C –*3* and 40°C –*4*. $C$=0.4 kg·m$^{-3}$, $\lambda$=632.8 nm, $\Theta$=90°.

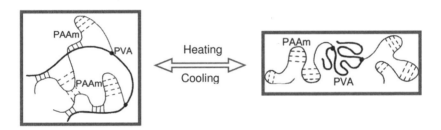

**Fig. 23**. Scheme of reversible conformational transition in a single PVA-*g*-PAAm macromolecule under the temperature effect.

copolymer macromolecules (**Fig. 23**) was designated as the *IntraPC* ↔ *segregated state* conformational transition. Its reversibility was proved in separate experiments.

The copolymer architecture influenced the temperature interval of the transition and the depth of the conformational alterations. The last one was evaluated by the $\Delta[\eta]$ value in the transition region. The lower the density and the higher length of the grafts, the greater conformational alterations (according to the $\Delta[\eta]$ value) were observed (**Fig. 21 a**). Additionally, the longer the grafted chains at constant their density, the lower the temperature of the transition onset was registered. Now, the phenomenon of the main chain collapse, initiated by deterioration in the solvent quality, under the action of different stimuli, is actively studied experimentally[156] and theoretically.[157] It should be added that in solutions of such an IntraPC forming graft copolymer as PVA-g-PVPD (poly-*N*-vinylpyrrolidone), no changes in [$\eta$] over a wide temperature range were observed.[154] The main reason for this is evidently the high thermostability of the hydrogen bonds between PVA and PVPD.

The influence of the hydrodynamic shear field on PVA-*g*-PAAm structures in aqueous solutions was studied[7,71,149] in order to explain: (i) the effect of the negative thixotropy (thickening) appeared in the semi-dilute copolymer solutions under slight stirring, (ii) the high hydrodynamic activity of the copolymers displayed in very dilute copolymer solutions in the turbulent flow regime (the value $W$ of the turbulent flow drag reduction attained to ~60% at $C$=0.02 kg·m$^{-3}$ and $Re$=5·10$^3$–5·10$^4$), and (iii) the resistance of the copolymer solutions (unlike PEO solutions) to degradation in turbulent flow.[158,159] Similar functional capabilities were shown also by some other IntraPC forming graft copolymers based on different polysaccharides and PAAm.[61,160]

Rheology experiments with PVA-*g*-PAAm1-3 series were carried out in a wide range of concentrations ($C$=0.5–5 kg·m$^{-3}$) involving dilute as well as semi-dilute solutions and also of the shear rate gradients ($j$=49-1312 s$^{-1}$). Two critical values $C'$ and $j^*$, correlated with each other, distinguishing the non-Newtonian (pseudoplastic) behavior of solutions from the viscoelastic one, and depending on the graft quantity $N$ (or the graft density), were established. Thus, two levels of alterations in the structures of all the copolymers under the action of the hydrodynamic shear field were found (**Fig. 24**). Indeed, at $C>C'$ and $j<j^*$ destruction of a network of pinnings, which appeared between the primary copolymer aggregates in semi-dilute solutions, took place (the case **a**), whilst at $C<C'$ and $j>j^*$ destruction of both primary aggregates and IntraPCs into separate macromolecules occurred (the case **b**). The later resulted in a sharp increase in the solution viscosity.

It was shown that all the structural transformations were fully reversible. The $C'$ values increased in the PVA-*g*-PAAm1-3 series from 2 to 4 kg·m$^{-3}$, thus demonstrating a higher resistance to the effect of the hydrodynamic shear field of the sample with smaller $N$ that is with a

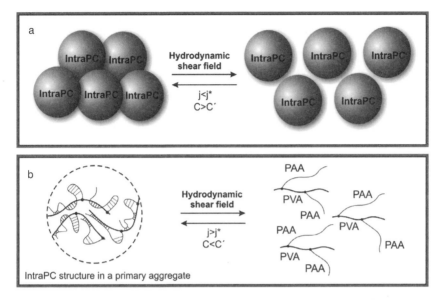

**Fig. 24.** The first (**a**) and second (**b**) levels of the PVA-g-PAAm structure destruction in aqueous solutions under the influence of the hydrodynamic shear field.

more developed H-bond system between PVA and PAAm. At the same time the largest value $j^*=243$ s$^{-1}$ was found in a dilute ($C=0.5$ kg·m$^{-3}$) solution of PVA-g-PAAm2. Note that the highest total number of H-bonds, such as *the main chain-the graft* and *the graft-the graft* distinguished this sample.

The reversibility of all the transitions was one of the key results. It allowed understanding the mechanism of resistance of PVA-g-PAAm solutions to degradation in the turbulent flow. Indeed, in solutions of PVA-g-PAAm, in contrast to those of PEO and PAAm, a special driving force was present, which was capable to restore the system to the initial state after stopping the hydrodynamic field effect. The thermodynamic affinity between the main and grafted chains was exactly this driving force.

Now we will review the behavior of other IntraPC forming graft copolymers in solution. Some studies,[62,161-164] dealt with the graft copolymers included classical pairs of the hydrogen-bonding polymer components, namely, PEO and PAAc or PMAAc. Taking into account the dependence of the graft copolymer properties, not only on the above mentioned parameters, but also on the graft distribution along the main chain, first of all, we will focus on the synthesis strategy. The PMAAc-*g*-PEO graft copolymers, with a polyacid main chain, and non-ionic grafts randomly distributed around the backbone, were obtained via coupling of methoxy-terminated MEPEG ($M_w$=5·10$^3$) with linear PMAAc, using dicyclohexyl-carbodiimide as a coupling agent.[162] The pH-sensitive state of the graft copolymers in dilute aqueous solutions was monitored by potentiometric titration, the PEO NMR spin-spin ($T_2$) relaxation, the pulsed gradient spin echo NMR and DLS. The results of potentiometric titration, presented as the dependence of $pK_a$ *vs* $\alpha$ according to the equation (17), showed that at $\alpha$<0.35 the $pK_a$ values for the graft copolymer turned out to be markedly higher than those for individual PMAAc and PMAAc/MEPEG blends. It means that such parameters as the number of hydrogen bonds between PMAAc and PEO chains and the value of the Gibbs free energy of complex formation, -$\Delta G^0$, are greater for the graft copolymer by contrast to the polymer blend. Note that we observed similar effects when we studied the pH-sensitivity of the InterPCs formed by PEO with different length, and the alternative copolymer of styrene and maleic acid, P(St-*alt*-MAc),[69] and used these potentiometric results for quantification of the intermolecular hydrogen bonds and determination of the -$\Delta G^0$ values in every InterPC.

The PMAAc-*g*-PEO macromolecules were in a collapsed state under acidic conditions because of IntraPC formation and underwent a coil expansion in the alkaline pH region, due to the IntraPC destruction initiated by the strong ionization of the main chain. At intermediate pH values two populations of PEO grafts were found: (i) the grafts bound with the PMAAc backbone, and (ii) those not participating in hydrogen-bonding.

A special kind of graft copolymers, based on the P(AAc-*co*-MA) main chains and PEO grafts with different length ($P_{n\ PEO}$=11 and 45), was synthesized by the "grafting through" method, namely, by copolymerization of AAc with PEOMA macromonomers via the RAFT technique, with dibenzyltrithiocarbonate as the chain transfer agent (CTA).[163] The reactivity ratios $r_{AAc}/r_{PEOMA}$ established in the process of copolymerization were very low (0.128 and 0.242). Due to this, and also as a result of the specific nature of CTA, the tapered triblock copolymers with central PAAc blocks and outer comb-like blocks were obtained. Three copolymer samples, such as $P(AAc_{356}\text{-}co\text{-}MA_{12})\text{-}g\text{-}PEO_{11}$ (**1**), $P(AAc_{657}\text{-}co\text{-}MA_{128})\text{-}g\text{-}PEO_{11}$ (**2**) and $P(AAc_{566}\text{-}co\text{-}MA_{24})\text{-}g\text{-}PEO_{45}$ (**3**) (here subscript near MA signifies the graft number $N$), distinguished one from another not only by the length of the main and grafted chains, but also by the EO/AAc molar ratio (0.37, 2.13 and 1.92 correspondingly), and the average effective graft density ($\sim$ 3, 16 and 4 grafts per 100 units of the main chain), were synthesized. That is why it was difficult to relate the data of DLS and TEM studies regarding pH-, salt-, and thermosensitivity of the copolymers with the parameters of their architecture. The $PMA_{76}\text{-}g\text{-}PEO_{11}$ "brush" (polymacromonomer) was synthesized and studied too.

It was established that the PMA-$g$-PEO "brush" was fully soluble in water in the entire pH range, unlike the P(AAc-$co$-MA)-$g$-PEO graft copolymers, which were insoluble at pH<5, poorly soluble at 5<pH<8 and soluble at pH>8. Such a behavior was in full agreement with the formation of IntraPCs in the copolymer macromolecules. In order to carry out the correct DLS measurements in the intermediate pH range, the copolymers were first dissolved in ethanol, in which hydrogen-bonding between the main chain and the grafts was minimized, and then pH-adjusted water was added in a drop-wise manner until stable self-aggregation was achieved and then final dialysis against the same pH-controlled water was performed. At pH=5 and $C$=16 kg·m$^{-3}$, the most developed spherical aggregates with $D_h$=321 nm were formed for the **1**st copolymer, while those in solutions of the **2**nd and **3**rd copolymers had smaller dimensions ($D_h$=31 and 95 nm). At pH=7, the $D_h$ value did not change for the **2**nd sample (31 nm), a bit increased for the **3**rd one (131 nm) and sharply decreased before 95 nm for the **1**st sample. Additionally, TEM images demonstrated the existence of dense spherical particles formed by **2**nd and **3**rd copolymers at pH=7 and hollow spheres (the vesicles) in the case of the **1**st one. The authors have attributed specific self-assembly of the **1**st $P(AAc_{356}\text{-}co\text{-}MA_{12})\text{-}g\text{-}PEO_{11}$ sample only to low EO/AAc molar ratio but the difference in other above-mentioned copolymer parameters must be also taken into consideration.

At pH=10, for all the copolymers, very large and loose aggregates were identified.

The temperature increase from 25 to 85°C at pH=7 did not influence the aggregate dimensions for the **1**st and **2**nd copolymers and led to some enhancement in $D_h$ for aggregates of the **3**rd copolymer up to a constant value of ~140 nm at $T$=60°C. The same behavior was observed for all the graft copolymers in the presence of salt (NaCl) in the 5<pH<8 range. Indeed, the $D_h$ value did not change until $C_{NaCL}$ was relatively low and sharply increased at $C_{NaCL} \geq 0.5$ mol·dm$^{-3}$, thus implying the beginning of macroscopic precipitation.

The effect of gelation in solutions of the graft copolymers composed of Pluronic (the triblock copolymer PEO-*b*-PPO-*b*-PEO with central poly(propylene oxide) block) main chain and PAAc grafts have been considered in two publications by Bromberg.[62,164] Grafting PAAc onto Pluronic was achieved by a "grafting to" method, using the transfer reactions in free radical polymerization of AAc initiated by ammonium persulfate in the presence of the triblock copolymer. One of the graft copolymer samples ($M_n$=3.13·10$^6$), with a well defined PEO$_{100}$-*b*-PPO$_{65}$-*b*-PEO$_{100}$ main chain, but unknown length and number of the grafts, was studied using the size exclusion chromatography (SEC), potentiometric titration, DSC and rheology (with determination of the storage (G′) and loss (G″) moduli). The influence of the copolymer concentration, solution pH, and salt additives on the gel point ($T_{gel}$) was studied.

The following main results were achieved. An increase in the copolymer concentration from 1 to 25 kg·m$^{-3}$ at pH=7 led to reduction in $T_{gel}$ from 27°C to 19°C. The effect of $T_{gel}$ decrease from 30°C to 25°C ($C$=10 kg·m$^{-3}$) was observed also upon increase in pH from 5.37 to 13.3. Addition of small portions of NaCl to the copolymer solution with $C$=10 kg·m$^{-3}$ enhanced the solution and gel viscoelasticity, while larger portions, such as 10–100 kg·m$^{-3}$ (at pH=7), gradually decreased the dynamic moduli and the $T_{gel}$ value. It was shown that the $T_{gel}$ values measured at pH=7 by the Winter-Chambon method (under condition of independence of the loss tangent on frequency) practically coincided in the entire concentration region with the temperatures of the copolymer aggregation onset, which were determined by DSC. The author used this to explain the gelation mechanism via hydrophobic micellization of the insoluble PPO blocks. These results were interpreted considering the existence of some equilibrium between intra- and inter-association of these blocks. The role of PAAc grafts consisted, to the author's opinion, only in the pH-dependent expansion or folding. Possible intra- and intermolecular complex formation between PEO and PAAc segments by hydrogen-bonding was not discussed at all. At the same time, there is at least one observation in the study, which contradicts this conception. This is the $pK_a$ *vs* $\alpha$ dependence calculated from the data of potentiometric titration. It demonstrates a stable compact (probably micellar) structure of the copolymer at $\alpha<0.35$ and a clear conformational transition in $0.35<\alpha<0.7$ range. If the conception proposed was the right one and if in the acidic pH region only the micelles with PPO-"core" and unbound PEO and PAAc segments in "corona" existed, the micellar destruction initiated by pH increase should not be accompanied by a sharp conformational transition. Similar conformational transitions occur when ionizable –COOH groups of some (co)polymer at low pH prove to be inside a compact structure with the lowered local dielectric permeability. It can be assumed that in the acidic pH region, either *core-shell-corona* micelles with the PPO-"core", the PEO·PAAc complexed "shell", and "corona" of the excessive PEO or PAAc segments (in dependence on the EO/AAc ratio), or the *core-corona* micelles with a mixed (PPO+PEO·PAAc)-"core", are formed. Moreover, it is clear that the gelation process developed in the intermediate pH range is not only due to inter-association of hydrophobic PPO blocks but also to inter-association of PEO and PAAc segments.

Separate pH- and thermosensitive properties of the IntraPC forming graft copolymers, based on the PAAc main chain and the grafts of different N-substituted polyacrylamides, have been also described.[81,165,166] The PAAc-*g*-PNIPAAm graft copolymers, with uniform grafts ($M_{nPNIPAAm}=2.2\cdot10^3$), randomly distributed along the PAAc chains, is one of the examples. These were synthesized by coupling PAAc with amino terminated PNIPAAm at PNIPAAm/PAAc= 20/80, 25/75, 30/70 and 50/50 weight ratios.[81] Using spectrophotometry at $\lambda=500$ nm, the temperatures of phase separation or the cloud points (CP) of the copolymer solutions ($C=2$ kg·m$^{-3}$) in citric-phosphate buffer were determined as a function of the solution pH. Some important effects were found in these studies: (i) independence of CPs on the initial PNIPAAm/PAAc ratio (within the above mentioned ratios) both at pH=4.0 and at pH=7.4, (ii) progressive decrease in CP values from 32°C to 16°C, with reduction in pH from 5 to 4, and (iii) essentially higher CP values for the corresponding P(AAc-*co*-NIPAAm) random copolymers. The reason for the first effect could not be easily found, taking into account the absence of real molecular characteristics of the copolymers in this work. Unlike this, the second and third effects were correctly interpreted by IntraPC formation in the copolymers and strengthening of the hydrogen-bonding between PAAc and PNIPAAm segments upon consequent protonation of ionized -COO⁻ groups.

In another work,[166] a well-characterized sample of the PAAc-*g*-PNIPAAm graft copolymer ($M_{PAAc}$=2.5·10$^5$, $M_{nPNIPAAm}$=2.3·10$^3$, PAAc/PNIPAAm=61/39) synthesized by the same coupling reaction was used for rheology and fluorescence studies in aqueous solution with $C$=2.5 kg·m$^{-3}$. A sharp decrease in both the reduced viscosity in the pH range from 9 to 4 and also in the known ratio $I_1/I_3$ of the pyrene probe from 1.73 at pH=6 to 1.4 at pH=4 have pointed on the formation of IntraPC with compact structure and hydrophobic areas.

In contrast to PNIPAAm having LCST near 32°C another N-substituted polyacrylamide such as PDMAAm is water-soluble over the wide temperature range from 5°C to 80°C. That is why investigations of the thermo- and pH-sensitivity of the IntraPC forming PAAc-*g*-PDMAAm graft copolymers, which were carried out by spectro-photometry and DLS, were of special interest.[165] Four copolymer samples with constant $M_{nPDMAAm}$~2·10$^4$ and variable content of DMAAm units in the copolymer, equal to 45, 56, 65 and 75 mol % (that is with growing graft density), were obtained by the "grafting through" method using free radical copolymerization of AAc with PDMAAm macromonomer, which was initiated by AIBN. The sample of random P(AAc-*co*-DMAAm) copolymer with a content of DMAAm units 48 mol % was also synthesized. The molecular weights of the random and graft copolymers were unknown. The properties of the copolymers were compared to those of PAAc/PDMAAm blend, which formed a stable InterPC at pH≤3.6, not dissociating even at high temperatures (up to 80°C).

Unlike the random copolymer, fully soluble within the range 10°C–70°C, and the polymer blend (InterPC), insoluble at the same temperatures, the PAAc-*g*-PDMAAm solution in distilled water ($C$=20 kg·m$^{-3}$) demonstrated the temperature-induced phase transition between friable hydrated and dense dehydrated states. An increase in the graft density (in the PDMAAm content) essentially influenced the temperature of phase separation (CP). In particular, the PAAc-*g*-PDMAAm(45) solution was turbid already at $T$=10°C, while PAAc-*g*-PDMAAm(75) solution was clear without any phase separation in the entire temperature range. The reason for this effect was evident, namely, an increase in the graft density resulted in steric hindrances, which created obstacles for realization of thermodynamic affinity between the main and grafted chains. The influence of the solution pH on CP was studied for the PAAc-*g*-PDMAAm(56) sample. The larger the pH (from 2.5 to 3.6) the higher the CP was observed. Finally, at pH=3.8, the copolymer solution became fully transparent at any temperature indicating destruction of the IntraPC. The more dilute solution ($C$=2 kg·m$^{-3}$) of the later copolymer in distilled water was probed by DLS at $T$=20.5°C. The bimodal size distribution, showing small individual macromolecules ($D_h$~10 nm) and their aggregates with $D_h$=141.3 nm, was observed. The aggregation dimensions remained unchanged up to $T$=27°C and then at higher temperatures sharply grew, implying the phase separation onset.

Two studies also, devoted to cellulose-based IntraPC forming graft copolymers where ionic PAAc chains played the role of grafts, should be mentioned.[167,168] The methylcellulose (MC)-*g*-PAAc graft copolymer with $M_{nMC}$=6.3·10$^4$ and the molar ratio MC/AAc=2/1 was synthesized[167] by a transfer reaction using free radical polymerization of AAc initiated by ammonium persulfate in solution of MC (the "grafting to" method). Formation of an IntraPC in the copolymer macromolecules, by hydrogen-bonding between –COOH groups of PAAc with oxygen atoms of MC, was confirmed by FTIR spectroscopy. A state of the copolymer macromolecules in dilute aqueous solutions ($C$<5 kg·m$^{-3}$) was characterized and dependences of the cloud points (CP) on the copolymer concentration, the solution pH, and ionic strength ($\mu$) were established. It was shown that IntraPC in the copolymer macromolecules existed only in the range $C$>2 kg·m$^{-3}$ (the solutions in pure distilled water). At lower concentrations, a gradual destruction of the H-bond system between MC and PAAc, due to the polyelectrolyte nature of the grafts, took place. In accordance with this fact the CP values at low copolymer concentration ($C<0.2$ kg·m$^{-3}$), turned out to be higher than those in solutions of pure MC, but at $C>0.2$ kg·m$^{-3}$, the CP values for the copolymer sharply decreased almost in 3 times by contrast with the data for MC. Dependences of CPs *vs* pH and μ, defined in the concentrated copolymer solution ($C=50$ kg·m$^{-3}$), had a natural view. At pH<3.5, very low CP values (~20°C) were observed but in the narrow 3.3<pH<3.5 interval they grew abruptly up to ~34°C and further did not change until pH=4.7. Obviously, at pH>3.3, the IntraPC destruction began. An increase in μ up to 0.01 caused a sharp CP reduction to 6°C and then at μ>0.01, this CP value remained practically constant.

Direct evidence for self-assembly of the IntraPC forming graft copolymers into spherical micelles, with a dense "core" and a friable "corona", was demonstrated in a study.[168] Two samples of the hydroxyethylcellulose (HEC)-*g*-PAAc copolymers with constant $M_{vHEC}=$ 9·10$^4$ and comparable $M_{vPAAc}=1.75\cdot10^4$ and 2.1·10$^4$, but different number of the grafts ($N\approx2$ and 3 for the **1**st and **2**nd sample, respectively), were obtained by the "grafting from" technique using the HEC/Ce$^{IV}$ redox system to initiate free radical polymerization of AAc (**Fig. 1**). DLS and TEM studies of the copolymer behavior in dilute aqueous solution ($C=1$ kg·m$^{-3}$) as a function of pH were carried out. Within the 4≤pH≤13.5 range, the macromolecules of both copolymers were in molecularly dispersed state with $D_h=50$–70 nm. As the pH value decreased to less than 4, an intense aggregation in the copolymer solutions, due to hydrogen-bonding between HEC and PAAc, developed, and the $D_h$ values attained by jump to 250 nm and 330 nm for the **1**st and **2**nd samples. TEM images showed the presence of the "hairy" *core-corona* type micelles in solution of the **1**st HEC-*g*-PAAc copolymer (pH=1.3) and the "crew-cut" micelles in solution of the **2**nd one. The HEC·PAAc hydrogen-bonded segments formed a dense micellar "core", while excessive hydrophilic parts of HEC were located in the "corona". The number of excessive HEC units in the **2**nd copolymer was lower, that is why only the "crew-cut" micelles formed in its solution.

## 7. The application spectrum of intramolecular polycomplexes

The block and graft copolymers forming IntraPCs can be considered first of all as "intelligent" stimuli-responsive polymer systems, which self-assembly, a phase state, and "functions" in solution, under different conditions, can be easily regulated by a careful selection of the interacting polymer partners, their length, and structural complementarity, and also their total macromolecular architecture. The IntraPCs are capable to bind different organic compounds, inorganic ions, colloid particles, and cells, but unlike InterPCs they cannot be destroyed to individual polymer components due to the existence of additional covalent bonds between them. The IntraPC aqueous solutions combine properties of InterPCs and amphiphilic block or graft copolymers, that is why they have many potential applications such as self-assembling nanocontainers for the drug and gene delivery,[133] and nanoreactors.[20] Furthermore, IntraPCs are capable to effectively regulate the stability of different colloidal systems as emulsifiers,[161,167] flocculants[61,169-173] and stabilizators.[174] One of these important colloid-chemical properties of IntraPCs, namely, their abnormally high flocculating ability and its reason is discussed in Chapter 8.

The IntraPCs in dilute aqueous solutions demonstrate a significant Thoms effect.[61,158-160] Therefore, they can be used as effective agents for the drag reduction of turbulent flow.

Semi-dilute solutions of IntraPCs, showing reversible thickening or even gelation under the effect of hydrodynamic shear field, temperature or pH changing,[62,121,149,164,175,176] can be widely used as displacement fluids in the oil recovery operations.

The cross-linked IntraPC forming block and graft copolymers are of a great interest as hydrogels dosage forms,[8,127] special nanoparticles for drug delivery,[168] super absorbents,[76] chemical and biosensors,[177,178] and finally as selective stimuli-responsive membranes with reversible mechano-chemical expansion and contraction.[74,75]

## References

1. I.M. Papisov and A.A. Litmanovich, in *Encyclopedia of Controlled Drug Delivery* (Wiley, New York, 1999), p. 397.

2. E. Bekturov and L. Bimendina, *J. Macromol. Sci., Part C: Polym. Rev.*, **37**, 501 (1997).
3. M. Jiang, M. Li, M. Xiang and H. Zhou, *Adv. Polym. Sci.*, **146**, 121 (1999).
4. V.A. Kabanov, *Russ. Chem. Rev.*, **74**, 3 (2005).
5. V.V. Khutoryanskiy, *Int. J. Pharm.*, **334**, 15 (2007).
6. T.B. Zheltonozhskaya, N.P. Melnik, S.G. Ostapchenko, D.A. Gavruschenko and B.V. Eremenko, *Ukr. Polym. J.*, **4**, 137 (1995).
7. T.B. Zheltonozhskaya, N.E. Zagdanskaya, O.V. Demchenko, L.N. Momot, N.M. Permyakova, V.G. Syromyatnikov and L.R. Kunitskaya, *Russ. Chem. Rev.*, **73**, 811 (2004).
8. A.M. Lowman and N.A. Peppas, *Macromolecules*, **30**, 4959 (1997).
9. S. Holappa, L. Kantonen, F.M. Winnik and H. Tenhu, *Macromolecules*, **37**, 7008 (2004).
10. V.A. Kabanov and I.M. Papisov, *Vysokomol. Soedyneniya. Ser. A.*, **21**, 243 (1979) (in Russian).
11. K. Abe, M. Koide and E. Tsuchida, *Macromolecules*, **10**, 1259 (1977).
12. V.Yu. Baranovsky, A.A. Litmanovich, I.M. Papisov and V.A. Kabanov, *Eur. Polym. J.*, **17**, 969 (1981).
13. J. Xie and P.Z. Dubin, in *Macromolecular Complexes in Chemistry and Biology. Ch. 15.* Ed. J. Bock (Springer-Verlag, Berlin, Heidelberg, 1994), p. 247.
14. J. Klier, A.B. Scranton and N.A. Peppas, *Macromolecules*, **23**, 4944 (1990).
15. A.B. Scranton, J. Klier and N.A. Peppas, *J. Polym. Sci., Part B: Polym. Phys.*, **29**, 211 (1991).
16. H. Cölfen, *Macromol. Rapid Commun.*, **22**, 219 (2001).
17. T. Goloub, A. de Keizer and M.A. Cohen Stuart, *Macromolecules*, **32**, 8441 (1999).
18. E.A. Bekturov, S.E. Kudaibergenov, R.E. Khamzamulina, V.A. Frolova, D.E. Nurgalieva, R.C. Schulz and J. Zöller, *Makromol. Chem., Rapid Commun.*, **13**, 225 (1992).
19. A.B. Lowe and C.L. McCormick, *Chem. Rev.*, **102**, 4177 (2002).
20. F. Liu and A. Eisenberg, *J. Am. Chem. Soc.*, **125**, 15059 (2003).
21. Q. Bo and Y. Zhao, *J. Polym. Sci., Part A: Polym. Chem.*, **44**, 1734 (2006).
22. N.P. Shusharina, E.B. Zhulina, A.V. Dobrynin and M. Rubinstein, *Macromolecules*, **38**, 8870 (2005).
23. Z. Wang and M. Rubinstein, *Macromolecules*, **39**, 5897 (2006).
24. C.-D. Vo, S.P. Armes, D.P. Randall, K. Sakai and S. Biggs, *Macromolecules*, **40**, 157 (2007).
25. J.-F. Gohy, S. Creutz, M. Garcia, B. Mahltig, M. Stamm and R. Jerôme, *Macromolecules*, **33**, 6378 (2000).
26. E.A. Bekturov, S.E. Kudaibergenov, V.A. Frolova, R.E. Khamzamulina, R.C. Schulz and J. Zöller, *Makromol. Chem., Rapid. Commun.*, **12**, 37 (1991).
27. S. Creutz, R. Jerôme, J.M.P. Kaptijn, A.W. van der Verf and J.M. Akkerman, *J. Coat. Technol.*, **70**, 41 (1998) (in Chinese).
28. C.S. Patrickios and E.N. Yamasaki, *Anal. Biochem.*, **231**, 82 (1995).
29. H. Walter, P. Müller-Buschbaum, J.S. Gutmann, C. Lorenz-Haas, C. Harrats, R. Jerôme and M. Stamm, *Langmuir*, **15**, 6984 (1999).

30. P. Uhlmann, N. Houbenov, N. Brenner, K. Grundke, S. Burkert and M. Stamm, *Langmuir*, **23**, 57 (2007).
31. N. Ayres, S.G. Boyes and W.J. Brittain, *Langmuir*, **23**, 182 (2007).
32. K. Yu, H. Wang, L. Xue and Y. Han, *Langmuir*, **23**, 1443 (2007).
33. A. Akinchina and P. Linse, *Langmuir*, **23**, 1465 (2007).
34. J.B. Webber, E.J. Wanless, V. Bütün, S.P. Armes and S. Biggs, *Nano Lett.*, **2**, 1307 (2002).
35. J.B. Webber, E.J. Wanless, S.P. Armes, Y. Tang, Y. Li and S. Biggs, *Adv. Mater.*, **16**, 1794 (2004).
36. G. Mino and S.A. Kaizerman, *J. Polym. Sci.: Polym. Chem. Ed.*, **31**, 242 (1958).
37. R.A. Wallace and D.G. Young, *J. Polym. Sci., Part A-1*, **4**, 1179 (1966).
38. S.A. Tashmukhamedov, P.P. Larin, R.S. Tillaev, Yu.T. Tashpulatov and Kh.U. Usmanov, *Vysokomol. Soedyneniya. Ser. A.*, **11**, 453 (1969) (in Russian).
39. D. Staszewska, *Makromol. Chem.*, **176**, 2481 (1975).
40. S.A. Tashmukhamedov, Sh.A. Azizov, A.Sh. Karabaev, A.T. Tsagaraev, R.S. Tillaev and Kh.U. Usmanov, *Vysokomol. Soedyneniya. Ser. A.*, **18**, 2171 (1976) (in Russian).
41. J. Wang, S.K. Varshney, R. Jerôme and Ph. Teyssie, *J. Polym. Sci., Part A: Polym. Chem.*, **30**, 2251 (1992).
42. H.-Q. Xie and D. Xie, *Prog. Polym. Sci.*, **24**, 275 (1999).
43. T.C. Chung, *Prog. Polym. Sci.*, **27**, 39 (2002).
44. G. Riess, *Prog. Polym. Sci.*, **28**, 1107 (2003).
45. H. Mori and A.H.E. Müller, *Prog. Polym. Sci.*, **28**, 1403 (2003).
46. T.B. Zheltonozhskaya, S.V. Fedorchuk and V.G. Syromyatnikov, *Russ. Chem. Rev.*, **76**, 731 (2007).
47. D.H. Solomon, E. Rizzardo and P. Cacioli, US Patent 4581429, 1985; *Chem. Abstr.*, **102**, 221335q (1985).
48. J.-S. Wang and K. Matyjaszewski, *Macromolecules*, **28**, 7572 (1995).
49. M. Kato, M. Kamigaito, M. Sawamoto and T. Hihashimura, *Macromolecules*, **28**, 1721 (1995).
50. J. Chiefari, Y.K. Chong, F. Ercole, J. Krstina, T.P. Le, R.T.A. Mayadunne, G.F. Meijs, G. Moad, C.L. Moad, E. Rizzardo and S.H. Thang, *Macromolecules*, **31**, 5559 (1998).
51. P. Corpart, D. Charmot, T. Bigdatti, S. Zard and D. Michelet, *Rhodia Chimie. Int. Appl.* WO 9858974; *Chem. Abstr.*, **130**, 82018 (1999).
52. C. Pugh and A.L. Kiste, *Prog. Polym. Sci.*, **22**, 601 (1997).
53. I.M. Papisov, *Vysokomol. Soedyneniya. Ser. A.*, **39**, 562 (1997) (in Russian).
54. S. Połowinski, *Prog. Polym. Sci.*, **27**, 537 (2002).
55. P. Aravindakshan, A. Bhatt and V.G. Kumar, *J. Appl. Polym. Sci.*, **66**, 397 (1997).
56. N.E. Zagdanskaya, T.B. Zheltonozhskaya and V.G. Syromyatnikov, *Voprosy Khim. Khim. Tekhnol.*, **3**, 53 (2002). (in Russian).
57. N. Permyakova, T. Zheltonozhskaya, O. Demchenko, L. Momot, S. Filipchenko, N. Zagdanskaya and V. Syromyatnikov, *Polish J. Chem.*, **76**, 1347 (2002).
58. T. Fang, L. Ji, J. Yu, L. Wang and S. Xu, *Polym. Bull.*, **29**, 71 (1992).

59. J. Gao, J. Yu, W. Wang, L. Chang and L. Tian, *J. Macromol. Sci. – Pure Appl. Chem.*, **A35**, 483 (1998).
60. X.-h. Peng, J.-p. Du, M.-j. Wang and Q. Lu, *Jingxi huagong=Fine Chem.*, **17**, 137 (2000) (in Chinese).
61. R.P. Singh, G.P. Karmakar, S.K. Rath, N.C. Karmakar, S.R. Pandey, T. Tripathy, J. Panda, K. Kanan, S.K. Jain and N.T. Lan, *Polym. Eng. Sci.*, **40**, 46 (2000).
62. L. Bromberg, *J. Phys. Chem. B*, **102**, 1956 (1998).
63. N.M. Permyakova, S.V. Fedorchuk, T.B. Zheltonozhskaya, N.E. Zagdanskaya and V.G. Syromyatnikov, *Voprosy Khim. Khim. Tekhnol.*, **8**, 108 (2007) (in Ukrainian).
64. L.N. Momot, T.B. Zheltonozhskaya, N.M. Permyakova, S.V. Fedorchuk and V.G. Syromyatnikov, *Macromol. Symp.*, **222**, 209 (2005).
65. C.-F. Huang, S.W. Kuo, F.J. Lin, C.-F. Wang, C.-J. Hung and F.-C. Chang, *Polymer*, **47**, 7060 (2006).
66. S.-W. Kuo, P.-H. Tung and F.-C. Chang, *Macromolecules*, **39**, 9388 (2006).
67. N.M. Permyakova, T.B. Zheltonozhskaya, S.V. Fedorchuk, N.E. Zagdanskaya and V.G. Syromyatnikov, *Mol. Cryst. Liq. Cryst.*, **468**, 405 (2007).
68. N. Permyakova, T. Zheltonozhskaya, V. Shilov, N. Zagdanskaya, L. Momot and V. Syromyatnikov, *Macromol. Symp.*, **222**, 135 (2005).
69. T.B. Zheltonozhskaya, L.N. Podolyak, B.V. Eremenko and I.A. Uskov, *Vysokomol. Soedyneniya. Ser. A.*, **29**, 2487 (1987) (in Russian).
70. T. Zheltonozhskaya, O. Demchenko, N. Kutsevol, L. Momot and V. Syrimyatnikov, in *Polymers of Special Applications* (Radom Techn. Univ., Radom, Poland, 2002), p. 242.
71. T.B. Zheltonozhskaya, Inter- and intramolecular polycomplexes stabilized by hydrogen bonds in the processes of flocculation and sorption, *Doctoral Thesis in Chemical Sciences* (Kiev National Taras Shevchenko University, Kiev, 2003) (in Ukrainian).
72. T. Zheltonozhskaya, O. Demchenko, I. Rakovich, J.-M. Guenet and V. Syromyatnikov, *Macromol. Symp.*, **203**, 173 (2003).
73. L.N. Momot, T.B. Zheltonozhskaya, T.V. Vitovetskaya, N.V. Strilchuk, L.R. Kunitskaya and V.G. Syromyatnikov, *Voprosy Khim. Khim. Tekhnol.*, **8**, 103 (2007).
74. T. Nonaka, T. Yoda and S. Kurihara, *J. Polym. Sci., Part A: Polym. Chem.*, **36**, 3097 (1998).
75. N. Kubota, T. Matsubara and Y. Eguchi, *J. Appl. Polym. Sci.*, **70**, 1027 (1998).
76. A. Suo, J. Qian, Y. Yao and W. Zhang, *J. Appl. Polym. Sci.*, **103**, 1382 (2007).
77. H.Q. Xie, J. Liu and D. Xie, *Eur. Polym. J.*, **25**, 1119 (1989).
78. Y. Zheng and G. Wan, *J. Appl. Polym. Sci.*, **57**, 623 (1995).
79. H. Katono, A. Maruyama, K. Sanui, N. Ogata, T. Okano and Y. Sakurai, *J. Control. Release*, **16**, 215 (1991).
80. T. Aoki, M. Kawashima, H. Katono, K. Sanui, N. Ogata, T. Okano and Y. Sakurai, *Macromolecules*, **27**, 947 (1994).
81. G. Chen and A.S. Hoffman, *Nature*, **373**, 49 (1995).
82. C.M. Hassan, F.J. Doyle III and N.A. Peppas, *Macromolecules*, **30**, 6166 (1997).
83. D.N. Robinson and N.A. Peppas, *Macromolecules*, **35**, 3668 (2002).

84. M. Coleman and P. Painter, *Prog. Polym. Sci.*, **20**, 1 (1995).
85. L. Zhu, M. Jiang, L. Liu, H. Zhou, L. Fan, Y. Zhang, Y.B. Zhang and C.J. Wu, *Macromol. Sci. Phys.*, **B37**, 805 (1998).
86. C. Wu, Y. Wu and R. Zhang, *Eur. Polym. J.*, **34**, 1261 (1998).
87. V.V. Khutoryanskiy, M.G. Cascone, L. Lazzery, N. Barbani, Z.S. Nurkeeva, G.A. Mun, A.B. Bitekenova and A.B. Dzhusupbekova, *Macromol. Biosci.*, **3**, 117 (2003).
88. J. Dai, S.H. Goh, S.Y. Lee and K.S. Siow, *Polym. J.*, **26**, 905 (1994).
89. T.G. Fox, *Bull. Am. Phys. Soc.*, **1**, 123 (1956).
90. M.M. Feldstein, *Polymer*, **42**, 7719 (2001).
91. S.W. Kuo and F.C. Chung, *Macromolecules*, **34**, 4089 (2001).
92. V.V. Khutoryanskiy, M.G. Cascone, L. Lazzery, Z.S. Nurkeeva, G.A. Mun and R.A. Mangazbaeva, *Polym. Int.*, **52**, 62 (2003).
93. M. Jiang, X. Qui, W. Qin and L. Fei, *Macromolecules*, **28**, 730 (1995).
94. M. Jiang, W. Chen and T. Yu, *Polymer*, **32**, 984 (1991).
95. X. Zhang, K. Takegoshi and K. Hikichi, *Polymer*, **33**, 718 (1992).
96. W. Herrera-Kao and M. Aguilar-Vega, *Polym. Bull.*, **42**, 449 (1999).
97. Z.S. Nurkeeva, G.A. Mun, A.V. Dubolazov and V.V. Khutoryanskiy, *Macromol. Biosci.*, **5**, 424 (2005).
98. V.V. Khutoryanskiy, A.V. Dubolazov, Z.S. Nurkeeva and G.A. Mun, *Langmuir*, **20**, 3785 (2004).
99. Yu.S. Lipatov, V.V. Shilov, Yu.P. Gomza, L.A. Bimendina and E.A. Bekturov, *Dopov. AN Ukr. Ser. B.*, 52 (1981) (in Ukrainian).
100. H. Feng, Z. Feng and Shen, L. *Polymer*, **34**, 2516 (1993).
101. L. Li, C.-M. Chan and L.-T. Weng, *Polymer*, **39**, 2355 (1998).
102. S.N. Cassu and M.I. Felisberti, *Polymer*, **40**, 4845 (1999).
103. K. Lewandowska, *Eur. Polym. J.*, **41**, 55 (2005).
104. V.V. Khutoryanskiy, M.G. Cascone, L. Lazzeri, N. Barbani, Z.S. Nurkeeva, G.A. Mun and A.V. Dubolazov, *Polym. Int.*, **53**, 307 (2004).
105. T. Nishi, *J. Macromol. Sci. – Phys.*, **B17**, 517 (1980).
106. L. Leibler, *Macromolecules*, **13**, 1602 (1980).
107. G.H. Fredrickson and F.S. Bates, *Annu, Rev. Mater. Sci.*, **26**, 501 (1996).
108. A.Yu. Grossberg and A.R. Khokhlov, *Statistical Physics of Macromolecules* (Nauka, Moscow, 1989) (in Russian).
109. K.S. Schweizer and J.G. Curro, *Adv. Polym. Sci.*, **116**, 319 (1994).
110. I. Hamley, *Block Copolymers* (Oxford Univ. Press., Oxford, 1999).
111. I. Hamley, *Adv. Polym. Sci.*, **148**, 113 (1999).
112. K. Ishizu and S. Uchida, *Prog. Polym. Sci.*, **24**, 1439 (1999).
113. I. Hamley and V. Castelletto, *Prog. Polym. Sci.*, **29**, 909 (2004).
114. L.-B. Feng, S.-X. Zhou, B. You and L.-M. Wu, *J. Appl. Polym. Sci.*, **103**, 1458 (2007).
115. P. Huang, J.X. Zheng, S. Leng, R.M. Van Horn, K.-U. Jeong, Y. Guo, R.P. Quirk, S.Z.D. Cheng, B. Lotz, E.L. Thomas and B.S. Hsiao, *Macromolecules*, **40**, 526 (2007).
116. N.M. Permyakova, T.B. Zheltonozhskaya, V.V. Shilov, N.E. Zagdanskaya, L.R.

Kunitskaya, V.G. Syromyatnikov and L.S. Kostenko, *Theor. Exper. Chem.*, **41**, 382 (2005).

117. T.B. Zheltonozhskaya, S.V. Fedorchuk, Yu.P. Gomza, N.M. Permyakova and V.G. Syromyatnikov, *Voprosy Khim. Khim. Tekhnol.*, **8**, 78 (2007). (in Russian).

118. V.P. Privalko and V.V. Novikov, *The Science of Heterogeneous Polymers. Structure and Thermophysical Properties* (John Wiley & Sons, Chichester, New York, Brisbane, Toronto, Singapore, 1995), p.127.

119. V.A. Bershtein and V.M. Egorov, *Differential Scanning Calorimetry in Physical Chemistry of Polymers* (Chemistry, Leningrad, 1980) (in Russian).

120. H.-Q. Xie, X.-H. Liu and J.-S. Guo, *Eur. Polym. J.*, **26**, 1195 (1990).

121. G.P. Karmakar and R.P. Singh, in *Proceedings of the SPE International Symposium on Oilfield Chemistry* (TX, Houston, 1997), p. 739.

122. S. Mishra, A. Panda and B.C. Singh, *J. Appl. Polym. Sci.*, **73**, 677 (1999).

123. O. Demchenko, N. Kutsevol, T. Zheltonozhskaya and V. Syromyatnikov, *Macromol. Symp.*, **166**, 117 (2001).

124. O. Demchenko, T. Zheltonozhskaya, J.-M. Guenet, S. Filipchenko and V. Syromyatnikov, *Macromol. Symp.*, **203**, 183 (2003).

125. V.V. Shilov, V.V. Shevchenko, P. Pissis, A. Kyritsin, G. Georgoussis, Yu.P. Gomza, S.D. Nesin and N.S. Klimenko, *J. Non-Cryst. Sol.*, **275**, 116 (2000).

126. J.T. Koberstein and L.M. Leung, *Macromolecules*, **25**, 6205 (1992).

127. N.A. Peppas and J. Klier, *J. Control. Release*, **16**, 203 (1991).

128. M.A. Lagutina, G.V. Rakova, N.V. Yarygina, S.A. Dubrovskii and K.S. Kazanskii, *Vysokomol. Soedyneniya, Ser. A*, **44**, 1295 (2002) (in Russian).

129. N.A. Plate, A.D. Litmanovich and O.V. Noa, *Macromolecular reactions* (Chemistry, Moscow, 1977) (in Russian).

130. Z. Tuzar and P. Kratochvil, in: *Surface and Colloid Science, Vol.15*. Ed. E. Matijevic (New York, Plenum Press, 1993), p. 1, chapter 1.

131. H. Shen, L. Zhang and A. Eisenberg, *J. Phys. Chem. B*, **101**, 4697 (1997).

132. P.L. Soo and A. Eisenberg, *J. Polym. Sci., Part B: Polym. Phys.*, **42**, 923 (2004).

133. A. Wittemann, T. Azzam and A. Eisenberg, *Langmuir*, **23**, 2224 (2007).

134. C. Wang, P. Ravi, K.C. Tam and L.H. Gan, *J. Phys. Chem., B*, **108**, 1621 (2004).

135. O. Terreau, C. Bartels and A. Eisenberg, *Langmuir*, **20**, 637 (2004).

136. E.B. Zhulina and T.M. Birshtein, *Vysokomol. Soedyneniya. Ser. A.*, **28**, 773 (1986) (in Russian).

137. I.W. Hamley, *The Physics of Block Copolymers*, 4. (Oxford Science Publication, Oxford, 1998).

138. P. Linse, in *Amphiphilic block copolymers: self-assembly and applications*. Eds. P. Alexandridis and B. Lindman (Elsevier, Amsterdam, 2000). p. 13.

139. J.-F. Gohy, S.K. Varshney and R. Jerôme, *Macromolecules*, **34**, 3361 (2001).

140. S. Holappa, M. Karesoja, J. Shan and H. Tenhu, *Macromolecules*, **35**, 4733 (2002).

141. V.E. Eskin, *Light Scattering by Polymer Solutions and Properties of Macromolecues* (Nauka, Leningrad, 1986) (in Russion).

142. A.H.E. Müller, X. Andre, C.M. Schilli and B. Charleux, *Pol. Mater. Sci. Eng.*, **91**, 252 (2004).

143. X. Andre, M. Zhang and A.X.E. Müller, *Macromol. Rapid. Commun.*, **26**, 558 (2005).
144. M. Koussathana, P. Lianos and G. Staikos, *Macromolecules*, **30**, 7798 (1997).
145. T.V. Burova, N.V. Grinberg, V.Ya. Grinberg, E.V. Kalinina, V.I. Lozinsky, V.O. Aseyev, S. Holappa, H. Tenhu and A.R. Khohlov, *Macromolecules*, **38**, 1292 (2005).
146. G. Li, L. Shi, Y. An, W. Zhang and R. Ma, *Polymer*, **47**, 4581 (2006).
147. G. Mountrichas and S. Pispas, *Macromolecules*, **39**, 4767 (2006).
148. O.S. Hernandez, G.M. Soliman and F.M. Winnik, *Polymer*, **48**, 921 (2007).
149. T.B. Zheltonozhskaya, O.V. Demchenko, N.V. Kutsevol, T.V. Vitovetskaya and V.G. Syromyatnikov, *Macromol. Symp.*, **166**, 255 (2001).
150. N. Zagdanskaya, L. Momot, T. Zheltonozhskaya, J.-M. Guenet and V. Syromyatnikov, *Macromol. Symp.*, **203**, 193 (2003).
151. N. Kutsevol, T. Zheltonozhskaya, N. Melnik, J.-M. Guenet and V. Syromyatnikov, *Macromol. Symp.*, **203**, 201 (2003).
152. D.J. Hemker and C.W. Frank, *Macromolecules*, **23**, 4404 (1990).
153. G.A. Mun, Z.S. Nurkeeva, V.V. Khutoryanskiy, V.A. Kan, A.D. Sergaziev and E.M. Shaikhutdinov, *Polym. Sci., Ser. B*, **43**, 289 (2001).
154. S.A. Tashmukhamedov, R.S. Tillaev, Kh.I. Akbarov and G. Khamrakulov, *Thermodynamical Properties of the Graft Copolymer Solutions* (FAN, Tashkent, (1989) (in Russian).
155. S.C. De Vos and M. Möller, *Makrromol. Chem., Macromol. Symp.*, **73**, 223 (1993).
156. H. Chen, J. Li, Y. Ding, G. Zhang, Q. Zhang and C. Wu, *Macromolecules*, **38**, 4403 (2005).
157. O.V. Borisov and E.B. Zhulina, *Macromolecules*, **38**, 2506 (2005).
158. N.P. Mel'nik, T.B. Zheltonozhskaya, L.N. Momot and I.A. Uskov, *Dokl. Akad. Nauk Ukr. SSR, Ser. B*, (11), 50 (1988) (in Russian).
159. N.P. Mel'nik, T.B. Zheltonozhskaya, L.N. Momot, I.A. Uskov, M.M. Matveenko, *Ukr. Khim. Zhurn.*, 56, 539 (1990) (in Russian).
160. S.R. Deshmukh, K. Sudhakar and R.P. Singh, *J. Appl. Polym. Sci.*, **43**, 1091 (1991).
161. A.M. Mathur, B. Drecher, A.B. Scranton and J. Klier, *Nature*, **392**, 367 (1998).
162. G.D. Poe, W.L. Jarrett, C.W. Scales and C.L. McCormick, *Macromolecules*, **37**, 2603 (2004).
163. E. Khousakoun, J.-F. Gohy and R. Jerôme, *Polymer*, **45**, 8303 (2004).
164. L. Bromberg, *J. Phys. Chem. B*, **102**, 10736 (1998).
165. T. Shibanuma, T. Aoki, K. Sanui, N. Ogata, A. Kikuchi, Y. Sakurai and T. Okano, *Macromolecules*, **33**, 444 (2000).
166. N. Chourdakis, G. Bokias and G. Staikos, *J. Appl. Polym. Sci.*, **92**, 3466 (2004).
167. R.A. Mangazbaeva, G.A. Mun, Z.S. Nurkeeva, E. Vecherkina and V.V. Khutoryanskiy, *Eurasian Chem. Tech. J.*, **7**, 219 (2005).
168. H. Dou, M. Jiang, H. Peng, D. Chen and Y. Hong, *Angew. Chem. Int. Ed.*, **42**, 1516 (2003).
169. S.K. Rath and R.P. Singh, *Colloids Surf. A. Physicochem. Eng. Asp.*, **139**, 129 (1998).

170. R.P. Singh, T. Tripathy, G.P. Karmakar, S.K. Rath, N.C. Karmakar, S.R. Pandey, K. Kannan, S.K. Jain and N.T. Lan, *Current Science*, **78**, 798 (2000).

171. T.B. Zheltonozhskaya, N.V. Kutsevol, L.N. Momot, N.P. Melnik and B.V. Eremenko, Ukrainian Patent, UA 23743 (2001); *Promislova Vlastnist'* (11), (2001).

172. T.B. Zheltonozhskaya, N.V. Kutsevol, L.N. Momot, O.O. Romankevich, V.G. Sytomyatnikov, V.Ya. Demchenko and V.M. Olenchenko, Ukrainian Patent, UA 29933 (2001); *Promislova Vlastnist'* (3), (2002).

173. N.E. Zagdanskaya, N.M. Permyakova, T.B. Zheltonozhskaya and S.V. Fedorchuk, Ukrainian Patent, UA 78649 (2007); *Promislova Vlastnist'* (4), (2007).

174. A.W.M. De Laat and H.F.M. Schoo, *Colloid Polym Sci.*, **276**, 176 (1998).

175. D. Hourdet, F. L'Alloret and R. Audebert, *Polymer*, **35**, 2624 (1994).

176. I. Kaur, S. Maheshwari, A. Gupta and B.N. Misra, *J. Appl. Polym. Sci.*, **58**, 835 (1995).

177. D.M. Disley, J. Blyth, D.C. Cullen, H.-X. You, S. Eapen and C.R. Lowe, *Biosensors & Bioelectronics*, **13**, 383 (1998).

178. O.V. Demchenko, T.B. Zheltonozhskaya, V.G. Syromyatnikov, N.V. Strelchuk, V.N. Yashchuk and V.Yu. Kudrya, *Functional Materials*, **7**, 711 (2000).

# CHAPTER 6

## INTERPOLYMER COMPLEXES OBTAINED BY TEMPLATE POLYMERIZATION

Stefan Połowiński

*Department of Physical Chemistry of Polymers,*
*Technical University of Lodz, Zeromskiego 116, 90-543 Lodz, Poland*
*stepol@mail.p.lodz.pl*

The current chapter outlines mechanisms of template polymerization, polycondensation and copolymerization. The mechanisms, illustrated with numerous examples, indicate a range of possibilities in synthesis of interpolymer complexes. Specific structure of products obtained by template polymerization as well as separation of the template and the daughter polymer is described. Literature in the field is also analyzed.

## 1. Introduction

Synthesis of polymers in which specific interaction between template macromolecules and growing chain takes place is called template or matrix polymerization. These interactions influence the structure of the polymerization product called usually a "daughter polymer" or "newborn polymer" and the kinetics of the process. Natural template processes such as self-replication of DNA or synthesis of proteins in living organisms are still inimitable in laboratory conditions. In this chapter, the synthesis of polypeptides or polynucleic acids is not considered, however, it is difficult to avoid some analogies between natural biological processes and template polymerization of simple synthetic polymers and copolymers.

The term "replica" polymerization was first used by Szwarc[1] and then was displaced by "template" polymerization. Since then, many articles, reviews[2-4] and chapters in encyclopaedias[5-8] have been published.

Template polymerization is a particular case of a more general group of processes such as polymerization in organized systems.[9] However, the term "template polymerization" usually refers to one phase systems in which monomer, and template is soluble in the same solvent. On the other hand, a compact interpolymer complex, gel or a precipitate of aggregated polycomplexes is formed as a product of the template process. It seems that it would be convenient not to limit template polymerization to processes running in liquids but include processes running in swollen gels, thin polymer films or on surfaces, but each case is specific, and a boundary is not sharp.

The template processes can be realized as template poly-condensation, polyaddition, ring-opening polymerization and ionic or radical polymerization.[10,11]

The influence of the template on the process and the product is usually called a "template effect" or "chain effect". The template effects can be expressed as: (a) kinetic effect -usually an enhancement of the reaction rate and change in kinetics equation; (b) molecular effect - influence on the molecular weight and molecular weight distribution; (c) effect on tacticity - the created polymer can have the complementary structure to the structure of the template used.

In the case of template copolymerization, the template effect deals with the sequence distribution of units. This effect is very important in biological synthesis, for instance in DNA replication.

Usually the system containing a monomer, solvent and a template is an initial state of template polymerization and two phenomena are important in the process - orientation and association of monomer units onto template and aggregation of polymerization product with the template.

The phenomena are related to the nature of forces operating in the system. It can be hydrogen bonds, dipole-dipole interactions, covalent bonding etc.

The first phenomenon can be considered in terms of a more general problem - interaction of a macromolecule in a mixed solvent.

The second phenomenon is very important and related to the structure of the polymerization product. In many cases, the products of template polymerization are interpolymer complexes with a specific, more regular structure.

In this chapter, polymerization of monomer units covalently bonded to a template will be omitted due to the fact that in this case as a product of polymerization ladder-type polymer is obtained, not an interpolymer complex.[11]

## 2. Radical template polymerization

The majority of papers published in the field of template polymerization deals with radical template polymerization. As in usual radical polymerization, template process is a chain reaction, and consists of initiation, propagation and termination. Examples of radical template polymerization of vinyl monomers are summarized in **Table 1**.

One of the earliest examples of radical template polymerization was described in Ref. 13. The authors suggest that polymerization of acrylic acid in the presence of poly(vinyl pyrrolidone) proceeds according to the template process. The mechanism of the polymerization is such that PVP molecules absorb the monomeric acrylic acid creating high local concentrations. On initiation, the acrylic acid molecules "zip-up" to produce poly(acrylic acid). According to the authors the reaction proceeds according to the scheme presented in **Fig. 1**.

Generalized kinetic model of radical-initiated template polymerizations were published by Tan and Alberda van Ekenstein.[32]

It is generally accepted that template polymerization can proceed according to two different mechanisms: type I (the "zip" mechanism) or type II (the "pick-up" mechanism). In the case of "type I", monomer units are connected to a template by strong forces (electrostatic, hydrogen bridges or even covalent bonds).

**Table 1**. Examples of vinyl monomers template polymerisation

| Monomer | Template | Solvent | Initiator | Reference |
|---|---|---|---|---|
| Acrylic acid | Poly(ethylene imine) | $H_2O$ | $K_2S_2O_8$ | 12 |
| Methacrylic acid | Poly(ethylene glycol) | $H_2O$ | $K_2S_2O_8$ | 13 |
| Methacrylic acid | Poly(ethylene glycol) | $H_2O$+50% methanol | $K_2S_2O_8$ | 14 |
| Methacrylic acid | Poly(ethylene glycol) | $H_2O$ | $Na_2S_2O_8^P$ | 15 |
| Methacrylic acid | Poly(vinyl pyrrolidone) | $H_2O$ | $UO_2SO_4^P$ | 16 |
| Methacrylic acid | Poly(vinyl pyrrolidone) | $H_2O$ | $K_2S_2O_8$ | 9 |
| Methacrylic acid | ionene | $H_2O$ | $K_2S_2O_8$ | 17 |
| p-styrene sulfonic acid | ionene | $H_2O$+ isopropanol | AIBN | 18 |
| Acrylic acid | Poly(vinyl pyrrolidone) | $H_2O$ | $K_2S_2O_8$ | 13 |
| Methacrylic acid | Poly(2-vinylpyridine) | DMF | TBPC* | 19,20,21 |
| Sodium acrylate | Poly(allylamine hydrochloride) | $H_2O$ | $K_2S_2O_8$ | 22 |
| Acrylic acid | chitozan | $H_2O$ | $K_2S_2O_8$ | 23 |
| 4- winyl pyridine | Poly(acrylic acid) | $H_2O$ | $K_2S_2O_8$ | 24,25,26 |
| N-vinyl imidazole | Poly(methacrylic amid) | $H_2O$ | AIBP** | 27,28,29 |
| p-styrenosulfonic acid | Ionene hydroxide | Water-propanol | AIBN | 18 |
| Dimethylaminoethyl methacrylate | Poly(acrylic amid) | Water-acetone | AIBN | 30 |
| N-vinyl pyrrolidone | Poly(methacrylic amid) | DMF | AIBN | 31 |

P- photoinitiation;    * t-butylcyclohexyl peroxydicarbonate;    ** 2,2` - azobis (2-aminopropane)

In the next step, polymerization proceeds along the template and daughter polymer forms a complex with the template. This is illustrated in the following scheme (**Fig. 2**), where T and M are interacting groups in the template and in the monomer respectively:

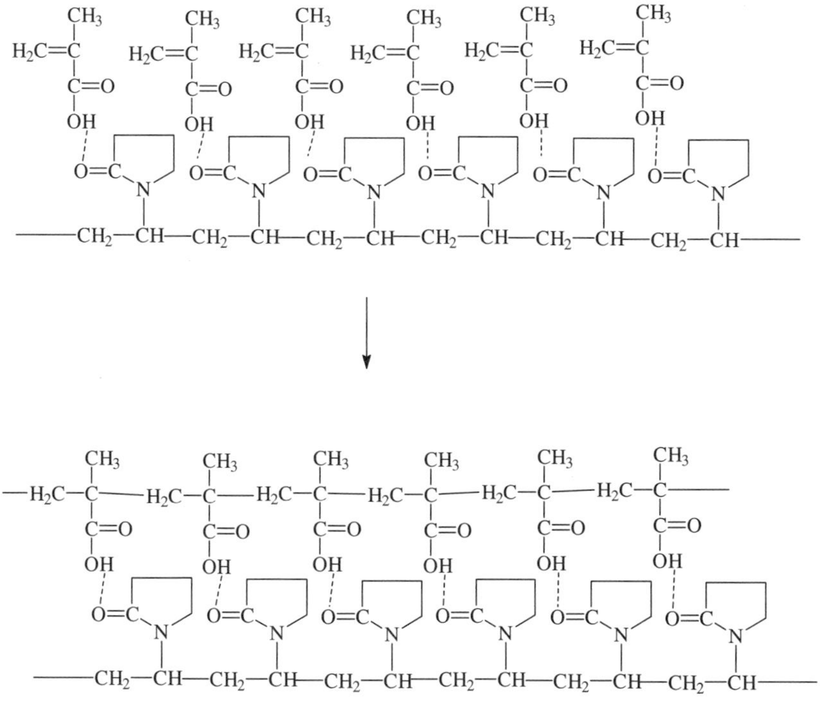

**Fig. 1.** Template polymerization of methacrylic acid onto poly(vinyl pyrrolidone).

Challa and Tan[3] rank among "template polymerization type I" the following examples: polymerization of acrylic acid in the presence of poly(ethylene imine) or ionenes, 4-vinylpyridine onto various polyacids and many others. In all mentioned cases, strong interaction between the monomer used and the template leads to the interpolymer complex.

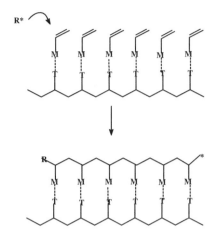

Fig.2. Schematic representation of "type-I" template polymerization.

For instance, the interaction between an ionene and monomer with sulfonic acid group leads to the strong complex on the basis of electrostatic forces, as demonstrated in **Fig. 3**.

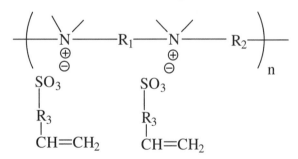

**Fig. 3**. Complex of a monomer with sulfonic acid groups with an ionene.

In the "type II" case, the monomer at the beginning of the reaction is "free" and polymerisation starts outside the template. When oligoradicals reach a critical length, complexation with the template occurs, and propagation proceeds along the template by adding monomer molecules from the surrounding solution. This is illustrated in **Fig. 4**.

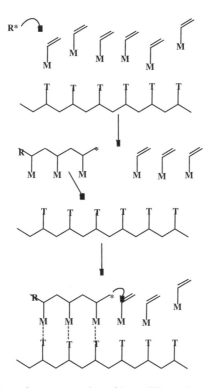

**Fig. 4.** Schematic representation of "type-II" template polymerization.

Challa and Tan[3] on the basis of kinetics examination suggest that polymerization of methacrylic acid in the presence of: poly(ethylene glycol), poly(L-lysine), isotactic poly(methylmethacrylate) and even poly(vinyl pyrrolidone) belongs to "type II case".

Typical examples of template reaction type II is the polymerization of acrylic or methacrylic acid onto poly(ethylene glycol).[13,14]

Where only a weak interaction takes place, as between poly(acrylic acid) and poly(ethylene glycol), the ability of a polymer-polymer complex to form will be related to the molecular weight of the two polymeric species. This is due to the fact that the interaction is reversible: Template + oligomer ↔ Complex.

When one of the reactants, normally the template is of high molecular weight and the other is a growing oligomer, the latter needs to reach a sufficiently high molecular size to allow the co-operative interactions along the chain to overcome the tendency of other forces such as Brownian motion to separate the chains.

The specific interaction between the complementary nucleic acids was used in template polymerization of vinyl monomers containing nucleic acid bases on the complementary template polymer.[33] The monomers used were the N-β-metacryloyloxyethyl derivatives, and templates with complementary groups shown in **Fig. 5**.

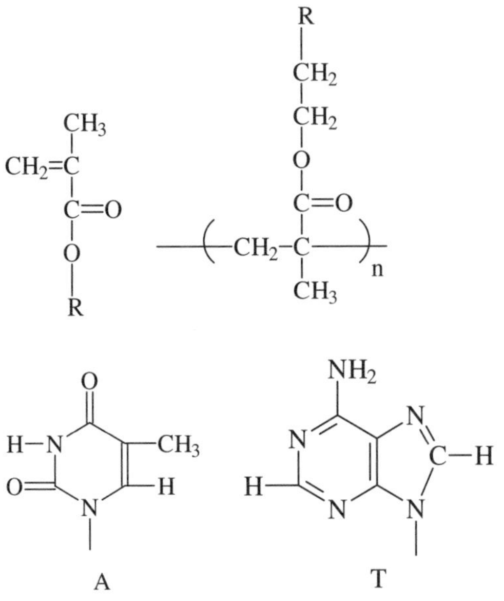

**Fig. 5**. Template polymerization of vinyl monomers containing nucleic acid bases. Monomers and templates; R- adenine (A) , uracil (U), or thymine (T) group.

It is well known that the interaction between such groups as thymine and adenine (**Fig. 6**) is very important in stabilization of double helix structure of DNA and adenine – uracil interaction in transcription process of RNA.

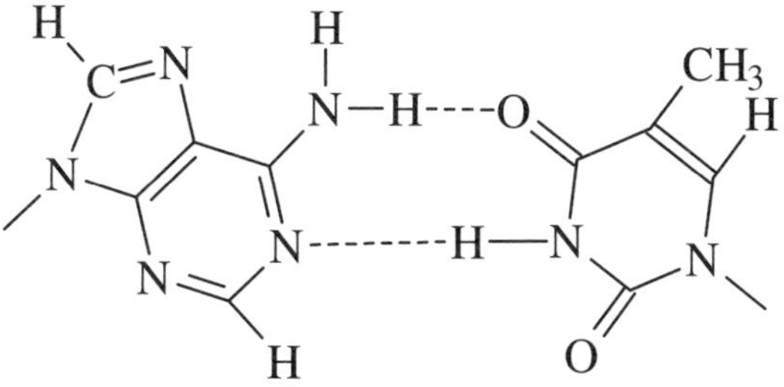

**Fig. 6.** Interaction between adenine and thymine group.

However, it was found by the authors that the interaction between adenine and uracil groups (**Fig. 7**) is not strong, therefore the monomer can not interact with the polymer, but the growing oligomer and the polymer containing uracil can form the complex with the template polymer containing adenine, as in typical "type II" template polymerization.

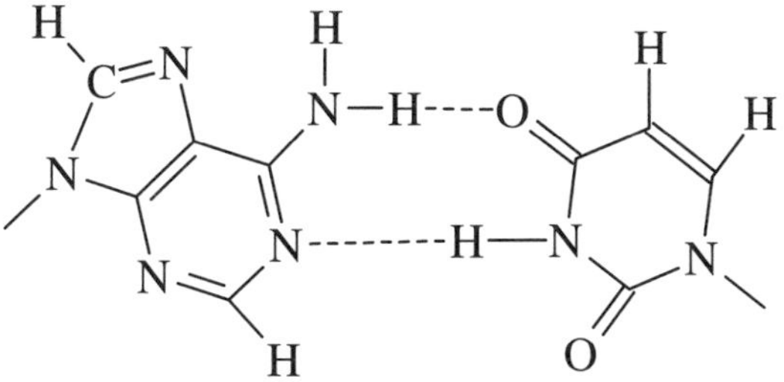

**Fig. 7.** Interaction between adenine and uracil groups.

The effects of the stereoregularity of the template polymer, polymerization temperature, and solvent used on the rate of polymerization were studied.[33] The authors discussed not only the influence of interaction between the monomer group and the group connected with the template, but also the intramolecular self-association of the bases along the polymer chain.

Another interesting example of a template "type II" polymerization was described by the authors.[34] They found that the radical polymerization of methyl methacrylate, if performed in the presence of isotactic poly(methyl methacrylate), proceeds according to the template mechanism. It was known that isotactic poly(methyl methacrylate) forms a stereocomplex with syndiotactic poly(methyl methacrylate). According to the authors, the driving force for stereoselectivity of this reaction is the occurrence of association between iso- and syndiotactic PMMA chains. It was found, that the complex formation depends on the template molecular weight, solvent type and temperature.

A very specific method of template polymerization by ring-opening polymerization was proposed by Bamford.[2,36] As a substrate, α-amino acid N-carboxy anhydrides with secondary amino groups was used (NCA). In the course of polymerization, NCA absorbs onto the polymer containing the amine group at the end of the macromolecule and a ring opening reaction takes place with a restoration of growing center and elimination of $CO_2$.

New type of ring-opening template polymerization and copolymerization was published.[36,37] The polymerization takes place according to scheme presented in **Fig. 8**. This polymerization differs from a conventional template polymerization as in this case, polymerization and separation take place simultaneously. It leads to both the daughter polymer and template polymer without chemical treatment after polymerization what is rather exceptional and it will be mentioned at the end of this chapter.

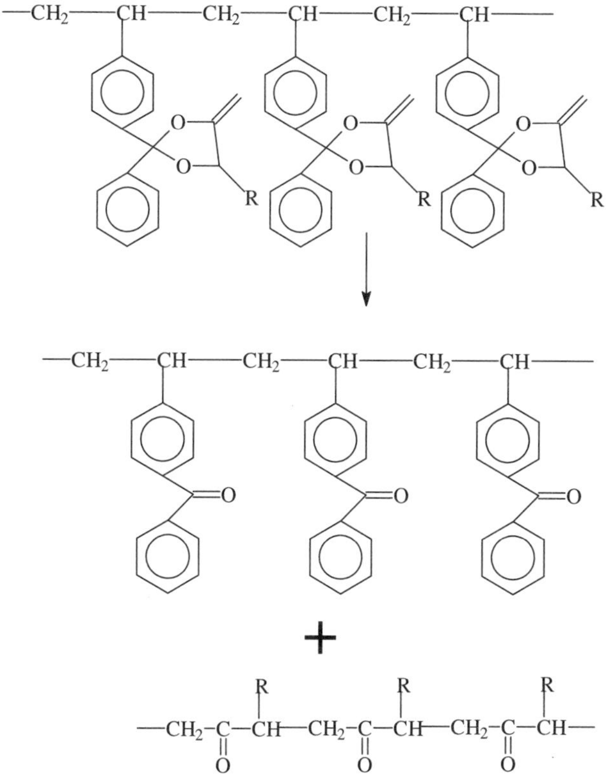

**Fig. 8**. Ring-opening template polymerization with simultaneous separation.

## 3. Template polycondensation

Template polycondensation is still not a very well studied process. This type of synthesis was revived by Korshak and Vasnev.[10]

Usually a polycondensation process takes place at a high temperature, which makes difficult to use a template for process control through interactions between the reaction substrates and the template. Even the processes carried out under milder conditions have been insufficiently studied in terms of process kinetics. However, the synthesis of polyamides by direct polyamidation of terephtalic acid with 4,4′-diaminodiphenylmethane in the presence of poly(ethylene oxide), poly(4-vinylpyridine) and poly(vinylpyrrolidone) has been described by Higashi.[38–41]

Direct polycondensation of aminoacids in the presence of poly(vinylpyrrolidone) and activating agents – triphenyl phosphate and LiCl has been also described by Higashi.[41]

The template polycondensation of urea and formaldehyde has been described by Papisov.[42] The author has suggested in his papers that the polycondensation of urea and formaldehyde on a poly(acrylic acid) template results in the formation of a complex with polyacid showing a specific structure being different than that of the complex obtained by mixing urea-formaldehyde resin with poly(acrylic acid).

If pH of the reaction mixture is lower than 3.7, a complex is formed according to the following reaction presented in **Fig. 9**.

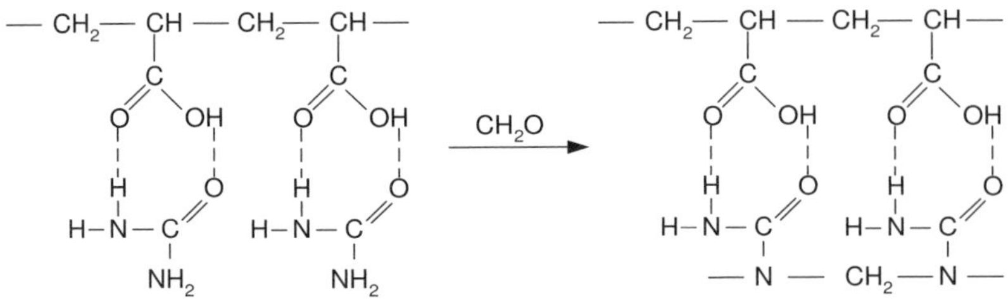

**Fig. 9**. Template polycondensation of urea-formaldehyde resin.

## 4. Template radical copolymerization

In the literature, there is limited information about the template copolymerization than about the template homopolymerization. However, the process seems to be very interesting on account of the fact that template can influence not only the molecular weight and tacticity of daughter polymer but also the composition and distribution of units in copolymer macromolecule. Only few different examples of template copolymerization can be considered. One of the examples deals with the system in which one comonomer (A) forms a complex with the template, while the other comonomer (B) reveals a weak interaction or is not interacting with the template at all. This case is illustrated in **Fig. 10**.

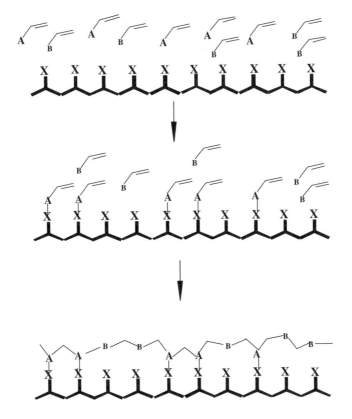

**Fig. 10.** Scheme of copolymerization in which only A-group of monomer is interacting with x-groups of template.

It is known, that the composition and the distribution of units in copolymerization are controlled mainly by propagation process. From this point of view, the equations have been formulated[43] and it was shown, how the reactivity ratios are dependent on the template concentration and individual reactivity rate constants of monomers taking part in the template copolymerization process. However, if the long critical lengths are necessary for adsorption of a growing macroradical onto the template any template effect can be destroyed.

Experimentally checked systems for such a case of template copolymerization are not very numerous, but all known results confirm the influence of template on the reactivity ratios of comonomers.[43-46]

The second case deals with the system when the template is a copolymer and template units X interact with monomer A and template units Y interact with monomer B. It is presented in **Fig 11**.

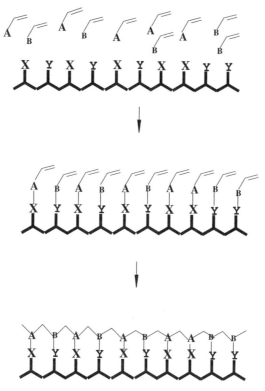

**Fig. 11**. Scheme of copolymerization in which A-group of monomer are interacting with X-and B-group with Y group of the template.

Literature in this field is quite limited. A very interesting template effect was described by Ref. 47. It was found that the copolymer formed in the early stages of the reaction can play a role of template for growing copolymer macromolecules.

## 5. Separation of the template and the daughter polymer

Many authors tried to separate and characterize the polymer obtained in a template process. However, there are two main problems, which makes the procedure difficult. The first is related to the strong bonding and the existence of chain entanglements between the component macromolecules, the second - to the secondary reactions in template polymerization.

Ferguson and Shah[13] used electrophoresis for the polymerization product separation in the polymerization of metacrylic acid onto poly(vinyl pyrolidone). Poly(methacrylic acid) ionizes in aqueous solution and PVP does not, so only PMM moves in the electric field. Using this method, the authors showed that graft copolymer is formed parallel to the template polymerization of acrylic acid. The same method applied to the product of template polymerization of acrylic acid onto poly(ethylene imine) leads to the conclusion that full separation of the complex components is possible.[12] The complex was soluble below pH 2.4 and above pH 9.9.

Klein[48] reported separation of poly(2-vinyl pyridine) from poly(methacrylic acid). The first step in separation of the template and the polymer is usually the solubilization of the template-polymer complex. The author reported that the poly(2-vinylpyridine)-poly(methacrylic acid) mixture could be separated in a strong acid medium. Klein claimed a complete separation of a polymer and a template. However, Blumstein and Kakivaya[18] reported that poly(2-vinylpyridine) and poly(acrylic acid) formed onto this template could not be properly separated by Klein's technique and remained contaminated with each other up to 15–20 wt. %. Separation of the template-formed complex poly(N-vinylimidazole) and poly(methacrylic acid)  by column chromatography using  $Al_2O_3$ appeared to be unsuccessful. Also precipitation of the components by different liquids led to incomplete separation of poly(N-vinyl imidazole) contaminated with up to 10% of poly(methacrylic acid).

Sato and co-authors[49] obtained polymer complex by template polymerization of maleic acid in the presence of poly(vinyl pyrrolidone). It was found that thermal stability of this complex is substantially different from those of template and daughter polymer mixture. In order to separate the components, the polymer complex was dispersed in benzene and excess of diazomethane was added. After methylation, products were analyzed by IR spectroscopy and gel permeation chromatography. However, about 30%–40% of probably grafted product was insoluble.

In order to separate the complex formed in polymerization of methacrylic acid in the presence of poly(vinylpyrrolidone), a similar method was applied.[50] The complex dispersed in THF was converted by diazomethane according to reaction presented in **Fig. 12**:

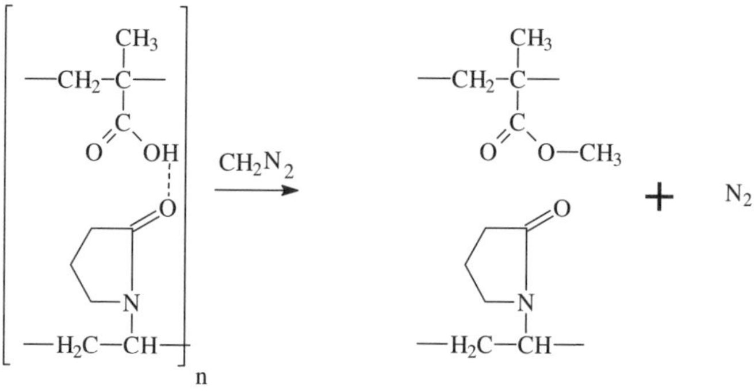

**Fig. 12**. Separation the template by methylation.

After separation, PVP remained insoluble in esterification medium and it was possible to analyze molecular weight and molecular weight distribution of poly(methacrylic acid) converted to PMM.

The separation of template polymerization products is additionally complicated by contamination of the main product by substances formed in the secondary reactions such as grafting, crosslinking, branching.

The problem of secondary reactions in template polymerization was discussed in Refs. 4, 8 and is related to the definition of template processes. However, some secondary reactions are definitely different from the template process. First of all, in many systems grafting takes place as it was mentioned in the case of polymerization studied by Ferguson and Shah.[13] Chain transfer to the monomer in polymerization of N-vinylimidazole in the presence of poly(methacrylic acid) and leads to branching or crosslinking.[27–29,51]

Litmanovich and co-workers found that the structure of urea-formaldehyde complex formed as a result of template polycondensation in the presence of poly(acrylic acid) depends on pH of the medium. Structure, swelling and stability of such complexes were described.[52]

# References

1. M. Szwarc, J. Polym. Sci., **13**.317 ( 1954).
2. C.H. Bamford, in "Developments in Polymerization", Ed. R. N. Haward (Applied Sci. Pub., London, 1979), p. 215.
3. G. Challa G. and Y.Y. Tan, Pure Appl. Chem. **53** . 627 (1981).
4. Połowiński S., Prog. Polym. Sci., **27**.537. (2002).
5. Y.Y. Tan, in *Comprehensive Polymer Science*, Eds. G. Allen and J.C. Bevington, (Pergamon Press, Oxford, 1989), Vol. 3, p. 245.
6. Y.Y. Tan and G. Challa, in *Encyclopedia of Polymer Science and Engineering*, Ed. Mark, Bikales, Overberger and Menges (Wiley, New York, 1989), Vol. 16, p. 554.
7. S. Połowiński, in *The Encyclopedia of Advanced Materials,* Eds., D. Bloor, R.J. Brook, M.C. Flemings, and S. Mahajan (Pergamon Press, Oxford, 1994), p. 2784.
8. S. Połowiński, in *Polymeric Materials Encyclopedia*, Ed. J.C. Salamone (CRC, Boca Raton 1996), Vol. 11, p. 8280.
9. N. Shavit and J. Cohen in *Polymerization of Organized Systems*, Ed. Elias H.G. (Gordon and Breach Sci. Pub., New York, 1977), p. 213.
10. V.V, Korshak and V.A. Vasnev (1989), in *Comprehensive Polymer Science*, Ed. G. Allen and J.C. Bevington (Pergamon Press, Oxford, 1989), Vol. 5, p. 160.
11. S. Połowiński, Template Polymerization; ChemTec Pub. Toronto-Scarborough. (1997).
12. J. Ferguson and S.A.O. Shah, Europ. Polym. J. **4** p. 611 (1968)
13. J. Ferguson and S.A.O. Shah, Europ. Polym. J. **4** p. 343 (1968)
14. I. M. Papisov, V.A. Kabanov, E. Osada, M. Leskano-Brito, J. Reimont, and A.N. Gvozdetskii, Vysokomol. Soj. **A14** p. 2462 (1972) (in Russian).
15. J. Matuszewska-Czerwik and S. Połowiński, Eur. Polym. J. **24** p. 791 (1988).
16. J. Matuszewska-Czerwik and S. Połowiński, Eur. Polym. J. **26** p. 549 (1990).
17. E. Tsuchida and Y. Osada, J. Polym. Sci., Polym. Chem. Ed., **13** p. 559 (1975).
18. A. Blumstein and S. Kakivaya in H.G. Elias ed., *Polymerization of Organized Systems* (Gordon and Brech Sci. Pub., 1977), p. 189.
19. J. Smid, Y.Y. Tan and G. Challa, European Polym J. **19** p. 853 (1983.)
20. J. Smid,Y.Y. Tan and G. Challa, European Polym J. **20** 1095 (1984).
21. J. Smid, J.C. Speelman, Y.Y. Tan and G. Challa, Europ. Polym. J. **21** p. 141 (1985).
22. P. Cerrai, G.D. Guerra, S. Maltini, and M. Tricoli, Macromol. Chem. Commun. **15** p. 983 (1994).
23. P. Cerrai, G.D. Guerra, M. Tricoli, S. Maltini, N. Barbani, and L. Petarca, Macromol. Chem. Phys. **197**, p. 3567 (2003).
24. V.A. Kabanov, K.V. Aliev, O.V. Kargina, T.J. Petrikeeva and V.A. Kargin, J. Polym. Sci. **C 16,** p. 1079 ( 1967).
25. V.A. Kabanov, O.V. Kargina and V.A. Petrovskaya, Vysokomol. Soed., **A13** pp. 348 (1971) (in Russian).

26. O.V. Kargina, L.A. Maszustina, V.J. Sverun, G.M. Lukovkin, V. P. Evdakov and V.A. Kabanov, Vysokomol. Soed., **A16** p. 1755 (in Russian).

27. H.T. Van de Grampel Y.Y. Tan and G. Challa, Makromol. Chem. Makromol.Symp. **21/22**, p. 83. (1988).

28. H.T. Van de Grampel, Y.Y. Tan and G. Challa, Macromolecules, **23** p. 5209 (1990).

29. H.T. Van de Grampel, Y Y. Tan and G. Challa, Macromolecules, **24**, p. 3767 (1991).

30. G.O.R. Alberda van Ekenstein, D.W. Koetsier and Y.Y. Tan, Europ. Polym. J. **17** p. 845 (1981).

31. Z.H. Abd-Ellatif, Polym. International **28** p. 301 (1992).

32. Y.Y. Tan. and G.O.R. Alberda van Ekenstein, Macromolecules, **24**, p. 1641 (1991).

33. M. Akashi, H. Takada, Y. Inaki and K. Takemoto, J. Polym. Sci., Polym. Chem. Ed., **17** p. 747 (1979).

34. R. Buter, Y.Y. Tan and G. Challa, (1972); J. Polym. Sci. **A1 10** p.1031 (1972).

35. D.G.H. Ballad and C.H. Bamford, Proc. Roy. Soc. **A236** p. 384 (1956).

36. J. Sugiyama, T. Yokozawa and T. Endo, J. Am. Chem. Soc. **115** p. 2041.(1993).

37. J. Sugiyama, T. Yokozawa, T. Endo, Macromolecules **27** p. 5536 (1994).

38. F. Higashi, and Y. Tagushi, J. Polym. Sci., Chem. Ed., **18** p. 2875.( 1980).

39. F. Higashi, M. Got and H. Kakinoki, J. Polym. Sci., Chem. Ed., **18** p. 851 (1980).

40. F. Higashi, Y. Nakano, M. Goto and H. Kakinoki, J. Polym. Sci., Chem. Ed., **18** p. 1099 (1980).

41. F. Higashi, K. Sano and H. Kakinoki, J. Polym. Sci., Chem. Ed., **18** p. 1841 (1980).

42. I.M. Papisov, O.E. Kuzovleva, S.V. Markov, and A.A. Litmanovich, Europ. Polym J. **20** p. 195 (1984).

43. S. Połowiński, J. Polym. Sci., Polym. Chem., Ed. **22** p. 2887 (1984).

44. S. Połowiński, Eur. Polym. J. **19** p. 679 (1983).

45. S. Połowiński, Acta Polym. **43** p. 99. (1992).

46. A. Chapiro, Eur. Polym. J. **25** p. 713 (1989).

47. A. Chapiro, J. Dulieu, Z. Mankowski and N. Schmitt, Eur. Polym. J. **25** p. 879 (1989) (in French).

48. C. Klein, Makromol. Chem. **161** p. 85 (1972) (in German).

49. T. Sato, K. Nemoto, S. Mori and T. Otsu, J. Macromol. Sci. Chem. **A13** p. 751 (1979).

50. J. Szumilewicz, S. Połowiński, J. Supera, Sci. Bull Lodz Tech. Univ. **701** p. 35 (1994).

51. H.T. Van de Grampel, Y.Y. Tan and G. Challa, Macromolecules **24** p. 3773 (1991).

52. A.A. Litmanovich, S.V. Markov and I.M. Papisov, Vysokomol. Soj, A28 p.127 (1986) ( in Russian).

# CHAPTER 7

# INTERPOLYMER COMPLEXES CONTAINING COPOLYMERS

Gina-Gabriela Bumbu[1*], John Eckelt[2], Cornelia Vasile[1]

[1] *"P.Poni" Institute of Macromolecular Chemistry, 41A Gr. Ghica Voda Alley, Ro 6600 Iasi, Romania*
*E-mail: bmgg@icmpp.ro*
[2] *University of Mainz, Institute of Physical Chemistry, Jakob Welder Weg 13, 55099 Mainz, Germany*

## 1. Introduction

The incorporation of small amounts of groups able to develop hydrogen-bonding interactions between the polymers in a blend is known as a method to enhance polymer-polymer miscibility.[1-4] When the number of the groups capable to participate in interpolymer H-bonding is large enough, a new structure with different physico-chemical properties known as interpolymer complex, IPC, is formed.[5-7]

For a long time, the majority of the experimental work was focused on IPCs formed through H-bonding interactions between synthetic complementary polymers like weak polyacids and Lewis polybases.[8-10] However, during the last decade, the researchers interest have been caught by the intermacromolecular associations between non-complementary polymers containing variable number of functional groups on the structural unit like polysaccharides[11-14] or random,[15,16] alternant,[17,18] graft-,[19] and block-copolymers[20] as they can better simulate the intermolecular interactions between natural complex macromolecules. In addition to H-bonding, hydrophobic interactions were found to contribute to the formation and stabilization of the IPCs.[21]

Generally, the formation and structure of interpolymer complexes depend on factors such as the chemical structure of both partners,[22] their chain lengths,[8,9,23] the environment like solvent[22,24] and temperature,[11] the feed composition, and the concentration.[24] Jiang *et al.* investigated the influence of the chemical structure and of the environment on the complexation between poly(styrene-*co*-vinylphenol) (STVPh) and poly(styrene-*co*-vinylpyridine) (STVPy). They found that the minimum required amount of functional groups for the formation of an H-bonding IPC depends on the used solvent. They discovered that while chloroform and butan-2-one are not interfering in the formation of the IPC through hydrogen bonding between STVPh and STVPy, tetrahydrofuran (THF) has a substantial influence on complexation, and the addition of *N, N* dimethylformamide (DMF) even causes decomplexation.[22] If the interpolymer complexation takes place in water, it is expected that the number of H-bonding required to form a stable IPC is much higher than in organic solvents due to competitive H-bonding of the polymers with the water molecules.[15] Using random copolymers as one of the partners of the IPC, one can notice different effects on the complexation process depending on the nature and the amount of the comonomer. Bokias *et al.* studied the influence of the nature and the amount of non-complexable groups ($x$) on the ability of an polyacid to form IPCs.[25] Both in the case of ionic and neutral non-complexable units, the presence of these groups in the copolymers weakened its ability to form IPCs. When the non-complexable units were charged, the weakening was faster and appeared at lower amounts of inert groups in the copolymer. For example, the complexation of poly(ethylene oxide) (PEO) with copolymers of poly(acrylic acid) (PAA) with poly(*N*, isopropylacryl amide) (PNIPAM) was inhibited at a molar content of PNIPAM of 29 mol % while for copolymers of PAA with vinylsulfonic acid, it was suppressed already at $x = 12$ mol %. The presence of hydrophobic non-complexable comonomers promoted the formation of the complex stabilizing its structure due to the hydrophobic forces.[16,17] Partially ionized PAA could be regarded as a random copolymer of acrylic acid units, able to participate to H-bonding with ionized acrylic acid units, unable to form H-bonding, that leads to the weakening of the complexation process. Iliopoulos *et al.*[26,27] observed a so called "structural defects" effect on the complex formation of partially ionized PAA with poly(oxyethylene) or poly(vinylpyrrolidone) (PVP). A weakening of the cohesion between the interacting polymer chains was expected due to the strong interaction of the anionic acrylate residues of the partially ionized PAA with water.[15] Bokias *et al.* stated that the limiting content of non-complexable ionic "structural defects" for the formation of H-bonding IPCs is *ca.* 10 mol % when linear copolymers are used,[28-30] while for graft copolymers hydrogen-bonding IPC formation is still very significant at this copolymer composition.[31] The authors assigned this large difference to the architecture of the used copolymers. In the case of the linear statistical copolymers, the charged units are distributed along the chain and act as strong structural defects inhibiting complexation, mostly by generating a stretching of the polyacid chain.[30] In the case of the graft copolymers, the charged units are placed in form of separate side chains, resulting in a much less pronounced stretching of the polyacid chain, permitting its complexation with polybases. This behaviour is further enhanced as the molecular mass of the side chains increases.[32]

The most common methods to investigate the complexation process and the IPC formation are viscometry, pH-metry, turbidimetry, Fourier transformed infrared spectroscopy (FTIR), and differential calorimetry (DSC). It is out of the purpose of this work but for the sake of clarity and better understanding of this chapter we will briefly present the basic principles of the above-mentioned methods.

It is well known that the comparison of the experimental specific, reduced or intrinsic viscosity of a polymer mixture with an "ideal value" can be used as a criterion for the compatibility or association of the two polymers in solution.[33,34] The ideal viscosity of a polymer mixture represents the viscosity of the system in the absence of any specific interactions between the two components. In the case of a solution of a mixture of two neutral polymers or of charged polymers in salt, the ideal values for the viscosity of the mixtures are calculated as the weight average of the viscosity values of the two components. When one of the partners is a polyelectrolyte, like a weak polyacid, the polyelectrolyte effect (the increase of the reduced viscosity of the polymer solution by decreasing concentration) has to be taken into account.[35] In such a case, the following equation has been proposed for the evaluation of the reduced "ideal viscosity" of the mixture:

$$\left( \eta_{sp} \Big/ c_{pol} \right)_m = w_1 \left( \eta_{sp,1} \Big/ c_1 \right) + w_2 \left( \eta_{sp,2} \Big/ c_{pol} \right) \tag{1}$$

where $\eta_{sp,1} \Big/ c_1$ is the reduced viscosity of the polyelectrolyte at concentration $c_1$, and $\eta_{sp,2} \Big/ c_{pol}$ is the reduced viscosity of the neutral polymer at the total polymer concentration in the system ($c_{pol}$).

The deviation of the experimental viscosity values from the "ideal" ones, is better put in evidence by the ratio between the experimental and the ideal intrinsic or reduced viscosity:

$$\eta_{red, r} = \eta_{red, exp} / \eta_{red, id} \tag{2}$$

Any deviations of the viscosity ratio from unity or of the experimental viscosity from the "ideal" one indicate interactions between the components; a negative deviation points out the formation of an interpolymer association with a compact structure while a positive deviation is an indication of a gel-like association, cf. **Fig. 1**.[35]

Another method to study the formation of H-bonded IPCs is potentiometry. In a system formed from a polyacid and a polybase, the deviation of the pH value of their mixed aqueous solution from that of the corresponding polyacid aqueous solution, at the same polyacid concentration, represents the contribution of H-bonding to the pH.[25,36] For any polyacid concentration the pH values of the polyacid/water system are lower than the pH values of the polyacid/polybase/water, cf. **Fig. 2**. This means that in mixture containing IPC, there are fewer free hydrogen ions than in a pure polyacid solution. Thus, the IPC are formed through H-bonding between non-dissociated groups of

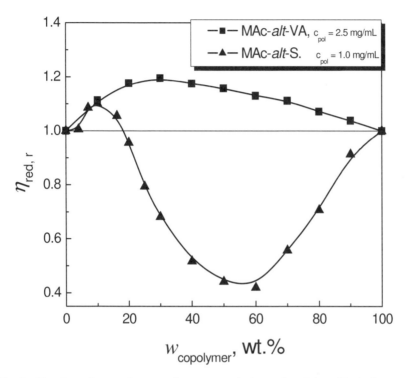

**Fig. 1.** The dependence of $\eta_{red,\,r}$ of aqueous solutions of maleic acid-vinyl acetate copolymer (MAc-*alt*-VA) / hydroxypropyl cellulose (HPC) and maleic acid-styrene (MAc-*alt*-S)/HPC mixtures, on the weight fraction of copolymer in the polymer mixture at T = 25 °C and constant $c_{pol}$.

polyacid and proton-acceptor groups of the polybase. The presence of a polybase in polyacid solutions produces a shift of the dissociation equilibrium of the polyacid to the non-dissociated form, due to the H-bonding between the proton-acceptor groups of the polybase and the non-dissociated, proton-donor COOH groups of the polyacid (see more details in chapter 3).

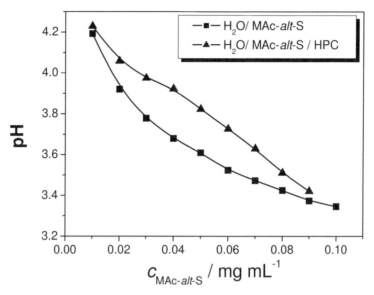

**Fig. 2.** The dependence of the pH of the MAc-*alt*-S solutions on polyelectrolyte concentration in presence of HPC, the complexation partner, and in its absence, at T = 25 °C. (Taken from Bumbu, G-G., Vasile, C., Eckelt, J. and Wolf, B.A.: Investigation of the Interpolymer Complex between Hydroxypropyl Cellulose and Maleic Acid-Styrene Copolymer, 1 Dilute Solutions Studies. Macromolecular Chemistry and Physics, 2004, 205, 1869-1876. Copyright Wiley-VCH Verlag GmbH & Co. KGaA. Reproduced with permission.

    A system containing IPC is sensitive to changes in pH, ionic strength, mixing ratio between the partners, and the temperature of the environment. The presence of H-bonded IPC in the system can be easily confirmed by turbidimetry as the solubility of the IPC in water varies with the above-mentioned parameters and therefore the optical density of the solution.[23,37,38] For example, upon a decrease in the pH a system containing an H-bonding IPC starts to undergo phase separation. The pH value at which this process occurs is called critical pH of complexation ($pH_{crit}$)[9] and it can be used to evaluate the complexation ability of the partners.[39-42] Below the $pH_{crit}$ value, the formation of IPCs with hydrophobic structure takes place and the polycomplex particles undergo further aggregation and precipitation. The complexation ability of a system and the stability of an IPC can be evaluated from the thermodynamic functions, e.g., enthalpy and entropy of complexation and the complexation constant. These parameters are mainly obtained from pH-metry. Detailed descriptions of these methods are the objective of chapter 3 in this book.

The glass transition temperature of a polymer mixture that presents specific interactions between the components can be described by the following equation[43]:

$$T_{gM} = w_1 T_{g1} + w_2 T_{g2} + q w_1 w_2 \tag{3}$$

where $w_1$ and $w_2$ are the weight fractions of the components of the blend, $T_{g1}$, $T_{g2}$, and $T_{gM}$ are the glass-transition temperatures of the components and of the mixture respectively, and $q$ is Kwei's constant. The $q w_1 w_2$ term is proportional to the number of specific interactions existent in the mixture and can be interpreted as the contribution of the hydrogen bonding. According to Painter *et al.*,[44] the $q$ parameter depends on the balance between the process of self-association and inter-associations in the blend. A high $q$ value is an indication of the presence of a large number of specific interactions in the system and reflects therefore the possibility of IPC formation.

In the FTIR spectra of IPCs, one can notice shifts of the characteristic bands of the functional groups of the polymers due to the specific interactions. The stoichiometry of the IPC can be calculated based on the ratio of the areas of the bands of associated and non-associated groups, cf. **Fig. 3**:

$$F_F^N = \frac{A_{free}}{A_{free} + \dfrac{a_{free}}{a_{assoc}} A_{assoc}} \tag{4}$$

where $A_{free}$ and $A_{assoc}$ are the areas of the bands of free and associated groups, respectively, and $a_{free}$ and $a_{assoc}$ are the absorptions of the two bands, assuming an absorption ratio of unity.

The selected bands depend on the chemical composition of the system, e.g., carbonyl (at 1737 cm$^{-1}$ and 1708 cm$^{-1}$),[45] carboxylic acid (1700 cm$^{-1}$, 1728 cm$^{-1}$) or amide I region (1640 cm$^{-1}$, 1656 cm$^{-1}$, 1680 cm$^{-1}$). In order to assess the free and hydrogen bonded groups, different approaches have been used, e.g., temperature dependence of the area of the bands, deconvolution of the overlapped bands by Gaussian and Lorentzian bandshapes, etc.

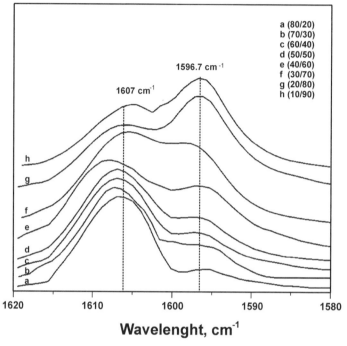

**Fig. 3**. Scale-expanded FTIR difference spectra (substraction of polystyrene) bands of the blend of poly(styrene-co-methacrylic acid), containing 29 mol % of methacrylic acid, SMA-29 with poly(styrene-co-4-vinylpyridine) containing 15 mol % of 4-vinylpyridine, S4VP-15, in the 1620-1580 cm-1 region recorded at room temperature. Reprinted with permission of John Wiley & Sons, Inc from Bennour, S., Metref, F. and Djadoun, S. Journal of Applied Polymer Science, 2005, 98, 806-811.

Painter and co-workers[46-48] developed an association model to describe the H-bonding in polymer mixtures. They included in the expression of the free energy of mixing an additional term that takes into account the contribution of hydrogen bonds ($\Delta G_H$). Using the association model the temperature-composition phase diagram and the enthalpy of mixing, $\Delta H_M$ can be calculated.

Moolman *et al.*[49] used molecular dynamic modelling to predict the optimum mixing ratio for the formation of an IPC between polyvinyl alcohol (PVOH) and a copolymer of maleic acid with methyl vinyl ether (MAc-*alt*-MVE). Different configurations of the polymer chains have been generated and placed in a cubic box using the amorphous-cell code based on the "self-avoiding" random-walk method of Theodorou and Suter[50] and on the Meirovitch scanning method.[51] The oligomers were equilibrated using both NVT and NPT (constant number of molecules, temperature, and volume, or pressure respectively) dynamic and the cohesive energy density (CED's) of the model were computed. From the CED's, the energies of mixing and Flory Huggins interaction parameter ($\chi$) values were estimated. They found negative $\chi$-values for all PVOH/MAc-*alt*-VME blends, indicating favourable interaction between the polymers with a minimum around 0.6–0.7 mass fraction of MAc-*alt*-VME. The results of molecular dynamics are in good agreement with the experimental viscosity results for blends at low pH.

This chapter reviews the work done in the field of IPC where at least one of the components is a copolymer. The copolymers used for IPC investigations can be random, alternant, graft or block-copolymers and can act as polybase or polyacid or both.

## 2. IPC containing the copolymer as a polybase

### 2.1. *Random copolymers*

Relatively limited number of studies deals with IPCs where the proton acceptor component (the polybase) is a copolymer.[15,16,39,40,52] Morawetz and Wang investigated the IPC formation between PAA and a co-polymer consisting of two hydrophilic nonionic monomers, N, N- dimethylacrylamide and acrylamide[15] with a content of acrylamide varying form 6 to 55 mol %. Both comonomers can form H-bonding and are therefore able to participate in the complexation process depending on the ionic strength and the pH of the medium. At lower pH (e.g. pH = 3), both comonomers can participate to the H-bonding and the authors found that the ratio of monomer units of PAA and copolymer in the complex is independent on the copolymer composition. At higher pH, pH = 4, the complexation abilities of PAA with the copolymer decrease dramatically as the acrylamide is not any longer able to participate in H-bond formation. For a content of 55 mol % acrylamide in the copolymer, no complex can be formed.

If one of the monomers forms hydrogen bonds and the other one does not but is relatively hydrophobic, two concurrent effects appear in the system: a decrease in the strength of the complexation abilities of the copolymer as the content of inactive monomer increases due to a disturbance in the complementarity[53] and an increase in the stability of the complex as the content of hydrophobic groups rises.

Kudaibergenov *et al.*[16] and Mun *et al.*[39,40] investigated the influence of the hydrophobic/hydrophilic balance of the copolymers of hydrophilic vinyl ethers of glycols with hydrophobic vinyl butyl ether (VBE) on their complexation abilities with PAA or poly(methacrylic acid) (PMAA). The copolymers of vinyl ether of ethylene glycol (VEEG) with a content of VBE between 8 and 28 mol % form IPCs both with PAA and PMAA.[16] and their complexation ability depends on the concentration of the polymer in the system.[39] An increase of PAA/VEEG-*co*-VBE concentration shifts the critical pH values to the higher region. The turbidity of the IPC solutions increases as the content of the hydrophobic VBE in the copolymer increases. The stoichiometry of the complexes, [PAA]:[VEEG-*co*-VBE] = 2:1 and [PMAA]:[VEEG-*co*-VBE] = 1:1, depends on the kind of the used polyacid but not on the copolymer composition.[16] In isopropanol, the composition of the IPCs of VEEG-*co*-VBE with PAA strongly depends on the VBE content in the copolymer. The increase of the VBE content shifts the composition of the IPCs from a stoichiometric to a non-stoichiometric ratio. The copolymers with VBE content higher than 29 mol % do not form IPCs in isopropanol. A complexation between a copolymer of acrylic acid with VBE (AA-*co*-VBE) and VEEG-*co*-VBE was only observed in *n*-butanol, but only for a VBE content in the AA-*co*-VBE lower than 20 mol %.[39] The complex based on PAA and VEEG-*co*-VBE with a content of VBE of 8 mol % is stable in aqueous solution that contains up to 7 vol. % DMF, 10 vol. % ethanol, or 17 vol. % 1,4-dioxane. The IR spectra of the complexes based on PMAA and poly(vinyl ethers) of glycols show a new band at 1724 cm$^{-1}$, assigned to the H-bonds between the carboxylic groups of PMMA and the hydroxyl groups of the copolymers is noticed.[16] Khutoryanskiy *et al.* showed that the presence of IPCs of PAA with VEEG-*co*-VBE in water/*n*-hexane emulsions increase significantly its stability. The copolymers with a higher hydrophobicity stabilized the emulsions more efficiently.[52]

The IPCs formed between PAA and copolymers of 2-hydroxy-ethylacrylate with a content of VBE varying between 2.4–9.4 mol % (HEA-*co*-VBE)[52] have relatively low critical pH values (within 2.9–3.3), just slightly higher than the pH$_{crit}$ of IPC between PAA and poly(2-hydroxyethylacrylate) (PHEA), indicating a low complexation ability of the partners. By increasing the content of hydrophobic VBE groups in the copolymer, the pH$_{crit}$ values increase proving the significance of hydrophobic interactions to the stabilization of the IPC. The complexes of HEA-*co*-VBE with PMAA - a more hydrophobic acid - exhibit much higher pH$_{crit}$ values (within 4.2–4.6) and was classified as a strongly complexing polymer pair.[42] At a content of PAA lower than the IPC stoichiometry, the phase separation temperature of the VBE-*co*-HEA/PAA system decreases while at significant excess of PAA in the system an increase in the phase separation temperature was noticed. The authors explained this behavior by the binding of extra PAA chains. The copolymer containing the highest amount of VBE (9.4 mol %) formed a turbid solution with PAA in a broad temperature range as long as the ratio between the [PAA]/[copolymer], r is lower than 1.7. When r is higher than 4.8, the binary system became soluble at low temperatures and phase separated only upon heating. These IPCs are sensitive both to the change of pH and temperature.

## 2.2. Graft-copolymers

H-bond interpolymer complexes between homopolymers are soluble within a narrow pH window.[54-57] Usually at pH values higher than 4–5, the amount of ionized sites (carboxylate groups) in the polyacid chain increases leading to the dissociation of the hydrogen-bonding of the interpolymer complexes while at pH lower than 3–3.5 associative phase separation takes place.[19] The extension of the solubility of the H-bonding IPCs in the low pH range could enlarge the spectrum of their potential applications. In order to broaden the solubility range of the IPC at low pH, Staikos and co-workers proposed the complexation either of a polyacid with a graft copolymer of neutral side chains on charged copolyacid backbone[19] or of a polybase with a graft copolymer of charged side chains on a weak polyacid.[31]

As polybase graft copolymer, the authors used copolymers of poly(*N,N*-dimethylacrylamide) (PDMAM) as side chains grafted on a backbone based on copolymers of acrylic acid (AA) with a content of 2-acrylamido-2-methylpropane sulfonic acid (AMPSA*x*), of either 36 mol % or 70 mol %, P(AA-*co*-AMPSA*x*). They noticed that the IPC between P(AA-*co*-AMPSA*x*)-*g*-PDMAM and PAA, with a nominal molecular mass of 90 kDa, are more soluble at lower pHs (e.g. 2.5) than the IPC based on neutral homopolymer PDMAM and PAA, their solutions being less turbid. The turbidity decreased with the amount of ionic groups (AMPSA) present in the copolymer. At pH lower than 3.75 the graft copolymer with a higher content of AMPSA, P(AA-*co*-AMPSA70)-*g*-PDMAM/PAA form an IPC with a compact structure, the degree of compactness increasing with the decrease of the pH. The authors reported that at pH = 2.5, the stoichiometry of the IPC appeared at a molar unit fraction of PAA in the mixture of 0.6. At low pH, H-bonding association between PAA and these graft copolymers led to the formation of core-corona nanoparticles, with a compact hydrophobic core, consisting of the H-bonding interpolymer complex formed between PAA and the PDMAM side chains of the graft copolymer, and an extended hydrophilic corona, consisting in the anionic backbone of the graft copolymer.[58]

At high polymer concentrations, 60 mg mL$^{-1}$ and a ratio between the aqueous mixtures of the partners P(AA-*co*-AMPSA70)-*g*-PDMAM/PAA of 1:1 (w/w), where PAA has a nominal molecular mass of 500 kDa, Staikos *et al.* remarked an increase in the viscosity of about 4 orders of magnitude as the pH decreased from 3.75 to 2.5. They assigned this effect to the architecture of the graft copolymer that combines the proton acceptor ability of the PDMAM side chains with the strongly hydrophilic character of its negatively charged backbone.[19] The noncomplexable negatively charged AMPSA groups did not allow the polyacid chain to adopt the appropriate conformation to associate with the nonionic polybase chain.[27-30,59]

## 2.3. Block copolymers

Du Prez *et al.* showed that the addition of small amounts of amino end-capped block copolymers of HO–PEO–PPO–PEO–OH to a mixture of HO–PEO–PPO–PEO–OH/PAA leads to viscous liquids showing pronounced shear-thickening that becomes stronger as the amount of added amino group increases. The authors explained this effect by the fact that the interpolymer complex formed between PEO blocks and PAA are further stabilized by the additional electrostatic interaction between the protonated amino-containing end groups and the corresponding counter anions in PAA. Mixtures of HO–PEO–PPO–PEO–OH end functionalised with short linear polyethylenimine chains with PAA were found to lead to the formation of solid precipitates.[20]

## 3. IPC based on polyacid copolymers in water

### 3.1. *Alternating or random copolymers*

In the majority of the studies involving polyacid copolymer as one of the partners in interpolymer complexation, the polyacid is a copolymer of maleic acid. The maleic acid copolymers are mainly obtained by the hydrolysis of their corresponding maleic anhydride copolymer. As the maleic anhydride has a very low reactivity constant it leads with some exception (e.g. the copolymer with acrylic acid) to alternating copolymers.

Vasile *et al.* investigated the complexation abilities of an alternating maleic acid–vinyl acetate copolymer (MAc-*alt*-VA) in aqueous dilute solution with polybases with different hydrophilic/hydrophobic balance e.g. poly(ethylene glycol) (PEG), polyacrylamide (PAM) or poly(*N*-

isopropylacrylamide) (PNIPAM).[18] They found that MAc-*alt*-VA does not form an IPC with PAM, the interactions between these to partners being very weak, but it forms strong hydrogen-bonding IPCs with a compact structure both with PEG and PNIPAM. The higher deviation of the ratio of intrinsic viscosities of the PNIPAM/MAc-*alt*-VA aqueous solutions from unity comparing with that of PEG/MAc-*alt*-VA indicates that PNIPAM/MAc-*alt*-VA forms a stronger IPC than PEG/MAc-*alt*-VA, and that the hydrophobic interaction has a very important contribution to the formation of the hydrogen-bonding interpolymer complexes. The stoichiometry of PEG/MAc-*alt*-VA was found to be 1:1 indicating that one polybase monomer unit interacts with one carboxylic group. For PNIPAM/MAc-*alt*-VA, the reported stoichiometry was 3:2. The complexation constant, $K_c$ calculated from potentiometric data were $K_{c,PEG/MAc-alt-VA} = 8 \pm 3$ mol $l^{-1}$ and $K_{c,PNIPAM/MAc-alt-VA} = (4 \pm 3) \times 10^2$ mol $l^{-1}$. The $K_c$ value of PNIPAM/MAc-*alt*-VA about two orders of magnitude higher than that of PEG/MAc-*alt*-VA was considered as a proof that PNIPAM forms a much stronger complex with MAc-*alt*-VA than PEG. This difference was attributed to the higher hydrophobicity of PNIPAM compared with PEG.[18]

The ability of MAc-*alt*-VA to form IPCs with polybases with a semirigid structure and many functional groups *per* monomeric unit as dextran (D), pullulan (P), and hydroxypropyl cellulose (HPC) was reported both in dilute and in semi-dilute concetration regime.[60] In the dilute concentration regime, $c_{pol} = 2.5$ mg mL$^{-1}$, MAc-*alt*-VA formed IPCs with expanded structures both with dextran, pullulan and HPC. The expanded structure was assigned both to the presence of a high number of carboxylate groups, manifesting repulsive interactions between each other and to the semi-rigid structure of the polysaccharides that leads to a sterical hindrance and less possibility to form a compact structure. Based on the intensity of the interactions between the components of the mixtures, the strength of the IPC was found to increase in the series: D/MAc-*alt*-VA < P/MAc-*alt*-VA < HPC/MAc-*alt*-VA. The weakest IPC is formed with dextran as dextran has a branched structure that is not sterically favorable for H-bonding formation and the strongest one is given by HPC where the presence of the hydrophobic propylenoxide side chains strengthen the complex through the hydrophobic forces.

In the semi-dilute concentration regime, $c_{pol} = 100$ mg mL$^{-1}$, due to the high cooperativity of the specific interaction, the polysaccharide/MAc-*alt*-VA mixtures became immiscible, phase separation occurred in blends containing HPC or pullulan. In semi-dilute solutions, the number of carboxylate groups is very small, therefore mainly the attractive interactions manifest in these systems. Also the number of molecules in the systems is higher leading to a higher number of H-bonding between the components. These two phenomena together give rise to a collapsed structure of the IPCs in the semi-dilute regime. The temperature has a different effect on the miscibility of the mixtures depending on the hydrophilicity/hydrophobicity balance. The miscibility of blends based on hydrophilic polysaccharide (pullulan and dextran) increases with increasing temperature while the miscibility of the mixtures containing hydrophobic polysaccharides (HPC) decreases by raising the temperature.[60] IR spectra of the films made of polysaccharide/MAc-*alt*-VA indicated the presence of both H-bonding and chemical linkages between of the partners, depending on the kind of the used polysaccharide and the ratio between the partners.[61-63]

The influence of the comonomers hydrophilicity/hydrophobicity on the ability of MAc-copolymers to form IPCs with HPC in dilute aqueous solutions, $c_{pol} = 2$ mg mL$^{-1}$, has been reported by Bumbu *et al.*[17] Copolymers of maleic acid with acrylic acid (MAc-*co*-AA, a highly hydrophilic comonomer), vinyl acetate (MAc-*alt*-VA, moderate hydrophilicity), and styrene (MAc-*alt*-S, hydrophobic) have been investigated. The authors found that independent on the type of the used comonomer IPCs with a compact structure form. Besides H-bonding, strong hydrophobic forces occur between HPC and MAc-*alt*-S, strengthening this IPC. The strength of the interpolymer interactions was estimated to increase in the order: HPC/MAc-*alt*-VA < HPC/MAc-*co*-AA < HPC/MAc-*alt*-S. After a certain concentration, called threshold concentration, the presence of the MAc-copolymers in a HPC aqueous solution leads to a significant decrease of its cloud point temperature. The threshold concentration depends on the hydrophobic/hydrophilic character of the MAc-copolymer. The lower the threshold concentration, the more hydrophobic the MAc-copolymer is. It was found that the IPCs of HPC with MAc-*alt*-VA or MAc-*co*-AA are water-soluble at pH higher than 3, while the IPC between HPC and MAc-*alt*-S is water-soluble at pH values above 4.5.

Bumbu *et al.* stated that the HPC/MAc-*alt*-S system exhibits a lower critical solution temperature (LCST) behavior[17] that depends on the total polymer concentration in the system. At a $c_{pol}$ = 1 mg mL$^{-1}$ the system presents a cloud point temperature of 54°C, for a content of MAc-*alt*-S of 10 wt. % in the HPC /MAc-*alt*-S mixture, while for MAc-*alt*-S higher than 10 wt.% the solutions are clear in the whole investigated temperature range (5°C–80°C). These data suggest that the system containing HPC and MAc-*alt*-S in a mixing ratio of 90:10 is still hydrophobic, but not as hydrophobic as the pure HPC, while for the other mixing ratios the system is hydrophilic; the IPC structure is probably organized such that the hydrophilic groups are orientated to the exterior.[11]

At $c_{pol}$ > 2 mg mL$^{-1}$ and polyacid concentration lower than 1 mg mL$^{-1}$, the cloud point temperatures of HPC/MAc-*alt*-S aqueous systems increase up to 55°C in comparison with that of pure HPC of 44°C. At higher polyacid concentrations, the HPC/MAc-*alt*-S aqueous system becomes cloudy and the IPC precipitates.[17] For a $c_{pol}$ = 5 mg mL$^{-1}$, an irreversible phase segregation upon heating of the homogeneous HPC/MAc-*alt*-S aqueous systems was reported. The segregation temperatures depend on the mixing ratio of the partners. By applying a constant shear rate ( $\dot{\gamma}$ = 50 s$^{-1}$) the phase segregation temperatures of the HPC/MAc-*alt*-S aqueous system is 10°C–15°C lower than for an un-sheared one. At $c_{pol}$ = 10 mg mL$^{-1}$, the phase separation was already noticed at room temperature when the two binary polymer solutions are mixed.[12]

The stoichiometry of the HPC/MAc-*alt*-S interpolymer complex, in mole units, was estimated as being HPC:MAc-S = 2:5[11] and it is independent of $c_{pol}$.[12] By raising the temperature, the complexation process is favored and therefore an increase in the IPC concentration and the apparent complexation constant (from 1.2 to 2.2x10$^6$ L$^{5/2}$·mol$^{-5/2}$), was observed. This behavior was assigned to the presence of the hydrophobic forces in the system. Potentiometric measurements yield an enthalpy of 32 kJ·mol$^{-1}$ and an entropy of 225 J·mol$^{-1}$·K$^{-1}$ for the complexation reaction between MAc-*alt*-S and HPC pointing out a thermodynamically favorable process for temperature above $-130°C$.[11]

The IPC formation between HPC and MAc-*alt*-S through the H-bonding was evidenced also by FTIR spectroscopy. During mixing, the C=O groups of the carboxylic moieties, which were involved in a self-associated hydrogen-bonding structure, are set free and shift to a higher wavenumber (1730 cm$^{-1}$–1732 cm$^{-1}$). The characteristic band of the ether groups of HPC is shifted to a lower wavenumber in the blends, e.g., the band of the side chain ether group is shifted from 1123 cm$^{-1}$ to 1118 cm$^{-1}$ and that of the ether of the glycosidic linkage from 1083 cm$^{-1}$ to 1077 cm$^{-1}$. This indicates their involvement in hydrogen bonding. The intermolecular hydrogen bonds form mostly between the OH from the carboxyl groups of MAc-*alt*-S and the oxygen from the ether groups of HPC.[12]

Heng *et al.* investigated both the complexation between poly(methyl vinyl ether-maleic acid) copolymer (MAc-*alt*-MVE) and poly(vinylpyrrolidone) (PVP K-90D) in aqueous solutions and the release profiles of sodium diclofenac from the composite films of PVP/MAc-*alt*-MVE prepared in the presence of N-methyl-2-pyrrolidone.[64] They found that MAc-*alt*-MVE and PVP form an IPC with a compact structure and a stoichiometry of 1:1. The small downshifts of about 2 cm$^{-1}$ in the overall bands in the Raman spectrum of the films prepared from mixture of 10 wt. % aqueous solutions of PVP and MAc-*alt*-MVE, in a weight ratio of 1:1 were attributed to the formation of intermolecular hydrogen bonds between the components. Shifts of the peak for the hydrogen bonded carbonyl group in comparison with the pure materials are affected by the strength and the number of the hydrogen bonds formed.[65] The authors reported that the PVP/MAc-*alt*-MVE film, where the ratio between the partners is equal with the complex stoichiometry, 1:1 has the highest bioadhesive property, followed by the films with a composition of 3:2, 2:3, 4:1, and 1:4. The bioadhesive property of the 3:2 film was similar to that of the 2:3

film. The drug release from the system with an increased amount of MAc-*alt*-MVE was found slower probably due to partial conversion of diclofenac sodium into its less soluble acid form and not to the presence of a large amount of complex in the system.[64]

Moolman *et al.* investigated both experimental and theoretical the interpolymer complexation between PVOH and MAc-*alt*-MVE (Gantrez S97)[49] as the formation of interpolymer complexes between PVOH and MAc-*alt*-MVE is the basis of a novel oxygen barrier technology for plastics packaging. For blends with unadjusted pH, the optimum blend ratio for interpolymer complexation appears at around 10–14 wt. % MAc-*alt*-MVE. [49] Lowering the pH leads to a shift in the optimal blend ratio towards higher MAc-*alt*-MVE content because of a lower degree of ionization of the carboxylic acid groups. This is possibly because of increased intra- and intermolecular hydrogen-bonding within MAc-*alt*-MVE at lower pH,[9] thus leaving fewer groups available for intermolecular H-bonding with PVOH. At pH ~1, the optimum blend ratio for complexation is around 70 wt % MAc-*alt*-MVE, which is in good agreement with predictions from molecular dynamics based on $\chi$-parameter values.

Radic *et al.* reported the formation of interpolymer complexes between poly(4-vinylpyridine) (P4Vpy) and maleic acid-ethylene (MAc-*alt*-E) copolymer.[66,67] Fourier transform infrared spectroscopic studies evidenced that the carboxylic groups of MAc-*alt*-E interact with the pyridine ring of P4VPy. The presence of the H-bonding interactions is indicated by the appearance of new bands at 1640 cm$^{-1}$ assigned to a ring-stretching vibration of the pyridinium cation interacting with the acid group and a broadening of the band assigned to the pyridinic ring at 1607 cm$^{-1}$.[67] The broadening of the band attributed to the stretching of the pyridinic ring reflects two contributions, one corresponding to the free pyridinic ring and the other to the perturbed ring as a consequence of the H-bonding interaction of the pyridinic nitrogen with the hydrogen of the polyacid. The authors calculated the inter-association constant, $K_A$ between MAc-E and P4VPy from FTIR data, based on the Association Model developed by Painter and Coleman.[46-48] They analysed one of the vibration bands of the pyridinic (C=C) double bond that consists of three spectral contributions: one due to the stretching vibration of the "free" pyridine ring at 1597 cm$^{-1}$, the second one at 1607 cm$^{-1}$ and the last one at about 1640 cm$^{-1}$, corresponding to the pyridinium cation. The area of the "free" band increases with temperature while that corresponding to the ionic band decreases. The fraction of "free" pyridine at each temperature decreases as the composition of MAc-*alt*-E increases. Using the association model they obtained an inter-association constant of 2557 at 298K and an enthalpy of the interpolymer complex formation of $\Delta H = -28$ J mol$^{-1}$, in relatively good agreement with those obtained from flow calorimetry of $-39.60$ J g$^{-1}$.

## 3.2. Graft-copolymers

Staikos *et al.* investigated the IPC formation at low pH between PEO and graft copolymers of poly(2-acrylamido-2-methylpropanesulfonic acid) (PAMPSA) on a PAA backbone (PAA-*g*-*x*PAMPSA*y*) where *x* is the weight percent of PAMPSA in the copolymer (being 14, 21, and 52 wt%) and *y* is the numeric molecular weight of PAMPSA ($M_n = 4.5$ kDa or 20kDa).[31] The negatively charged PAMPSA side chains are expected to provide sufficient hydrophilicity[68] to the IPC formed between PEO and the PAA backbone of PAA-*g*-*x*PAMPSA*y* to keep the complex soluble even at low pH.

The authors found that the IPC formed between PEG and PAA-*g*-*x*PAMPSA*y* are more soluble at pH $< 3.75$[31] than the IPC of PEG/PAA;[69] the turbidity of aqueous solution of PEG/PAA-*g*-*x*PAMPSA*y* at pH $< 3.75$ is substantially low. The formation of IPCs through H-bonding with PEG is inhibited when *x* reaches 50 wt % for PAMPSA with $M_n = 4.5$ kDa. When the copolymer with the higher molecular mass PAMPSA side chains ($M_n = 20$ kDa) is involved, the H-bonding interpolymer complexation with PEG takes place at pH 2.75, even for $x = 50$ wt %.[30]

It was reported that the increase of the ionic strength of the solution (e.g. in 0.1 M NaCl solutions) favours the H-bonding interpolymer complexation.[14,70] For example, Staikos *et al.* found that the mixtures of PEG with PAA-*g*-14PAMPSA4.5 and PAA-*g*-21PAMPSA4.5 form soluble hydrogen-bonding IPCs in pure buffer solutions, but in the presence of 0.1 M NaCl they exhibit considerable turbidity, dependent on the PAMPSA content of the graft copolymer. The mixture PAA-*g*-52PAMPSA4.5/PEG remains almost clear for pH-values below 3.75, both in pure buffer and in 0.1 M NaCl solutions.

At pH lower than 3.5, PAA-*g*-*x*PAMPSA4.5 copolymers form H-bonded IPCs with PEO with a compact structure. Viscometric measurements indicate a stoichiometry of 1:1 for the formed H-bonded IPCs.[31]

## 4. IPC in organic solvents

Traditionally in the complexation studies, each polymeric unit has a functional group able to participate in H-bonding. However, there is a large number of papers in the literature reporting about interpolymer complexes between polymers that do not possess functional groups able to participate in H-bonding on each monomeric unit.[71-75] The H-bonding density in the blends of these polymers is controllable by adjusting either the content of proton donor or proton acceptors groups in the corresponding copolymers so that an interpolymer complex can be formed. Apart from the above-mentioned IPCs that form in water and are destroyed by adding organic solvents, the copolymers that do not have functional groups on each monomeric unit can form complexes only in organic solvents. Their ability to form IPCs is strictly related to the content of functional groups able to participate to H-bonding in direct correlation with the quality of the solvent used.

Djadoun *et al.* studied both in solution and solid state the complexation behavior of various mixtures of poly(styrene-*co*-methacrylic acid) (SMA) containing 12 or 29 mol % of methacrylic acid units (SMA-12 and SMA-29) with poly(styrene-*co*-4-vinylpyridine) containing 6 or 15 mol % of 4-vinylpyridine (S4VP-6 and S4VP-15).[24] They found that S4VP-6 formed H-bonding interpolymer complexes with SMA-12 and SMA-29 in butan-2-one, but not in THF. The higher value of Kwei's constant ($q$) for the SMA-29/S4VP-15 system in butan-2-one being higher ($q = 40$) than in THF ($q = 27$) confirmed that H-bonding specific interactions are stronger in butan-2-one than in THF and provided the evidence of IPC formation in the first mentioned solvent. By increasing the content of the methacrylic acid in SMA-copolymers, the IPC becomes stronger; the negative deviation of the viscosity form the ideality increases. A minimum in the reduced viscosity of both SMA-12/S4VP-6 and SMA-29/S4VP-6 mixtures was observed at approximately 1:1 ratio. The authors noticed in the IR spectra of SMA/S4VP both modifications in the intensity of the bands around $1737 \text{ cm}^{-1}$–$1740 \text{ cm}^{-1}$ and $1698 \text{ cm}^{-1}$, attributed to free and associated carboxylic groups, respectively, and the apparition of a new band of increasing intensity at around $1723 \text{ cm}^{-1}$, assigned to the free carboxylic groups. The intensity of the free carbonyl groups increases as the content of pyridine groups within the S4VP increases. The bands associated to the free pyridine ring at $1597 \text{ cm}^{-1}$ shifted to higher wavenumber, $1607 \text{ cm}^{-1}$ due to their involvement in H-bonding.

The miscibility or complexation of poly(styrene-*co*-methacrylic acid) with poly(ethyl methacrylate) (PEMA), poly(2-(N,N-dimethylamino) ethyl methacrylate) (PMAD), and poly{styrene-*co*-[2-(*N,N*-dimethylamino) ethyl methacrylate]} containing 3, 12, or 21 mol % of 2-(*N,N*-dimethylamino) (SMAD-3, SMAD-12 and SMAD-21) respectively have been reported.[73] It was shown that for a content of methacrylic acid of about 22 mol % in the SMA, SMA-22, the copolymer form IPCs in butan-2-one that precipitate with PMAD, SMAD-12, and SMAD-21. The strongest complex appeared in the PMAD/SMA-22 system and the authors explained it by the presence of amine-carboxylic acid interactions besides the ester - carboxylic acid interactions that could occur in PEMA/SMA-22 and PMAD/SMA-22 systems.[73] In THF, the minimum content of functional groups required for the interpolymer complexation was higher than in butan-2-one; interpolymer complexation between SMA-22 and SMAD-12 appears only in the presence of an excess of SMAD-12. The complex stoichiometry of polymeric systems called *fixed mean stoichiometry* (FMS)[26] varies with the quantity of functional groups within the copolymers and gradually approached a 1:1 ratio when the amount of functional groups was increased.[6,22] The value of the FMS for these systems in butan-2-one and THF varied with the content of MAD in the SMAD copolymer and equalled the 1:1 ratio for the PMAD/SMA-22 one.[73] The positive deviations of the glass transition temperatures of PEMA/SMA-22, PMAD/SMA-22, and SMAD-21/SMA-22 blends from the average one in all composition range confirmed the presence of specific interactions within these binary systems. The authors reported a value of 32 for the Kwei's constant for the PEMA/SMA-22 system.[73] The decrease of the content of the MAD comonomer within the SMAD copolymer from 12 to 3 mol % led to the immiscibility of SMAD-3/SMA-22 blends.

Jiang *et al.* reported that an IPC forms in toluene between poly(methyl methacrylate) (PMMA) and poly{styrene-*co*-[p-(2,2,2-trifluoro-1-hydroxy-1-trifluoromethyl)ethyl-α-methylstyrene]} (PS(OH)) when the p-(2,2,2-trifluoro-1-hydroxy-1-trifluoromethyl)ethyl-α-methyl-styrene (HFMS) content in PS(OH) reached 8 mol %.[6,10] In the 50 wt % / 50 wt % mixtures of polystyrene-*b*-poly(methyl methacrylate), PS-*b*-PMMA with PS(OH), the complexation takes place when the content of PS(OH) in mixtures is higher than 8 mol % or the content of HFMS in PS(OH) is higher than 8 mol %. The IPC formation was evidenced by the presence of a single peak in dynamic light scattering (DLS), the peaks associated with the component polymers disappearing. Using laser light scattering (LLS) and non-radiative energy transfer fluorospectroscopy, it was shown that PS-*b*-PMMA and PS(OH) form in toluene stable core-shell structures with the core and shell, respectively, made of the insoluble PMMA/PS(OH) complexes and the soluble PS blocks as long as the HFMS content of PS(OH) is higher than 8 mol %. As the HFMS content of PS(OH) increases from 8 to 49 mol %, the hydrodynamic radius of the complex ($<R_h>$) decreases from ca. 100 nm to 70 nm while the average molecular weight of the complex, $M_w$ increase from 2.8 to $4.9 \times 10^4$ kDa. Laser light scattering studies indicated that the molar mass and the core density of the complex micelles increased, but the size decreased as the HFMS content increased.[75] The authors observed that when the ratio of PS-*b*-PMMA to PS(OH) is very asymmetric, i.e., 7:1 and 1:7, not all the polymer chains are involved in IPC. LLS showed in addition to the micelle peak, a small peak located at *ca.* 4 nm that indicates the existence of the individual chains in the system. For the ratio of the partners ranging between 1:2 and 2:1, nearly all the polymer chains participated in the complexation. At a ratio between PS-*b*-PMMA and PS(OH) around 2:1, $<R_h>$ had its lowest value, implying a maximum complexation. The authors noticed that at a fixed ratio between the partners, dilution had nearly no effect on already formed micelles, indicating that the structure was kinetically frozen and the chains do not undergo any spatial rearrangement by dilution. However, the size and the molar mass of the complexes decreased as the polymer concentration in the initial solutions decreases, indicating that the complexation was a diffusion-controlled process and the micellization process may not reach the thermodynamic equilibrium.[75]

Poly(styrene-*co*-vinylphenol) (STVPh) and poly(styrene-*co*-vinylpyridine) (STVPy) blends can form an interpolymer complex due to hydrogen bonding in tetrahydrofuran, butan-2-one, and chloroform, provided that the contents of the hydroxyl and pyridyl reach certain levels.[22] The number of functional groups in the copolymers needed to form aggregates depends on the proton-accepting ability of the solvent. In THF, for OH contents in STVPh higher than 30 mol %, IPCs with compact structure appear between STVPh and STVPy25. By increasing the OH content the strength of the complex increases and the blend solutions turn turbid. The fixed mean stoichiometry of the STVPh/STVPy25 blends in THF varies, as those of PS(OH) and poly(methyl methacrylate) in toluene,[6] with the content of hydroxyl in STVPh and gradually approaches 1:1 ratio as the hydroxyl content reaches 50 mol % or more. In butan-2-one, which has little effect on complexation, 6 mol % of OH in STVPh leads to complexation with STVPy25. However, for vinylpyridine (VPy) content in STVPy higher than 75 mol % even at a lower content of OH in STVPh e.g. 22 mol % a complexation in THF was put in evidence. The greater the VPy content in STVPy is, the smaller the OH content in STVPh is needed to form interpolymer complex aggregates in THF.[22] In chloroform, no interpolymer complexation occurs when the OH content in STVPh is less than 3 mol %. The lowest OH content (6 mol %) in STVPh required to form complexes with STVPy50 in chloroform is lower than that required in THF (>22 mol %). The $<R_h>$ distributions in STVPh50/STVPy blends depend on the VPy content in STVPy. The size of the complex is usually one order of magnitude larger than the individual polymer coils, e.g., *ca.* 80 nm for STVPh50/STVPy25 blend solution and has a relatively narrow distribution.

The minimum OH content in STVPh required to form interpolymer complex with poly(ethyl methacrylate) (PEMA) is 9 mol % in toluene and 22 mol % in 1-nitropropane.[76] Complete decomplexation of the STVPh/PEMA complexes in toluene can be realized by adding a small amount of tetrahydrofuran. The higher the OH content in STVPh, the more THF (e.g. up to about 0.4 mol $L^{-1}$ for 17 % mol OH) is needed to destroy the STVPh/PEMA complexes. The size of the complex depends on both hydroxyl content in STVPh and blend composition and substantially increases from about $10^2$ nm for the blends with large excess of STVPh up to about $10^3$ nm for the blend having a 1:1 base ratio.

Goh *et al.*[77] reported that poly(4-vinylphenol) (PVPh) and poly(styrene-*co*-vinylpyridine) (STVPy) with a content of VPy of 70 % formed IPCs in ethanol on all the compositional range but not in DMF. The complexation ability between poly(styrene-*co*-allyl alcohol) with a content of hydroxyl groups of 4.5 or 6.5 wt % (SAA4.5 and SAA6.5) and poly(*N,N*-dimethylacrylamide) (PDMA) and poly(*N*-methyl-*N*-vinylacetamide) (PMVAc) in methyl ethyl ketone (MEK), THF, and DMF were investigated.[78,79] In MEK, SAA4.5 and SAA6.5 form interpolymer complexes with PDMA, when the content of PDMA in the mixture is lower than 50 % and 75 %, respectively. No complexes appeared in THF and DMF. PMVAc form IPC with SAA in MEK at all mixing ratio, but in THF complexation occurs only when the mixture in rich in SAA. The compositions of the complexes were reported in the range of 50-70 mol % for SAA4.5/PMVAc system and 55-74 mol % for SAA6.5/PMVAc mixture.

Qi *et al.* studied the complexation of poly(styrene-co-octyl acrylate-co-acrylic acid) as proton donating polymer (PDP) and poly(styrene-co-octyl acrylate-co-4-vinylpyridine) as proton accepting polymer (PAP).[80] The complex stoichiometry, PDP: PAP is relatively insensitive to the composition of PAP, but it increases with the decrease in the content of octyl acrylate in the PDP and gradually approaches 0.5 when the PDP does not contain octyl acrylate. The authors concluded that the dependence of the complex stoichiometry on chain composition of the PDP is the result of the combination of steric hindrance of side chains of octyl acrylate units in the PDP with intra-association of carboxyls.

Kato *et al.* investigated the complexation between poly(4-vinylpyridine) (P4VP) and sequence-ordered ethylene (E) methacrylic acid (MAA) copolymers e.g. alternating, MAc-*alt*-E or periodic (MAc-2E with MMA:E = 1:2) structures, in organic solvents.[81] Both copolymers form insoluble IPC in methanol (0.1M of each functional group) with a stoichiometry between carboxyl and pyridyl groups of *ca.* 1 : 1 molar ratio. In the IR spectra of the mixtures, two new broad bands at 2515 cm$^{-1}$ and 1930 cm$^{-1}$ appear, ascribed to the strong hydrogen bonds between carboxyl and pyridyl groups,[82] and a shoulder around 1720 cm$^{-1}$ assigned to the carbonyl group set free as a consequence of the complexation between the carboxyl proton and the pyridyl nitrogen atom. The experimental T$_g$ values of the IPCs were found *ca.* 20°C–30°C above the calculated ones.

Based on the variation of the extinction coefficient of pyridyl groups (at 255 nm) in the UV spectra of the mixture of the MAc-E copolymers and P4VP in dilute methanol solutions ($10^{-4}$ M) the complexation constants were calculated: K$_{c, \text{MAc-E/P4VP}}$ ~ 5 x $10^2$ and K$_{c, \text{MAc-2E/P4VP}}$ ~ 2 x $10^2$.[81] The authors explained the smaller complexation constant of MAc-2E/P4VP in comparison with MAc-E/P4VP based on the space between neighboring carboxyl groups on the acid polymer chains. The carboxyl groups of MAc-E, and MAc-2E appear on every 4, and 6 main-chain carbon atoms, respectively, while the complementarily binding pyridyl groups in polymer P4VP appear on every 2 carbon atoms. Consequently, the pairing between carboxyl and pyridyl groups, and therefore the complexation becomes more difficult in the MAc-2E/P4VP. The presence of ethylene units in the copolymers increased the flexibility of the main-chain and decreased the steric hindrance induced by the tacticity. Thus, the smallest K$_c$ value in acid polymer MAc-2E/P4VP can be also ascribed to the polymer sequence effect.[81] These complexes can be stabilized also by ionic interaction.[83]

## References

1. L. G. Parada, L. C. Cesteros, E. Meaurio and I. Katime, *Polymer*, **39**, 1019 (1998).
2. D. Li and J. Brisson, *Polymer*, **39**, 801 (1998).
3. S. Y. Liu, G. Z. Zhang and M. Jiang, *Polymer*, **40**, 5449 (1999).
4. Y. H. Wang, G. R. Qi, H. Y. Peng and S. L. Yang, *Polymer*, **43**, 2811 (2002).
5. X. P. Qui and M. Jiang, *Polymer*, **36**, 3601 (1995).
6. X. P. Qiu and M. Jiang, *Polymer*, **35**, 5084 (1994).
7. M. Jiang, X. P. Qiu, W. Qin and L. Fei, *Macromolecules*, **28**, 730 (1995).
8. E. A. Bekturov and L. A. Bimendina, *Adv. Polym. Sci.*, **41**, 99 (1981).
9. E. Tsuchida and K. Abe, *Adv. Polym. Sci.*, **45**, 1 (1982).
10. M. Jiang, M. Li, M. L. Xiang and H. Zhou, *Adv. Polym. Sci.*, **146**, 121 (1999).
11. G. G. Bumbu, C. Vasile, J. Eckelt and B. A. Wolf, *Macromol. Chem. Physic.*, **205**, 1869 (2004).
12. G. G. Bumbu, J. Eckelt, B. A. Wolf and C. Vasile, *Macromol. Biosci.*, **5**, 936 (2005).
13. V. V. Khutoryanskiy, M. G. Cascone, L. Lazzeri, Z. S. Nurkeeva, G. A. Mun and R. A. Mangazbaeva, *Polym. Int.*, **52**, 62 (2003).
14. V. V. Khutoryanskiy, A. V. Dubolazov, Z. S. Nurkeeva and G. A. Mun, *Polym. Sci. Ser. B*, **45**, 89 (2003).
15. Y. Wang and H. Morawetz, *Macromolecules*, **22**, 164 (1989).
16. S. E. Kudaibergenov, Z. S. Nurkeeva, G. A. Mun, B. B. Ermukhambetova and A. T. Akbauova, *Macromol. Chem. Physic.*, **196**, 2203 (1995).
17. G. G. Bumbu, C. Vasile, G. C. Chitanu and G. Staikos, *Macromol. Chem. Physic.*, **206**, 540 (2005).
18. C. Vasile, G. G. Bumbu, Y. Mylonas, I. Cojocaru and G. Staikos, *Polym. Int.*, **52**, 1887 (2003).
19. M. Sotiropoulou, G. Bokias and G. Staikos, *Macromolecules*, **36**, 1349 (2003).
20. Y. Wang, E. J. Goethals and F. E. Du Prez, *Macromol. Chem. Physic.*, **205**, 1774 (2004).
21. G. Staikos, K. Karayanni and Y. Mylonas, *Macromol. Chem. Physic.*, **198**, 2905 (1997).
22. M. L. Xiang, M. Jiang, Y. B. Zhang, C. Wu and L. X. Feng, *Macromolecules*, **30**, 2313 (1997).
23. T. Ikawa, K. Abe, K. Honda and E. Tsuchida, *J. Polym. Sci., Part A: Polym. Chem.*, **13**, 1505 (1975).
24. S. Bennour, F. Metref and S. Djadoun, *J. Appl. Polym. Sci.*, **98**, 806 (2005).
25. G. Bokias and G. Staikos, *Recent Res Devel Macromol Res*, **4**, 247 (1999).
26. I. Iliopoulos, J. L. Halary and R. Audebert, *J. Polym. Sci., Part A: Polym. Chem.*, **26**, 275 (1988).
27. I. Iliopoulos and R. Audebert, *Eur. Polym. J.*, **24**, 171 (1988).
28. J. J. Heyward and K. P. Ghiggino, *Macromolecules*, **22**, 1159 (1989).
29. I. Iliopoulos and R. Audebert, *Macromolecules*, **24**, 2566 (1991).

30. G. Bokias, G. Staikos, I. Iliopoulos and R. Audebert, *Macromolecules*, **27** , 427 (1994).
31. P. Ivopoulos, M. Sotiropoulou, G. Bokias and G. Staikos, *Langmuir*, **22**, 9181 (2006).
32. A. A. Litmanovich, I. M. Papisov and V. A. Kabanov, *Eur. Polym. J.*, **17**, 981 (1981).
33. G. R. Williams and B. Wright, *J. Polym. Sci. Part A*, **3**, 3885 (1965).
34. D. Staszewska and M. Bohdanecky, *Eur. Polym. J.*, **17**, 245 (1981).
35. G. Staikos, G. Bokias and C. Tsitsilianis, *J. Appl. Polym. Sci.*, **48**, 215 (1993).
36. D. J. Hemker, V. Garza and C. W. Frank, *Macromolecules*, **23**, 4411 (1990).
37. E. Tsuchida and S. Takeoka, in *Macromolecular Complexes in Chemistry and Biology*, Eds. P. Dubin, J. Block, R.M. Davis, D.N. Schulz and C. Thies (Springer, Berlin, 1994), p. 183.
38. M. Koussathana, P. Lianos and G. Staikos, *Macromolecules*, **30**, 7798 (1997).
39. G. A. Mun, Z. S. Nurkeeva, V. V. Khutoryanskiy and A. B. Bitekenova, *Macromol. Rapid Comm.*, **21**, 381 (2000).
40. G. A. Mun, Z. S. Nurkeeva, V. V. Khutoryanskiy and A. D. Sergaziyev, *Colloid Polym. Sci.*, **280**, 282 (2002).
41. V. V. Khutoryanskiy, A. V. Dubolazov, Z. S. Nurkeeva and G. A. Mun, *Langmuir*, **20**, 3785 (2004).
42. V. V. Khutoryanskiy, G. A. Mun, Z. S. Nurkeeva and A. V. Dubolazov, *Polym. Int.*, **53**, 1382 (2004).
43. T. K. Kwei, *J. Polym. Sci. Polym. Lett. Ed.*, **22**, 307 (1984).
44. P. C. Painter, J. F. Graf and M. M. Coleman, *Macromolecules*, **24**, 5630 (1991).
45. Y. Xu, P. C. Painter and M. M. Coleman, *Makromol. Chem., Macromol. Symp.*, **51**, 61 (1991).
46. J. Y. Lee, P. C. Painter and M. M. Coleman, *Macromolecules*, **21**, 346 (1988).
47. M.M. Coleman, J.F. Graft, P.C. Painter, in *Specific Interactions and Miscibility of Polymer Blends*, Eds. M.M. Coleman, J.F. Graft, P.C. Painter (Technomic, Lancaster, 1991) p. 221.
48. M.M. Coleman, J.F. Graft, P.C. Painter, in *Specific Interactions and Miscibility of Polymer Blends*, Eds. M.M. Coleman, J.F. Graft, P.C. Painter (Technomic, Lancaster, 1991) p. 309.
49. F. S. Moolman, M. Meunier, P. W. Labuschagne and P. A. Truter, *Polymer*, **46**, 6192 (2005).
50. D. N. Theodorou and U. W. Suter, *Macromolecules*, **18**, 1467 (1985).
51. H. Meirovitch, *J. Chem. Phys.*, **79**, 502 (1983).
52. G. A. Mun, Z. S. Nurkeeva, G. T. Akhmetkalieva, S. N. Shmakov, V. V. Khutoryanskiy, S. C. Lee and K. Park, *J. Polym. Sci., Part B: Polym. Phys.*, **44**, 195 (2006).
53. S. E. Kudaibergenov, Z. S. Nurkeeva, G. A. Mun and V. V. Khutoryanskii, *Vysokomol. Soedin.*, **40**, 1541 (1998).
54. A. D. Antipina, V. Yu. Baranovskii, I. M. Papisov and V. A. Kabanov, *Polym. Sci. Ser. A*, **14**, 1047 (1972).
55. I. Iliopoulos and R. Audebert, *Polymer Bulletin*, **13**, 171 (1985).
56. T. Petrova , I. Rashkov, V. Baranovsky and G. Borisov, *Eur. Polym. J.*, **27** , 189 (1991).

57. K. Sivadasan, P. Somasundaran and N. J. Turro, *Colloid Polym. Sci.*, **269**, 131 (1991).

58. M. Sotiropoulou, J. Oberdisse and G. Staikos, *Macromolecules*, **39**, 3065 (2006).

59. H. T. Oyama, D. J. Hemker and C. W. Frank, *Macromolecules*, **22**, 1255 (1989).

60. G.G. Bumbu, and C. Vasile, in *New Trends in Natural and Synthetic Polymer Science*, Eds. C. Vasile and G.E. Zaikov (Nova Science Publishers, Inc., New York, 2006) p. 33.

61. G. G. Bumbu, C. Vasile, G. C. Chitanu and A. Carpov, *J. Appl. Polym. Sci.*, **86**, 1782 (2002).

62. G. G. Bumbu, C. Vasile, G. C. Chitanu and A. Carpov, *Polym. Degrad. Stab.*, **72**, 99 (2001).

63. G. G. Bumbu, C. Vasile, M. C. Popescu, H. Darie, G. C. Chitanu, G. Singurel and A. Carpov, *J. Appl. Polym. Sci.*, **88**, 2585 (2003).

64. J. S. Hao, L. W. Chan, Z. X. Shen and P. W. S. Heng, *Pharm. Dev. Technol.*, **9**, 379 (2004).

65. L. S. Taylor and G. Zografi, *J. Pharm. Sci.*, **87**, 1615 (1998).

66. V. Villar, A. Opazo, L. Gargallo, D. Radic, H. Rios and E. Clavijo, *Int. J. Polym. Anal. Ch.*, **4**, 333 (1998).

67. V. Villar, L. Irusta, M. J. Fernandez-Berridi, J. J. Iruin, M. Iriarte, L. Gargallo and D. Radic, *Thermochim. Acta*, **402**, 209 (2003).

68. H. Dautzenberg, W. Jaeger, J. Kotz, B. Philipp, C. Seidel and D. Stscherbina, in *Polyelectrolytes: Formation Characterization and Application* (Hanser, Munich, 1994).

69. F. E. Bailey Jr., R. D. Lundberg and R. W. Callard, *J. Polym. Sci. Part A*, **2**, 845 (1964).

70. V. A. Prevysh, B. C. Wang and R. J. Spontak, *Colloid Polym. Sci.*, **274**, 532 (1996).

71. M. C. Piton and A. Natansohn , *Macromolecules*, **28**, 1598 (1995).

72. M. C. Pascu, Gh. Popa, M. Grigoras and C. Vasile, *Synthetic Polym. J.*, **2**, 158 (1998).

73. N. Abdellaouli and S. Djadoun, *J. Appl. Polym. Sci.*, **98**, 658 (2005).

74. M. Jiang, *Macromol. Symp.*, **118**, 377 (1997).

75. S. Y. Liu, H. Zhu, H. Y. Zhao, M. Jiang and C. Wu, *Langmuir*, **16**, 3712 (2000).

76. M. L. Xiang, M. Jiang, Y. B. Zhang and C. Wu, *Macromolecules*, **30**, 5339 (1997).

77. J. Dai, S. H. Goh, S. Y. Lee and K. S. Siow, *Polym. J.*, **26**, 905 (1994).

78. J. Dai, S. H. Goh, S. Y. Lee and K. S. Siow, *Polymer*, **35**, 2174 (1994).

79. J. Dai, S. H. Goh, S. Y. Lee and K. S. Siow, *Polymer*, **34**, 4314 (1993).

80. Y. H. Wang, G. R. Qi, H. L. Li and S. L. Yang, *Eur. Polym. J.*, **38**, 1391 (2002).

81. Y. Inai, S. I. Kato, T. Hirabayashi and K. Yokota, *J. Polym. Sci., Part A: Polym. Chem.*, **34**, 2341 (1996).

82. J. Y. Lee, P. C. Painter and M. M. Coleman, *Macromolecules*, **21**, 954 (1988).

83. I. F. Piérola, M. Cáceres, P. Cáceres, M. A. Castellanos and J. Nuñez, *Eur. Polym. J.*, **24**, 895 (1988).

# CHAPTER 8

# INTER- AND INTRAMOLECULAR POLYCOMPLEXES IN POLYDISPERSED COLLOIDAL SYSTEMS

Tatyana Zheltonozhskaya[*], Nataliya Permyakova, Boris Eremenko

*Department of Macromolecular Chemistry, Faculty of Chemistry,*
*Kiev National Taras Shevchenko University*
*Vladimirskaya St. 64, 01033, Kiev, Ukraine*
*[*]E-mail: zheltonozhskaya@ukr.net*

Problems of stability in polymer-colloidal systems are discussed from the viewpoints of classical colloidal chemistry, macromolecular chemistry and statistical physics. The special situation in the systems containing polydispersed colloidal particles is highlighted. The behavior of the InterPC (PSMA+PEO) composed of poly(styrene-*alt*-maleic acid) and poly(ethylene oxide), and also of the IntraPC (PVA-PAAm) formed in the poly(vinyl alcohol)-*graft*-polyacrylamide copolymers in mono- and polydispersed silica/water suspensions is considered in order to explain the fact of a principally higher flocculating capability of polycomplexes as compared with individual polymer components. The approaches of colloidal and macromolecular chemistry are used in conjunction in these studies. The development of complicated non-equilibrium competitive processes in the ternary systems containing chemically complementary polymer components and polydispersed colloidal particles is demonstrated. The existence of an additional thermodynamic factor of attraction between large micrometer-scale particles covered by adsorption layers of polycomplexes and between triple polymer-colloidal complexes formed by both polymer components and nanoparticles is shown.

## 1. Introduction

Application of inter- and intramolecular polycomplexes (Inter- and IntraPC) as high efficient regulators of stability and structure formation of disperse systems[1-13] and also as functional coatings and membranes on solid surfaces[14-17] requires a knowledge of the processes occurring in multicomponent polymer-colloidal systems due to competitive interactions between their components. Until recently, the systematic studies in the field were practically absent. The main attention was paid to competitive interactions in multicomponent polymer systems[18-20] and systems consisting of chemically complementary polymers and micelles of surfactants.[21] From another aspect, a competitive adsorption of the blends of non-interacting (immiscible) polymers on the solid/liquid interfaces was studied.[22,23] Based on a profound analogy between both the processes of macromolecular connection in solutions and of polymer adsorption on a solid/liquid interface, our group has begun for the first time the detailed investigations of competitive interactions in ternary polymer-colloidal systems containing hydrogen bonded Inter- or IntraPC and also mono- or polydispersed colloidal particles. One of the driving forces of these investigations was to explain the effect of a principally higher flocculating capability of both types of polycomplexes compared to individual components and some known flocculating agents, that was established in two real polydispersed colloidal systems such as kaolinite clay/water and coagulant (aluminum sulfate)/natural water suspensions.[24-29] Interestingly, a similar phenomenon of extensive flocculation was revealed as an undesirable secondary effect in ternary polymer-colloidal systems upon development of multilayer-coated particles using the layer-by-layer technique.[15,30]

## 2. Stability of polymer-colloidal systems from some points of view

The problems of colloidal particles stability to aggregation and sedimentation in polymer solutions were studied in the frame· of classical colloidal chemistry, macromolecular chemistry, and statistical physics.

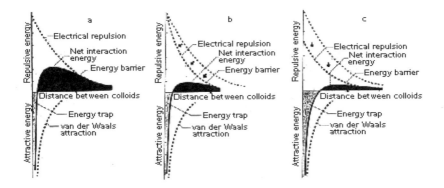

**Fig. 1.** The main interparticle forces according to DLVO theory (**a**), and the effects of reducing an energy barrier by shrinking the electric double layer (**b**) and lowering a particle charge (**c**).

*Colloidal chemistry approaches.* The main approach of classical colloidal chemistry is based on the influence of polymer adsorption on the interaction energy between colloidal particles. This is mostly correct in the case where the colloidal particle dimensions are essentially higher than those of the macromolecules in the dispersion medium, $R_p >> R_{polym}$. According to DLVO theory,[31,32] the stability or aggregation of monodispersed colloidal particles with the electric double layer on their surfaces in a polymer-free solution is determined by the balance between two forces such as *electrostatic repulsion* and *van der Waals attraction*:

$$V(h) = V_E(h) + V_{VDV}(h) \qquad (1)$$

Here $V_E(h)$ and $V_{VDV}(h)$ are the electrostatic and van der Waals interaction potentials but $h$ is a surface-to-surface distance between two colloidal particles. Changing in the overall interaction potential $V$ *vs* $h$ for particles with $R_p < 1$ μm shows (**Fig. 8.1(a)**) a primary (proximal) potential minimum and an energy barrier, of which the height can be essentially reduced by adding a low-molecular weight electrolyte, that results in shrinking of the electric double layer (**b**), or in decrease of the interfacial charge (**c**).

For micrometer-scale particles, the secondary (distal) potential minimum (it is not shown) appears due to increase in the $V_{VDV}(h)$ contribution. It leads to fixing (aggregation) the micrometer-scale particles on significant distances one from another even under existence of a high energy barrier. Later, a third force, such as *steric repulsion between interfacial adsorption layers of solvent* was found[33,34] but its contribution (an additional term $V_S(h)$ in the equation (1)) turned out to be weak enough as compared with electrostatic repulsion.

Polymer adsorption can change the $V_E(h)$ term and essentially increase the $V_S(h)$ contribution into the overall interaction potential of particles that depends on the properties of macromolecules and interfaces. Adsorption of an oppositely charged polyelectrolyte leads to decrease of the interfacial charge up to its neutralization and often to overcharging of an interface. At the same time the interfacial charge grows, when both adsorbed polyelectrolytes and interface have the same sign of charges.[32] Nonionic polymers either do not change the interfacial charge under adsorption[35] or reduce it due to the formation of hydrogen bonds and a definite orientation of polymer dipoles in the adsorption layer.[36,37] The potential of *steric repulsion between the adsorption polymer layers* of two drawing together colloidal particles consists of three terms[38]:

$$V_S(h) = V_V(h) + V_M(h) + V_{El}(h) \qquad (2)$$

Here $V_V(h)$, $V_M(h)$ and $V_{El}(h)$ are the repulsion potentials conditioned by: (i) the volume restrictions for "loops" and "tails" in overlapping adsorption layers, (ii) the osmotic contribution related to the difference in the chemical potentials of a dispersion medium in a shrinking layer between particles and in a volume, and (iii) the polymer chain elasticity. The higher molecular weight and concentration of polymer in the adsorption layer the higher the $V_S(h)$ contribution. Scaling theory predicts that there are three regions in the adsorbed polymer profile normal to the interface.[39] In the proximal region, the segment density is constant, while in the central and distal regions, the density decays with a power law and an exponential form, respectively. Modern experimental studies of the steric repulsion between polymer-coated

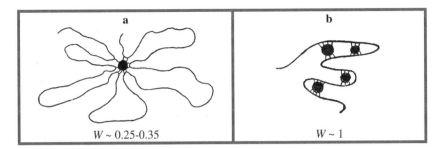

**Fig. 2.** Schematic representation of improbable (**a**) and most probable (**b**) building of PCC in the polymer/nanoparticles systems.

colloidal particles, which were carried out by SANS (small-angle neutron scattering), showed that adsorbed polymer layer in a good solvent partially collapses when the layers approach one another.[40]

In the mixtures, either of colloidal particles with non-adsorbing polymer,[41-44] or the particles coated by saturated interfacial polymer layers with abundant free polymer in a dispersion medium,[30] an additional force of the *depletion attraction between particles* (the additional $V_D(h)$ term in formula (1)) must be taken into consideration. This force has an entropic origin. Indeed, when a surface-to-surface distance becomes smaller than the hydrodynamic diameter of the macromolecules, they do not penetrate into the space between particles due to a considerable loss in entropy. As a result, the osmotic pressure stipulated by a difference in the chemical potentials of the macromolecules in a volume and between particles emerges and the particles attract one another. The depletion attraction is especially strong in the case of large particles and high polymer concentrations. Finally, a separate contribution of *hydrophobic attraction* between polymer-coated particles in aqueous medium should also be included, when a considerable interface hydrophobization upon polymer adsorption takes place.[34]

*Macromolecular chemistry approaches.* Significant progress in comprehension of the stability of polymer-colloidal systems with nanometer-scale monodispersed particles (the case $R_p \ll R_{polym}$) was achieved in the limits of macromolecular chemistry. On the basis of experimental[45-48] and theoretical[49-51] studies, a new conception about *cooperative polymer-colloidal reactions* and *polymer-colloidal complexes* (PCC) or multiplets instead of the traditional notions about polymer adsorption on small particles, was developed. It was shown that long macromolecule cannot form the classical adsorption layer around a small colloidal particle (**Fig. 2(a)**) because of a very low probability ($W$) of this process.[49] In reality, a macromolecule "collects" many particles in a PCC by a cooperative mechanism (**b**) that leads to a considerable gain in entropy of the system. From such a point of view, stability or phase separation in the polymer/nanoparticles systems in some solvent is mainly determined by PCC solubility or insolubility. It should be noted that the well-known La Mer's concept about the bridging mechanism of flocculation[52] could be most correctly used just in the case $R_p<<R_{polym}$. At the same time, if PCC is destroyed under the effect of external stimuli (solution pH, temperature and so forth), other kinds of phase separation in the system due to either immiscibility of polymer and particles[53] or possibly the depletion attraction between particles[44] can be observed.

One more important problem raised in the limits of macromolecular chemistry was *the influence of dispersed particles on a phase state of polymer blends in a solution*.[54] It was shown that highly dispersed filler (Aerosil) could improve[55] or deteriorate[56] the polymer compatibility in ternary polymer-colloidal systems. The following factors: (i) competitive adsorption of polymers at the interface, and (ii) alterations in molecular-weight distributions of polymers in solution caused by selective adsorption of the higher molecular-weight polymer fractions played an important role in these processes. Both effects were observed at high enough concentrations of components.

*Modern approach of statistical physics.* A quantitative theory of the binary systems, such as homopolymer/nanometer-scale monodispersed particles, was suggested recently in the limits of statistical physics.[57-60] The effect of a solvent was taken into consideration by variation of the polymer/particle interaction energy. This theory has comprehended practically all phenomena found in the polymer/small particles mixtures. Three situations in the system, namely, full absence of polymer/particle interaction, and also existence of weak and strong interactions between the components, were analyzed. The main conclusions are following. If *the polymer/particle interaction is absent*, even a small polymer additive sharply influences the distribution of the small particles in the space. An increase of the polymer fraction in the system leads to aggregation of the particles due to the effect of an entropy factor (the depletion attraction), but at $\rho_{polym}>0.8$ (here $\rho_{polym}$ is the number-average density of the polymer units in the system) the particles form a fluid-like structure even in the case where their average density is low. If *the polymer/particle interaction is weak*, two special temperatures $T_1^0$ (upper) and $T_2^0$ (lower), at which a state of particles surrounded by polymer segments resembles "ideal gas", appear. At $T<T_2^0$ and $T>T_1^0$, the polymer additive causes particle attraction and aggregation, but in the region $T_2^0<T<T_1^0$, the state of "absolute stability" of the system exists. In this region, compressibility of a disperse component proves to be lower than that in the system of individual particles, therefore the effect of particle stabilization by polymer is realized. Finally, in the case of *strong polymer/particle interaction,* the regime typical for systems near critical conditions arises. Long-active correlations emerge in a disperse subsystem but polymer chains turn out to be under conditions of poor solvent. Such a behavior was shown to be related only to polymer influence but not to attraction between particles. Thus, polymer additive causes a sharp strengthening of the aggregation (flocculation) processes under the condition of strong polymer/particle interactions. As a result, macroscopic areas enriched by dispersed particles appear. Here, separate nanoparticles are coated by dense polymer layers and connected by bridging between polymer segments.

Polydispersed colloidal systems, in contrast to monodispersed ones, demonstrate a very complicated behavior, even in the absence of polymers. There are only few mainly theoretical studies in the field.

*Polymer-free polydispersed colloidal systems.* One of the aspects of the polydispersity influence on aggregation and gravity-driven sedimentation processes was previously analyzed.[61] The following forces were shown to affect two differing in size and weight particles, which formed an aggregate through a liquid layer with thickness $h$ (the aggregation in a distal potential minimum) during sedimentation:

$$F(h) = \sum_i F_i + F_g + F_f = F_E + F_{VDV} + F_S + F_f^* \qquad (3)$$

Here $\sum F_i = F_E + F_{VDV} + F_S$ is a balance of the electrostatic, van der Waals, and steric forces between particles, while $F_g$ and $F_f$ are the gravity and friction forces acting on aggregates of two particles. The friction force $F_f^*$, which is the resulting force for $F_g$ and $F_f$, acts mainly as a repulsion force and causes the aggregate destruction. Only in one case, when a small particle is placed for sedimentation exactly under a large particle, the $F_f^*$ force promotes to a particle attraction. The higher polydispersity, the greater $F_f^*$ contribution in overall $F(h)$ force, that is the lower the probability of particle aggregation. In order to confirm this conception, the author has presented experimental data for polydispersed quartz suspension ($R_p = 1 \div 18$ µm), which has shown a high stability to aggregation during sedimentation in a centrifuge.

Interaction of particles different in size and electric double layer was considered in a previous study.[62] It was shown that approach of two non-identical particles with the same sign of potential on their interfaces caused first an increase in the electrostatic repulsion between the particles up to a maximum. But at small distances, the effect of overcharging a particle with lower interfacial potential took place and the particles were attracted.

Small particles in a polydispersed colloidal system can promote aggregation and sedimentation of large particles inducing additional depletion attraction between them.[63-65] This phenomenon has the same entropy origin as the particle depletion attraction initiated by non-adsorbing polymers. But overall effect of small particles on the "gas-liquid" and "gas-solid" phase transitions proved to be non-trivial even in the simplest bidispersed suspension (under the presence of both a weak van der Waals and a depletion attraction); that essentially depended on the volume fractions of small and large particles, and their size asymmetry.[66]

Fundamental review about the polydispersity impact on the phase state of colloidal, polymeric, and polymer-colloidal systems from a point of view of statistical physics was recently published.[67] There, theoretical approaches for predicting polydispersed phase equilibria were analyzed. Phase diagrams for monodispersed and bidispersed colloidal suspensions were compared. For a monodispersed suspension, a phase diagram in the coordinates: the free energy density $f$ vs the particle density $\rho$ ($\rho = N/V$, where $N$ is the particle quantity in a certain volume), was shown to be a curve with two minimums corresponding to the compositions of both coexisting phases. Essentially more complicated phase diagram in the coordinates: $T$-$\rho_1$-$\rho_2$ (the subscripts 1 and 2 designate the first and second kind of particles) was found in a simplest bidispersed suspension. Its analysis resulted in the conclusion that appearance of critical points of arbitrary order is a new feature in the phase behavior of polydispersed systems. A density distribution and temperature specify such critical points; their defining property is that, at those parameters, a single phase separates (upon lowering $T$, for example) into $n$ infinitesimally phases.

Based on the capability of small (nanometer-scale) particles to regulate interactions between the large (micrometer-scale) ones, a new nanoparticle engineering has been developed.[68] It was shown theoretically and experimentally[69,70] that highly charged nanoparticles can form a dynamic nanoparticle halos around negligibly charged microspheres in a binary suspension due to weak electrostatic attraction. Such "haloing" occurs in a definite region of nanoparticle volume fractions $\phi^L < \phi_{nano} < \phi^U$ (here $\phi^L$ and $\phi^U$ are the lower and upper critical nanoparticle volume fractions), thus ensuring transition from a colloidal gel to a stable fluid and further to a very dense colloidal gel upon increase in nanoparticle concentration in the system. Using binary suspensions in the region $\phi^L < \phi_{nano} < \phi^U$, one can design three-dimensional (3-D) colloidal crystals via gravitation settling of microspheres into a nanoparticle array.[71]

*Polydispersed colloidal systems with polymers.* At present, a state of polymer-colloidal systems composed of polydispersed particles and even alone polymer component remains practically not studied. That was pointed out also in review.[67] Its author referred to only one study,[72] which suggested that an intricate-and largely unexplored-phase diagram topologies may occur for fully polydispersed polymer-colloidal system.

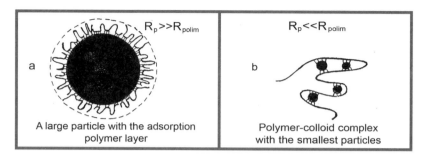

**Fig. 3.** Schematic picture of competitive polymer interaction with large (**a**) and small (**b**) particles in polydispersed polymer-colloidal systems.

To recent time, there was still limited information about the situation in polydispersed colloidal systems containing two polymers, especially interacting with each other. This fact, and also the abnormally high flocculating ability of Inter- and IntraPCs stimulated our group to begin versatile investigations in the field.

The behavior of two hydrogen-bonded polycomplexes such as InterPC (PSMA+PEO) based on poly(styrene-*alt*-maleic acid) and poly(ethylene oxide) and also IntraPC (PVA·PAAm) formed in the poly(vinyl alcohol)-*graft*-polyacrylamide copolymers in model poly-dispersed silica (SiO$_2$)/water suspensions has been studied. The polydispersity of colloidal particles must be one of the reasons for development of complicated competitive interactions in polymer-colloidal systems (**Fig. 3**). That is why both ternary polymer-colloidal systems SiO$_2$+PSMA+PEO and SiO$_2$+PVA-$g$-PAAm were examined in three cases: (i) the dispersed phase had a broad distribution of particles size (Aerosil), (ii) the SiO$_2$ particles were monodispersed (quartz) and their size essentially exceeded that one for polycomplexes (PC) and separate macromolecules, $R_p \gg R_{polym(PC)}$, and (iii) the particles were monodispersed (SiO$_2$ sol) but $R_p \ll R_{polym(PC)}$. The first case modeled competitive interactions, which develop in real ternary systems with Inter- or IntraPC and polydispersed colloidal particles, in particular, in *the natural water+coagulant+polycomplex flocculant system* arising in the process of water purification on waterworks. The other cases modeled two utmost situations in the ternary systems. Approaches of macromolecular chemistry and classical colloidal chemistry have been united for the first time in these studies.

All the components in the ternary systems were capable to interact with each other via hydrogen bonding. Indeed, the hydrogen bond structures between PEO or PVA and $SiO_2$ surface were well known from literature. The H-bonds between PVA main chain and PAAm grafts in the IntraPC forming PVA-*g*-PAAm graft copolymers were shown in Chapter 5 (**Fig. 6**), but the H-bond structures between PSMA and PEO, and also between PSMA or PAAm and $SiO_2$ surface as established by IR spectroscopy[73-75] are presented in **Figs. 4–6**. Now we will discuss the main results of these studies.

### 3. Behavior of Inter- and IntraPC in polydispersed suspensions

The Aerosil/water suspensions are typical polydispersed colloidal systems formed by aggregation of primary $SiO_2$ nanoparticles due to hydrogen-bonding and condensation of the surface silanolic groups.[76] Here, a very wide distribution of particle dimensions from ~10–20 nm for primary particles to 100–250 nm for small aggregates and then to 1–20 $\mu$m for large aggregates exists.[77] That is why a special strategy was chosen for studying the situation in the Aerosil/water suspension in the presence of one or two polymers. Indeed, the suspension of Aerosil (its specific surface $S_{sp}=1.75 \cdot 10^5$ m$^3 \cdot$kg$^{-1}$) with a constant concentration $C_{SiO_2}=9.26$ kg$\cdot$m$^{-3}$ was mixed with a solution of one of the polymers for a definite time. Then (after 24 hours) if it was necessary, a solution of another polymer was added and mixing was continued for the necessary time.

Further, sediment was separated off by centrifugation ($\omega$=6000 rot$\cdot$min$^{-1}$) and the supernatant was studied by different methods. The most supernatants contained ~2–3 weight percent of stable primary silica particles with $R_p$=28 nm as measured by static light scattering.

**Situation in $SiO_2$(Aerosil)+PSMA+PEO system.** In order to choose the system composition correctly and to interpret the results obtained, the following binary systems were preliminarily characterized.

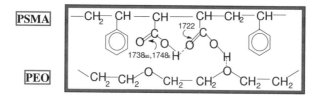

**Fig. 4.** Hydrogen bonds in the InterPC (PSMA+PEO) with participation of oxygen atoms of PEO and cyclic dimmers formed by adjacent –COOH groups in PSMA chains.[73]

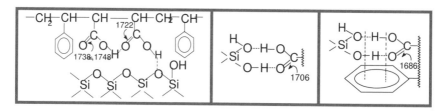

**Fig. 5.** Three kinds of hydrogen bonds between $SiO_2$ surface and PSMA chains with participation of silanolic and –COOH groups.[74]

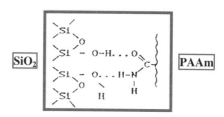

**Fig. 6.** Structure of hydrogen bonds between $SiO_2$ surface and PAAm chains with participation of silanolic and primary amide groups.[75]

**1) InterPC (PSMA+PEO).** The InterPC formation between a narrow fraction of PSMA ($M_v = 2.34 \cdot 10^5$) and two PEO samples with $M_v = 4 \cdot 10^4$ (PEO1) and $6.08 \cdot 10^5$ (PEO2) was controlled, according to the studies,[78,79] by measuring the PSMA solution pH upon addition of PEO solution[80] (**Fig. 7**). Note that PSMA is a polydibasic acid and the values of $pK_a^o$ (a negative logarithm of the characteristic dissociation constant) for its primary and secondary -COOH groups are essentially different ($pK_1°=2.4$, $pK_2°=7.0$, $\Delta pK=4.6$) due to formation of stable cyclic dimers in every copolymer unit (**Fig. 4**). That is why all the secondary – COOH groups of PSMA in the considered region pH<7 were fully uncharged.

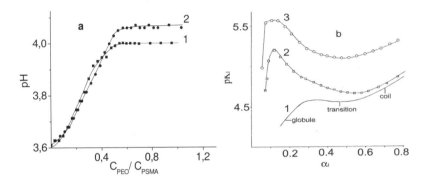

**Fig. 7.** Solution pH of PSMA *vs* relative concentration of PEO (**a**) for PEO1 –*1* and PEO2 -*2*, and also the pK$_1$ value for primary –COOH groups *vs* their dissociation degree (**b**) for PSMA –*1* and InterPCs with PEO1 –*2* and PEO2 –*3*. C$_{PSMA}$= 0.84 kg·m$^{-3}$.

The characteristic compositions ($\varphi_{char}$) for InterPCs were found on inflection points of curves 1 and 2 in **Fig. 7(a)**. Taking into account participation of primary –COOH groups of the polyacid simultaneously in both equilibriums of dissociation and H-bonding, and using a well-known method[81] and data of **Fig. 7(a)**, the fraction $\theta$ of PSMA units bound with PEO in corresponding InterPC was established:

$$C_b = C_{PSMA}^0 - C_{PSMA} \; ; \qquad \theta = C_b / C_{PSMA}^0 \qquad (4)$$

Here $C^0_{PSMA}$ is the initial concentration of PSMA units in a solution at a definite pH, $C_{PSMA}$ is the concentration of unbound PSMA units in the InterPC at the same pH, $C_b$ is the concentration of PSMA units bound with PEO.

Additionally, the Gibbs free energy of InterPC formation was determined based on the data of potentiometric titration of InterPCs at

**Table 1.** The main characteristics for intermolecular polycomplexes.

| InterPC | Solvent | $\varphi_{char}$, $w_{PEO}/w_{PSMA}$ | $\eta_{sp}/C_{PSMA}$, $m^3 \cdot kg^{-1}$ | $\theta$, % | $-\Delta G^0_{InterPC}$, $kJ \cdot mol^{-1}$ |
|---|---|---|---|---|---|
| PSMA+PEO1 | $H_2O$ | 0.38 | 0.007 | 44.0 | 1.34 |
| PSMA+PEO2 | | 0.49 | 0.021 | 66.9 | 1.43 |
| PSMA+PEO1 | 0.01 N NaCl | 0.40 | 0.001 | 49.7 | 0.92 |

$\eta_{sp}/C_{PSMA}$: specific viscosity of InterPC ($\varphi=\varphi_{char}$) reduced to PSMA concentration

$\varphi=\varphi_{char}$ and individual PSMA by NaOH, which are presented in **Fig. 7(b)** in the coordinates of equation (17) from Chapter 5. The PSMA macrochains underwent the *globule-coil type* conformational transition due to the ionization of primary –COOH groups (that resulted in appearance of three characteristic parts in curve 1), whilst the InterPC particles were fully destroyed in this process. Using the Leyte and Mandel approach,[82] the Gibbs free energies for the transitions $PSMA_{globule} \rightarrow PSMA_{coil}$, $\Delta G^\circ_{conf}$, and $InterPC_{globule} \rightarrow PSMA_{coil}$ (the InterPC destruction), $\Delta G^\circ_{destr <InterPC>}$, were calculated using the $pK_1=f(\alpha_1)$ curves. Then the Gibbs free energy of InterPC formation was found as a difference[80]:

$$- \Delta G^o_{InterPC} = -(\Delta G^o_{destr <InterPC>} - \Delta G^o_{conf}) \qquad (5)$$

All the parameters for the InterPCs are summarized in **Table 1**. The values of $\varphi_{char}$, $\theta$, and $-\Delta G^\circ_{InterPC}$ increased with growth of PEO length, thus implying an enhancement in stability of the InterPC structure. At the same time the energy of polymer interaction markedly lowered in the presence of low-molecular-weight electrolyte (NaCl). The InterPC particles had a very compact structure in solution that was reflected by very low values of the reduced viscosity (**Table 1**).

These studies allowed choosing the ratios between PEO1 or PEO2 and PSMA in the ternary systems; they were equal to $\varphi_{char}$ of InterPCs.

**2) Binary system $SiO_2$(Aerosil)+PEO.** The equilibrium adsorption of PEO with different molecular weights onto Aerosil (the contact time is 24 hours) and its effect on interfacial charge was studied in detail earlier.[36,83,84] Adsorption isotherms demonstrated high affinity of PEO to silica surface and an increase of the PEO maximum adsorption with increasing $M_{vPEO}$ (from $a_{max}$=0.33 for PEO1 to 0.39 mg·m$^{-2}$ for PEO2). The PEO adsorption reduced a negative interfacial charge of silica particles, and also a number of primary silica particles in supernatants.

Using these data, the ratios PEO1/SiO$_2$ and PEO2/SiO$_2$ in the ternary systems were selected. They corresponded to nearly 70% of the $a_{max}$ value to make sure that all the nonionic polymer macromolecules introduced in the Aerosil suspension were fully adsorbed at the interface and absent in solution, whilst the adsorption layer was close to saturation. Thus, the whole composition of every ternary system was determined.

**Table 2.** Parameters for the binary polymer-colloidal system with polydispersed silica.

| System | Solvent | $t_c$, | $\beta$, | $\eta_{sp}/C_{PSMA}$, m$^3$·kg$^{-1}$ | | $\theta$, | $-\Delta G°_{PCC}$, |
|---|---|---|---|---|---|---|---|
| | | min or h | % | Solution | Super natant | % | kJ·mol$^{-1}$ |
| SiO$_2$+PSMA | H$_2$O | 5 | 7.4 | 2.79 | 3.06 | 62 | 1.03 |
| | | 60 | 9.8 | 2.83 | 3.30 | 66 | 1.09 |
| | | 24 h | 7.8 | 2.80 | 2.74 | 68 | 1.14 |
| | 0.01 N NaCl | 5 | 15.0 | 0.48 | 0.63 | 76 | 1.52 |
| | | 60 | 11.7 | 0.47 | 0.60 | 79 | 1.54 |
| | | 24 h | 14.2 | 0.48 | 0.62 | 80 | 1.66 |

$t_c$ : The contact time of PSMA with SiO$_2$.
$\beta$ : The PSMA weight fraction adsorbed on aggregated Aerosil particles (in sediment).
$\eta_{sp}/C_{PSMA}$ : The PSMA reduced viscosity in solution and corresponding supernatant.
$\theta$ : The fraction of PSMA units bound in PCC.
$-\Delta G°_{PCC}$ : The Gibbs free energy of PCC formation.

**3) Binary system SiO$_2$(Aerosil)+PSMA.** This system was explored under the same conditions and at the same ratio between PSMA and SiO$_2$ as the ternary systems. The main purpose was to establish the influence of the contact time of PSMA with SiO$_2$ and of the low-molecular-weight electrolyte (NaCl). The following main results (**Figs. 5** and **8(a)**, and also **Table 2**) were obtained investigating the supernatants by potentiometric titration, viscometry and IR spectroscopy.[74,80,85,86] Only an insignificant part of PSMA was adsorbed by large aggregated SiO$_2$ particles and proved to be in sediment (see the $\beta$ parameter in **Table 2**). The larger fraction of PSMA remained in solution and formed a stable polymer-colloidal complex (PCC) with primary silica nanoparticles. Such a

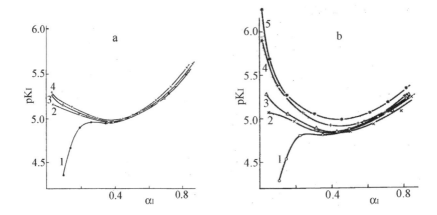

**Fig. 8.** Examples of the $pK_1$ *vs* $\alpha_1$ dependences for PSMA in water *–1* (**a,b**) and in supernatants of the binary system $SiO_2$+PSMA (**a**) stirred for 5 min *–2*, 60 min *–3*, and 24 hours *–4*, the ternary system ($SiO_2$+PEO1)+PSMA (**b**) after mixing PSMA with ($SiO_2$+ PEO1) for 5 *–2* and 60 min *–3*, and also of the ternary system ($SiO_2$+PSMA)+PEO1 (**b**) after mixing PEO1 with ($SiO_2$+PSMA) for 5 *–4* and 60 min *–5*. $C^0_{PSMA}$=0.9 kg·m$^{-3}$, *T*=25 °C. Analogous results were obtained upon using PEO2 instead PEO1, and in aqueous-salt (0.01 N NaCl) solution.

conclusion was confirmed by: (i) destruction of the initial compact structure of PSMA at once after its minimum contact with silica particles (**Fig. 8(a)**, curve 2), (ii) increase in the reduced viscosity of PSMA in supernatants as compared to pure polyacid solutions of equal concentration (**Table 2**), and (iii) identification of hydrogen bonds between PSMA and $SiO_2$ (**Fig. 5**).

It was shown[85] that dissociation of silanolic groups of Aerosil ($pK_a^o$=6.0) was fully suppressed in the pH region, where dissociation of essentially stronger primary carboxylic groups of PSMA occurred. That allowed determination of such PCC parameters as $\theta$ and $-\Delta G^o_{PCC}$ using the same approaches as in the InterPC (PSMA+PEO) studies (see above), and the $pK_1$=$f(\alpha_1)$ curves of **Fig. 8(a)**, and also the following relationship:

$$- \Delta G^o_{PCC} = -(\Delta G^o_{destr\, <PCC>} - \Delta G^o_{conf}) \qquad (6)$$

In the formula (6), $\Delta G^o_{destr\, <PCC>}$ is the Gibbs free energy for the transition PCC→PSMA$_{coil}$ that is for PCC dissociation. Thus, only a small enhancement in $\theta$ and $-\Delta G°_{PCC}$ values with time for PCCs formed in aqueous and aqueous-salt medium was found (**Table 2**). This fact indicates a high rate of PCC formation. At the same time, PCC demonstrated essentially a more stable and compact structure in the presence of NaCl due to evidently screening effect of the low-molecular-weight electrolyte.[86] It should be emphasized that the interaction energy between PSMA and $SiO_2$ was lower than that between both polymers in aqueous medium and vice versa it was proved to be higher than the interaction energy between PSMA and PEO in aqueous-salt solution (**Tables 1, 2**).

**Table 3.** Parameters of the ternary polymer-colloidal systems with polydispersed silica.

| System | Solvent | $t_c$, min or h | $\beta$, % | $\eta_{sp}/C_{PSMA}$, m³·kg⁻¹ | | $\theta$, % | $-\Delta G°_{PCC}$, kJ·mol⁻¹ |
| --- | --- | --- | --- | --- | --- | --- | --- |
| | | | | Solution | Super natant | | |
| ($SiO_2$+PEO1) +PSMA | $H_2O$ | 5 | 10.8 | 2.85 | 2.95 | 57 | 1.16 |
| | | 60 | 13.2 | 2.89 | 3.13 | 67 | 1.26 |
| | 0.01 N NaCl | 5 | 43.4 | 0.55 | 0.54 | 29 | 0.62 |
| | | 60 | 50.4 | 0.58 | 0.41 | 30 | 0.95 |
| | | 24 h | 54.3 | 0.62 | 0.47 | 35 | 1.57 |
| ($SiO_2$+PEO2) +PSMA | $H_2O$ | 5 | 11.7 | 3.06 | 3.05 | 66 | 0.97 |
| | | 60 | 19.7 | 3.23 | 3.10 | 69 | 1.19 |
| | | 24 h | 29.4 | 3.46 | 3.12 | 72 | 1.22 |
| ($SiO_2$+PSMA) +PEO1 | $H_2O$ | 5 | 11.5 | 2.87 | 0.16 | 84 | 1.70 |
| | | 60 | 11.3 | 2.86 | 0.26 | 90 | 1.75 |
| | 0.01 N NaCl | 5 | 29.5 | 0.51 | 0.01 | 74 | 1.87 |
| | | 60 | 28.0 | 0.51 | 0.01 | 72 | 1.82 |
| | | 24 h | 33.3 | 0.52 | 0.01 | 70 | 1.86 |
| ($SiO_2$+PSMA) +PEO2 | $H_2O$ | 5 | 15.8 | 3.15 | 0.76 | 82 | 1.84 |
| | | 60 | 15.3 | 3.14 | 0.88 | 87 | 1.82 |
| $SiO_2$+ (PSMA+PEO1) | $H_2O$ | 24 h | 3.4 | - | - | 79 | 1.76 |
| $SiO_2$+ (PSMA+PEO2) | $H_2O$ | 24 h | 13.6 | - | - | 81 | 2.26 |

$t_c$ : The contact time of a second polymer with the blend of $SiO_2$ and a first polymer or InterPC with $SiO_2$.

Now, we will consider the situation with the ternary polymer-colloidal system, which is partially reflected in **Fig. 8(b)** and **Table 3**. Such factors as (i) the order of the component addition, (ii) the time of their contact, (iii) the PEO molecular weight, and (iv) the presence of low-molecular-weight electrolyte affected the final state of the system $SiO_2$+PSMA+PEO with polydispersed colloidal particles considerably.[28,74,80,85,86] Due to the two first factors, this system could not be considered as one in equilibrium.

In the case of PSMA addition after preliminary adsorption of PEO on $SiO_2$ surface (this case was designated as ($SiO_2$+PEO)+PSMA in **Table 3**) polyacid actively bound to the aggregated Aerosil particles and proved to be in the sediment. The fraction $\beta$ of PSMA in the sediment essentially grew with increase in the contact time and $M_{vPEO}$, and also in the aqueous-salt medium unlike the aqueous one.

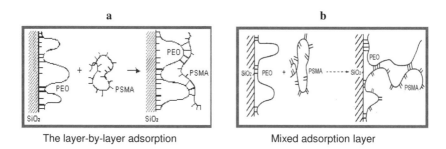

a                                              b

The layer-by-layer adsorption              Mixed adsorption layer

**Fig. 9.** Schemes illustrating PSMA adsorption on aggregated $SiO_2$ particles covered by the PEO adsorption layer in aqueous (a) and aqueous-salt (b) mediums.

Moreover, that parameter was greater in the ternary system than in the $SiO_2$+PSMA binary system under the same conditions (**Table 2**). Analysis of these data, which was carried out from the points of view: (1) full filling of a surface of aggregated $SiO_2$ particles with PEO chains before PSMA introduction, 2) zero desorption of PEO from $SiO_2$ surface during PSMA adsorption (absence of PEO in supernatants was confirmed by a special test), 3) increasing in size of "loops" and "tails" of PEO in the adsorption layer with increase in $M_{vPEO}$ and also in the presence of $Na^+$ ions in solution, allowed considering the process of PSMA adsorption as InterPC formation on the surface of aggregated $SiO_2$ particles (**Fig. 9**). But based on various values of the $\beta$ parameter and various correlations of the Gibbs interaction energies of PSMA with

PEO and $SiO_2$ in water and 0.01 N NaCl solution, essentially different structure of complexed adsorption layer in a given media was proposed (**Fig. 9(a)**, **(b)**).

The other PSMA fraction, which remained in solution, formed steady PCCs with silica nanoparticles (**Fig. 8(b)**, curves 2, 3). The main characteristics of these PCCs under equilibrium conditions (after mixing PSMA with the $SiO_2$+PEO blend for 24 hours) were practically the same as in the binary system $SiO_2$+PSMA in the corresponding medium (**Tables 2, 3**).

A state of the ternary system changed sharply, when PEO was added to the $SiO_2$+PSMA mixture (this case was designated as ($SiO_2$+PSMA)+ PEO in **Table 3**). Indeed, the most part of both polymers remained in the solution in the form of a compact triple PCC composed of both polymers and silica nanoparticles. The following results, namely, low numbers of the parameter β (**Table 3**), and sharp changing in the $pK_1=f(\alpha_1)$ curves (**Fig. 8(b)**, curves 4,5), moreover, appearance of a noticeable opalescence in the supernatants and significant decrease in their reduced viscosity, and finally remarkably higher $\theta$ and $-\Delta G^\circ$ values in this case, by contrast with binary systems PSMA+PEO and $SiO_2$+PSMA, have led to this important conclusion. According to the $-\Delta G^\circ_{PCC}$ values in **Table 3**, stability of the triple PCC was enhanced with increase in $M_{vPEO}$ and in the presence of salt.

An analogous situation, that is the formation of a steady triple PCC with smallest silica nanoparticles, took place, when the initially formed InterPC (PSMA+PEO) was added to a polydispersed Aerosil suspension (designation $SiO_2$+(PSMA+PEO) in **Table 3**). The highest $-\Delta G^\circ_{PCC}$ values were found just in this case, especially upon using longer PEO2.

It should be noted that the $SiO_2$+PSMA binary system and the $SiO_2$+(PSMA+PEO) ternary system were studied further in a wide range of polymer concentrations ($C_{PSMA}$=0.1–1.2 kg·m$^{-3}$ and $C_{InterPC}$= 0.14–1.67 kg·m$^{-3}$). But the picture observed was the same: (i) small adsorption of PSMA or InterPC onto aggregated $SiO_2$ particles (2–3% in the first system and 3–5 % in the second one), and (ii) formation of double or triple PCC in solution. These results led to a new outlook on the so-called "negative adsorption"[87,88] of some polymers, in particular of

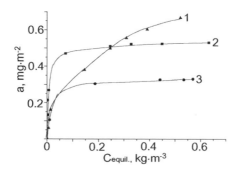

**Fig. 10.** Adsorption isotherms of PVA-*g*-PAAm1 *–1*, PVA-*g*-PAAm2 *–2* and PVA-*g*-PAAm3 *–3* onto Aerosil measured by interferometry. T=25 °C. $M_{vPVA}=8\cdot10^4$; $M_{vPAAm}\sim$ $1\cdot10^5$; N=25, 31 and 49 for the 1st, 2nd and 3rd copolymer sample, respectively.

poly(acrylic acid) (PAAc) onto polydispersed adsorbents. In fact, the adsorption value could not be correctly determined due to the appearance of a stable PCC in the solution.

Thus, in the ternary systems including two interacting polymers (InterPC) and polydispersed colloidal particles no equilibrium processes of competitive interactions can be developed. The final state of these systems is determined by the order and time of a component mixing, the nature and characteristics of polymers and particles, and also by correlation of the interaction energies of the component pairs in a given medium.

**$SiO_2$(Aerosil)+PVA-*g*-PAAm system.** Equilibrium adsorption of the IntraPC forming graft copolymers, having comparable length but different quantity of the grafts, onto Aerosil (**Fig. 10**) was measured by interferometry and viscometry.[28,75,89] Close $a_{max}$ values were obtained by both methods for all the samples that suggested a narrow molecular-weight distribution of the graft copolymers. The $a_{max}$ values regularly lowered with increase in the graft number $N$ that was explained by improvement in the graft copolymer solubility in water (reduction in the aggregation number $Z$) with $N$ increase (see Chapter 5, Sec. 6). It was shown that adsorption of individual macromolecules of the graft copolymers occurred only in the region $C\leq0.1$ kg·m$^{-3}$. The adsorption isotherms demonstrated a high affinity between PVA-$g$-PAAm and the surface of the aggregated $SiO_2$ particles. But the isotherm 1 was different from the other ones due to the strongest aggregation of PVA-$g$-PAAm1 macromolecules in solution. It was established that all supernatants did not contain silica nanoparticles, thus implying their full connection with the copolymer macromolecules and going over into the sediment. This fact has confirmed the high flocculating ability of the given graft copolymers with respect to colloidal particles of different size.[26]

In order to compare the adsorption ability of the graft copolymers with that of individual components, the adsorption of PVA ($M_v$=8·10⁴) and PAAm ($M_v$=2.72·10⁶, hydrolysis degree ~11%) on Aerosil was studied too. For PVA, the $a_{max}$ value determined by both above-mentioned methods was very small (~0.08 mg·m⁻²), but for PAAm that could not be found at all because of larger numbers of reduced viscosity and higher readings on an interferometer scale for supernatants (after adsorption) than for initial PAAm solutions (before adsorption). In fact, we observed the above-mentioned phenomenon of "negative adsorption" and showed that it can be attributed to the formation of PCC ($SiO_2$+PAAm) in solution. It is clear that the stability of PCC to sedimentation was ensured by the presence of ionic –COOH groups on PAAm chains.

Potentiometric titration of the Aerosil suspension in the PVA-$g$-PAAm solution ($C$=0.8 kg·m⁻³), and also of a free copolymer solution and of the copolymer-free suspension has revealed the nonionic character of the graft copolymers and the zero effect of their adsorption on the interfacial charge.

## 4. Inter- and IntraPC in monodispersed silica suspensions

The following part of our investigations was focused on two utmost cases, possible in considered ternary systems, namely, when $R_p$>>$R_{polym}$ (PC) and $R_p$<<$R_{polym}$ (PC). An influence of Inter- and IntraPC on the $\zeta$-potential and stability of quartz suspensions to aggregation was probed in the first case, whilst the complex formation between Inter- or IntraPC and silica sol particles was examined in the second one. Narrow fractions

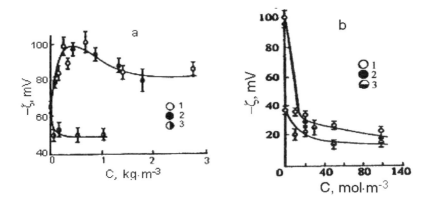

**Fig. 11.** Electrokinetic potential of quartz *vs* polymer (**a**) and $NaNO_3$ (**b**) concentration for PSMA $-1$ (**a,b**), InterPC $-2$ (**a,b**), PEO1 $-3$ (**a**), and polymer-free suspension $-3$ (**b**). $C_{SiO2}$=3.75 kg·m$^{-3}$, $C_{PSMA}$ in $NaNO_3$ solutions was 0.67 kg·m$^{-3}$, pH~4.4.

of quartz suspension were given off by sedimentation but monodispersed $SiO_2$ sol was obtained of the Aerosil suspension by a double centrifugation at $\omega$=6000 rot·min$^{-1}$.

**$SiO_2$+PSMA+PEO system at $R_p \gg R_{polym}$ (PC).** Equilibrium values of $\zeta$-potential of a quartz suspension ($R_p$=3–7 $\mu$m) in solutions of separate polymers and InterPC after stirring for 24 hours were determined using microelectrophoresis experiments.[28,90] The influence of PSMA and InterPC in a wide concentration range on $\zeta$-potential of quartz particles (**Fig. 11(a)**) was shown to be analogous to that of other polyelectrolytes, which have the same sign of charge as the interface. The absolute values of the negative $\zeta$-potential first grew up to a maximum (curves 1,2) due to increase in adsorption of negatively charged PSMA or InterPC at silica/water interface. Then, their decrease was observed, which was initiated by shrinking of the electric double layer under the effect of a growing ionic strength.. The analysis showed that negative $\zeta$-potential of quartz particles in solutions of PSMA and InterPC was ensured solely by dissociation of primary –COOH groups of the polyacid in the adsorption layer.

It was evident that, within experimental error, the curves 1 and 2 coincided in spite of essentially smaller charge of the InterPC particles unlike PSMA macromolecules (**Fig. 7(b)**, **Table 1**). Taking into account this fact, and also the stability of the InterPC during adsorption, it was concluded about the formation of a markedly thicker and denser adsorption layer of polycomplex on the quartz surface compared to the polyacid. $\zeta$-Potential of quartz in PEO1 solutions decreased (**Fig. 11(a)**, curve 3) that was conditioned by diminishing the interfacial charge upon PEO1 adsorption and by special orientation of PEO dipoles in the adsorption layer.[36,83,84]

It was clear that the stability of quartz particles to aggregation and sedimentation was ensured mainly by electrostatic repulsion. But the behavior of the same particles covered by adsorption layers of PSMA, PEO1 or InterPC under conditions of significant weakening of this factor was of a special interest. That is why electrostatic repulsion was decreased by addition of a low-molecular-weight electrolyte ($NaNO_3$) in the quartz suspension after its mixing with the corresponding polymer solution for 24 hours.[90] A sharp decrease in $\zeta$-potential of quartz particles in solutions of PSMA and InterPC under the effect of increasing $NaNO_3$ concentration is shown in **Fig. 11(b)**. A difference in the polymer effect on the aggregation stability of the quartz suspension under such conditions is reflected in **Fig. 12(a)**. An essentially lower value of the electrolyte coagulation threshold corresponding to the inflection point of the curves in **Fig. 12(a)** was found in the solution of InterPC ($10 \ mol_{NaNO3} \cdot m^{-3}$) by contrast with that in PSMA ($90 \ mol_{NaNO3} \cdot m^{-3}$) and PEO1 ($40 \ mol_{NaNO3} \cdot m^{-3}$) solutions. Thus, an abnormally high destabilizing activity of InterPC with respect to the large colloidal particles was fully confirmed.

Interestingly, a final state of the given ternary system did not depend on the order of the component addition and the time of their contact, thus suggesting its equilibrium character. This fact was established in separate experiments. Indeed, $\zeta$-potential constant values of quartz particles near $-100$ mV and the coagulation threshold near $10 \ mol_{NaNO3} \cdot m^{-3}$ were obtained in 60 mins and 24 hours after addition of PSMA, PEO1 or InterPC to the systems ($SiO_2$+PEO1), ($SiO_2$+PSMA), and $SiO_2$, respectively.

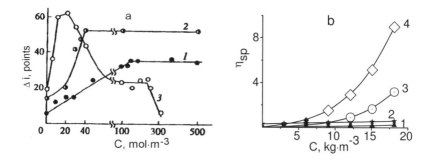

**Fig. 12.** Changes (**a**) in the light transmission ($\lambda$=480 nm, T=20°C) of quartz suspension in solutions of PSMA (C=0.67 kg·m$^{-3}$) –*1*, PEO1 (C=0.26 kg·m$^{-3}$) –*2*, and InterPC –*3* in 5 min after onset of its sedimentation *vs* NaNO$_3$ concentration, and (**b**) in the specific viscosity (T=25°C) of silica sol suspension in pure water –*1* and in solutions of PEO1 (C=0.27 kg·m$^{-3}$) –*2*, PSMA (C=0.70 kg·m$^{-3}$) –*3*, and InterPC –*4 vs* SiO$_2$ concentration. Large points designate a phase separation in the polymer-SiO$_2$ sol system.

**Table 4.** Parameters of polymer-colloidal complexes in binary and ternary systems.

| System | $\varphi$, $w_{SiO2}/w_{PSMA}$ | $n$ | $\theta$, % | $-\Delta G°_{PCC}$, kJ·mol$^{-1}$ |
|---|---|---|---|---|
| SiO$_2$+PSMA | 12,0 | 0.35 | 32 | 0.27 |
| | 25,9 | 0.76 | 72 | 0.72 |
| (SiO$_2$+PEO1)+PSMA | 12,0 | 0.35 | 45 | 0.78 |
| (SiO$_2$+PSMA)+PEO1 | 12,0 | 0.35 | 46 | 0.78 |
| SiO$_2$+(PSMA+PEO1) | 12,0 | 0.35 | 34 | 0.71 |
| SiO$_2$+(PSMA+PEO2) | 12,0 | 0.35 | 55 | 1.30 |

$n$ : The number of SiO$_2$ sol particles with $R_p$=11.2 nm per one PSMA macromolecule.
*Note* : The ratio PEO1/PSMA corresponded to $\varphi_{char}$ for InterPC (Table 1); the contact time of every polymer or InterPC with silica sol was equal to 24 h.

**SiO$_2$+PSMA+PEO system at $R_p$<<$R_{polym}$ (PC).** Spontaneous polymer-colloidal reactions between one or two polymers and monodispersed silica sol particles ($R_p$=11.2 nm according to static light scattering) were studied by potentiometric titration and viscometry[28,74,80] analogously to the studies of supernatants in the same system but with polydispersed Aerosil (Sec. 3). The influence of the order of the components addition and of the molecular weight of PEO, $M_{vPEO}$, was tested. The $pK_1=f(\alpha_1)$ curves calculated for the $SiO_2+PSMA$ binary system demonstrated a gradual destruction of the PSMA compact structure when the quantity of $SiO_2$ sol particles was increased. That was accompanied by growth in PCC stability (**Table 4**) and confirmed once again a strong interaction of PSMA with silica nanoparticles. For the triple PCC both $\theta$ and $-\Delta G^o$ parameters proved to be essentially larger than those for the double PCC. Additionally, they were enhanced with increasing $M_{vPEO}$, but they did not depend on the way of the triple PCC formation (**Table 4**). Therefore, this ternary system was thermodynamically in equilibrium.

It was revealed that at $\phi=12.0$ $w_{SiO2}/w_{PSMA}$ ($C_{SiO2}=7.9$ $kg \cdot m^{-3}$) the ternary system (including PEO1), unlike the corresponding binary systems, verged on phase separation. Such a separation took place with further, even insignificant, increase in the sol content in the system and resulted in a sharp increase of the reduced viscosity (**Fig. 12(b)**). This effect was a clear display for an abnormally high flocculating capability of InterPC, also with respect to silica nanoparticles. Now, we will consider only some results concerning the other systems.

**$SiO_2+PVA$-$g$-$PAAm$ system at $R_p>>R_{polym}$ (PC).** The effect of the IntraPC forming graft copolymers with different graft numbers (their parameters are shown below **Fig. 10**) and also individual components on the $\zeta$-potential of quartz with $R_p=70–110$ $\mu m$ was established[28,89] by the streaming potential method. The studies were carried out in such a concentration region (**Fig. 13(a)**) where adsorption of individual PVA-$g$-PAAm chains should occur (see Sec. 3). The influence of the partially hydrolyzed PAAm (curve 1) was analogous to that of the negatively charged PSMA and InterPC (**Fig. 11(a)**). Nonionic PVA caused a certain decrease in $\zeta$-potential (curve 2). But adsorption of the graft copolymers stimulated even more the $\zeta$-potential decrease (curves 3,4), which was the greatest among the similar effects of other nonionic water-soluble homopolymers.

Using the values of $\zeta$-potential and adsorption of PVA-$g$-PAAm2,3 or PVA, and taking into account a zero influence of the copolymers (and

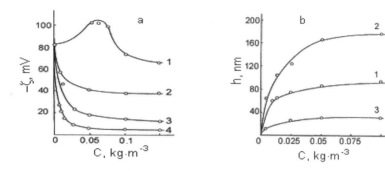

**Fig. 13.** Electrokinetic potential of quartz (**a**) and thickness of adsorption layer (**b**) *vs* polymer concentration for PAAm ($M_v$=2.72·10[6], the hydrolysis degree ~11 mol %)–*1* (**a**), PVA ($M_v$=8·10[4]) –*2*(**a**),*3*(**b**), PVA-*g*-PAAm2 –*3*(**a**),*1*(**b**), and PVA-*g*-PAAm3 – *4*(**a**), *2*(**b**).

PVA) on the interfacial charge of the silica particles (Sec. 3), the electrokinetic thickness of every adsorption layer formed by individual polymer chains was calculated[89] (**Fig. 13(b)**). Analysis of these data showed that strong $\zeta$-potential decrease in solutions of the graft copolymers was related to the appearance of thick adsorption layers on the surface of quartz particles and shifting of the shear plane into the solution.

**SiO$_2$+PVA-*g*-PAAm system at $R_p$<<$R_{polym\ (PC)}$.** Complex formation of the graft copolymers with 2 and 5 silica sol particles ($R_p$=11.2 nm) was studied[91,92] using viscometry, static light scattering and benzene solubilization. Compact PCC particles, more stable to dilution than initial IntraPCs in the copolymer macromolecules, were revealed in the PVA-*g*-PAAm/SiO$_2$ sol mixtures by the first method (the example is shown in **Fig. 14(a)**). Further, the $\Delta_v$ and $\Delta_u$ depolarization coefficients, characterizing optical homogeneity and dimension (form) of scattering particles (Chapter 5, Sec. 6) as well as the optical anisotropy parameter $\delta^2$ were determined for the copolymers and PCC by the second method (**Table 5**).

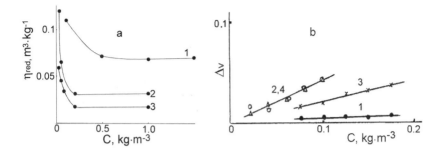

**Fig. 14.** Reduced viscosity (**a**) and the $\Delta_v$ depolarization coefficient (**b**) *vs* polymer concentration for PVA-*g*-PAAm2 *–1* (**a**) and triple PCC formed by PVA-*g*-PAAm2 with two *–2* (**a**), *1* (**b**) and five silica sol particles *–3* (**a**), *2* (**b**), and also by PVA-*g*-PAAm3 with two *–3* (**b**) and five SiO$_2$ sol particles *–4* (**b**). T=25°C.

Moreover, a considerable benzene solubilization in pure copolymer solutions (Chapter 5, Sec. 6) and SiO$_2$ sol suspension, unlike a zero effect in PCC solutions, were fixed by the third method.

Thus, PCC formation in solutions of PVA-*g*-PAAm2 (with smaller graft density) led only to a sharp decrease in the reduced viscosity, while the $\Delta_v$, $\Delta_u$, and $\delta^{\circ}$ parameters practically did not change. This fact indicates the high stability of the IntraPC structure in the graft copolymer macromolecules, in contrast to their aggregates, which were destroyed in the polymer-colloidal reaction. Probably, silica nanoparticles easily penetrated into the hydrophilic cavities of PVA-*g*-PAAm2 and interacted there with unbound segments of PVA and PAAm without IntraPC destruction. The interaction of PVA-*g*-PAAm3, having a larger graft density, with SiO$_2$ nanoparticles was more complicated. It was accompanied not only by a sharp reduced viscosity decrease but also by a significant reduction in all the optical parameters (**Table 5**). Therefore, both processes, the aggregate and the IntraPC destruction, resulting in the formation of essentially more compact and dense PCC particles, took place in the given polymer-colloidal reaction.

## 5. Possible reasons for the high flocculating ability of polycomplexes

Let us summarize the main results in accordance with the modern notions about the stability of polymer-colloidal systems (Sec. 2). The high destabilizing activity of the IntraPC forming graft copolymers with respect to large silica particles (and large particles of kaolinite clay or hydrolyzed aluminum sulfate in real polydispersed suspensions[26,27]) could be explained, at first sight, only by the formation of a thick adsorption layer, which shifts sharply the shear plane into the solution and lowers $\zeta$-potential of the particles. In this case, a strong decrease in the electrostatic repulsion contribution $V_E(h)$ in the overall interaction potential $V(h)$ of the particles evidently surpassed an increase in the steric repulsion contribution $V_S(h)$, related to a thick adsorption layer. At the same time, an abnormally high destabilizing activity of the InterPC (PSMA+PEO) could not be understood within traditional notions and required another interpretation. Indeed, the InterPC and PSMA adsorption led to an equal enhancement of the $\zeta$-potential of quartz (attributed to arising of a thicker interfacial layer of weakly charged InterPC particles) and to equal rate of the $\zeta$-potential decrease upon NaNO$_3$ addition (**Fig. 11**). Under such conditions, an increase in the quartz particle stability in the InterPC solutions, due to the growth of the $V_S(h)$ contribution, could be expected.

**Table 5.** Hydrodynamic and optical parameters for PCC based on graft copolymers.

| Copolymer | $n$ | $\Delta\eta_{red}$, m$^3$·kg$^{-1}$ | $\Delta_v^{\,o}$ | $\Delta_u^{\,o}$ | $\delta^2$ |
|:---:|:---:|:---:|:---:|:---:|:---:|
| | 0 | - | 0.006 | 0.030 | 0.052 |
| PVA-$g$-PAAm2 | 2 | 0.54 | 0.003 | 0.032 | 0.055 |
| | 5 | 0.74 | 0.003 | 0.030 | 0.052 |
| | 0 | - | 0.020 | 0.049 | 0.087 |
| PVA-$g$-PAAm3 | 2 | 0.58 | 0.001 | 0.024 | 0.042 |
| | 5 | 0.77 | 0.001 | 0.027 | 0.046 |

$n$ : The number of silica sol particles per one macromolecule of the graft copolymer.
$\Delta\eta_{red}$ : Falling of the PVA-$g$-PAAm reduced viscosity initiated by PCC formation.
$\Delta_v^{\,o}$ and $\Delta_u^{\,o}$ : The depolarization coefficients extrapolated to $C_{PVA-g-PAAm}=0$.
$\delta^2$ : The parameter of optical anisotropy.

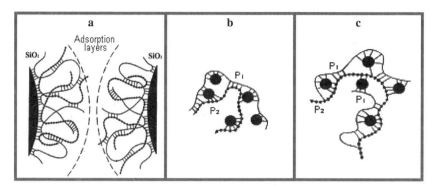

**Fig. 15.** Schemes illustrating drawn together large particles covered by adsorption layers of InterPC (**a**) and also PCC formed by Inter- (**b**) or IntraPC (**c**) with nanoparticles.

The real situation was reverse: the stability of the quartz suspension (the coagulation threshold) fell sharply under the effect of the InterPC. Moreover, the particle aggregation developed at high $\zeta$-potential, such as $-35$ mV (**Fig. 11(b)**). This result could not be also explained by additional contributions of hydrophobic or depletion interactions, thus implying the existence of another unknown factor of particles attraction. In our opinion, the existence of unbound segments of both components on the outer border of the mixed adsorption layer of Inter- or IntraPC (**Fig. 15(a)**) initiates the hydrogen bonding interactions between the adsorption layers of the different colloidal particles and ensures, in such a way, an additional thermodynamic factor of the large particles attraction.

In this context, a higher flocculating ability of the Inter- and IntraPC, as compared to the separate components, with respect to silica nanoparticles (and smallest particles of kaolinite clay or hydrolyzed aluminum sulfate in real suspensions[24-27]) could be attributed to (i) the formation of stable triple PCCs (**Fig. 15(b), (c)**) with reduced solubility in water, and (ii) effect of an additional thermodynamic factor of attraction between the PCC particles.

It should be added that IntraPCs with covalently bound chemically complementary partners are more promising flocculating agents than InterPCs, because they cannot be destroyed in any competitive processes. Possessing a practically universal binding ability, these flocculants are capable to clear water on waterworks simultaneously from pollutants not only of natural (colloidal particles, humic and fulvic acids[93,94]) but also of anthropogenic origin (in particular, phenol[95,96] and radionuclides $^{137}Cs$ and $^{90}Sr$ [97,98]).

## References

1. D.B. Brawn and Y.Z. Fan, Europ. Patent Appl. 0055489 A1 (1981).
2. G.P. Medvedev, E.A. Evelson, V.V. Khrabrov, V.I. Kurlyankina, V.A. Molotkov and S.I. Klenin, USSR Patent 940459 A (1980).
3. T. Lindström and G. Glad-Nordmark, *J. Colloid Interface Sci.*, **97**, 62 (1984).
4. V.I. Matsokin, A.S. Yuschenko, T.I. Gryshanova, G.M. Kurdyumov, D.A. Topchiev, E.V. Shurupov, A.B. Zezin, V.A. Kasayikin, N.N. Bushuev, M.I. Novikov and T.V. Solovyeva, USSR Patent 1401813 A1 (1986).
5. A.F. Nikolaev, V.G. Shibalovich, V.M. Bondarenko, G.P. Perina, B.A. Barbanel and M.B. Lyubimov, USSR Patent 1204576 A (1986).
6. R. Stephen, Phuong-Thao-Luong, U.S. Patent 4906386 (1990).
7. V.A. Kabanov, A.B. Zezin, V.A. Kasaikin, A.A. Yaroslavov and D.A. Topchiev, *Russ. Chem. Rev.*, **60**, 288 (1991).
8. J. Kötz and S. Kosmella, *J. Colloid Interface Sci.*, **168**, 505 (1994).
9. S.K. Rath and R.P. Singh, *J. Appl. Polym. Sci.*, **66**, 1721 (1997).
10. H.D. Bijsterbosch, M.A. Cohen Stuart and G.J. Fleer, *J. Colloid Interface Sci.*, **210**, 37 (1999).
11. T.V. Shevchenko, T.A. Krasnova and O.I. Korshunova, *Chemical Industry*, (11), 36 (2000) (in Russian).
12. T.B. Zheltonozhskaya and V.G. Syromyatnikov, in: *Scientific Transactions. Chemical Faculty. Kiev National Taras Shevchenko University. V.14* (Pedagogics, Kiev, 2004), p. 152 (in Ukrainian).
13. D.R. Biswald and R.P. Singh, *J. Appl. Polym. Sci.*, **102**, 1000 (2006).
14. H. Walter, P. Müller-Buschbaum, J.S. Gutmann, C. Lorenz-Haas, C. Harrats, R. Jerôme and M. Stamm, *Langmuir*, **15**, 6984 (1999).
15. G. Decher and J.B. Schlenoff, Eds., *Multilayer Thin Films: Sequential Assembly of Nanocomposite Materials* (Wiley-VCH, Weinheim, Germany, 2003).
16. E. Kharlampieva, S.A. Sukhishvili, *J. Macromol. Sci., Part C: Polym. Rev.*, **46**, 377 (2006).
17. S.E. Morgan, P. Jones, A.S. Lamont, A. Heidenreich and C.L. McCormick, *Langmuir*, **23**, 230 (2007).
18. V.Yu. Baranovskiy, G.S. Georgiev and V.A. Kabanov, *Vysokomol. Soedyneniya. Ser. A.*, **31**, 486 (1989) (in Russian).
19. A.V. Kabanov, T.K. Bronich, V.A. Kabanov, K. Yu and A. Eisenberg, *Macromolecules*, **29**, 6797 (1996).
20. V.A. Kabanov, *Russ. Chem. Rev.*, **74**, 3 (2005).
21. T.A. Korobko, V.A. Izumrudov and A.B. Zezin, *Polym. Sci.*, **35**, 71 (1993).

22. Yu.S. Lipatov, in: *Encyclopedia of Surface and Colloid Science. V.1*, Ed. A.T. Hubbard (Marcel Dekker, New York, 2002), p. 558.

23. Y.S. Lipatov, V.N. Blyznyuk, T.T. Todosiychuk, V.N. Chornaya, R.K. Katumenu and V.D. Konovalyuk, *Colloid Polym. Sci.*, **284**, 893 (2006).

24. T.B. Zheltonozhskaya, N.V. Kutsevol, L.N. Momot, B.V. Eremenko and V.G. Syromyatnikov, Ukrainian Patent, UA 17242 A (1997).

25. T.B. Zheltonozhskaya, L.N. Momot, N.V. Kutsevol, V.G. Syromyatnikov, V.Ya Demchenko and V.N. Olenchenko, Ukrainian Patent, UA 17814 A (1997); *Promislova Vlastnist'* (5), (1997).

26. T.B. Zheltonozhskaya, N.V. Kutsevol, L.N. Momot, N.P. Melnik and B.V. Eremenko, Ukrainian Patent, UA 23743 (2001); *Promislova Vlastnist'* (11), (2001).

27. T.B. Zheltonozhskaya, N.V. Kutsevol, L.N. Momot, O.O. Romankevich, V.G. Sytomyatnikov, V.Ya. Demchenko and V.M. Olenchenko, Ukrainian Patent, UA 29933 (2001); *Promislova Vlastnist'* (3), (2002).

28. T.B. Zheltonozhskaya, Inter- and intramolecular polycomplexes stabilized by hydrogen bonds in the processes of flocculation and sorption, *Doctoral Thesis in Chemical Sciences* (Kiev National Taras Shevchenko University, Kiev, 2003) (in Ukrainian).

29. N.E. Zagdanskaya, N.M. Permyakova, T.B. Zheltonozhskaya and S.V. Fedorchuk, Ukrainian Patent, UA 78649 (2007); *Promislova Vlastnist'* (4), (2007).

30. D.J. McClements, *Langmuir*, **21**, 9777 (2005).

31. B.V. Deryagin, *Theory of Stability of Colloids and Thin Films* (Science, Moscow, 1986) (in Russian).

32. J.N. Israelachvili, *Intermolecular and Surface Forces* (Academic Press, London, 1992).

33. R.G. Horn and D.T. Smith, *Non-Cryst. Solids*, **120**, 72 (1990).

34. N.V. Churaev, *Russ. Chem. Rev.*, **73**, 25 (2004).

35. V.A. Myagchenkov, S. Baran, E.A. Bekturov and G.V. Bulidorova, *Polyacrylamide Flocculants* (Kazanskiy State Technological University, Kazan, 1998) (in Russian).

36. B.V. Eremenko, Adsorption of nonionic polymers and electro-surface properties of dispersed systems, *Doctoral Thesis in Chemical Sciences* (Kiev National Taras Shevchenko University, Kiev, 1982) (in Russian).

37. O.V. Akopova and B.V. Eremenko, *Colloid J.*, **56**, 5 (1994).

38. F.T. Hesselink, A. Vrij and J.T.G. Overbeek, *J. Phys. Chem.*, **75**, 2094 (1971).

39. G.J. Fleer, M.A. Cohen Stuart, J.M.H.M. Scheutjens, T. Cosgrove and B. Vincent, in *Polymers at Interfaces* (Chapman & Hall, London, 1993).

40. D. Qiu, T. Cosgrove and A.M. Howe, *Langmuir*, **23**, 475 (2007).

41. S. Asakura and F. Oosava, *J. Polym. Sci.*, **33**, 183 (1958).

42. X.Y.T. Narayanan, P. Tong, J.S. Huang, M.Y. Liu, B.L. Carvalho and L.J. Fetters, *Phys. Rev. E.* **54**, 6500 (1996).

43. A. Weiss, K.D. Hörner and M. Ballauff, *J. Colloid Interface Sci.*, **213**, 417 (1999).

44. D. Velegol, G.L. Holtzer, A.F. Radović-Moreno and J.D. Cuppett, *Langmuir*, **23**, 1275 (2007).

45. V.A. Kasaikin, N.V. Pavlova, L.N. Ermakova, A.B. Zezin and V.A. Kabanov, *Colloid. J.*, **48**, 452 (1986) (in Russian).

46. R.I. Kaluzhnaya, Kh.Kh. Khulchaev, V.A. Kasaikin, A.B. Zezin and V.A. Kabanov, *Vysokomol. Soedyneniya*, **36**, 257 (1994) (in Russian).
47. B. Cabane, K. Wong, P. Lindner and F. Lafuma, *J. Reol.*, **41**, 531 (1997).
48. W. Nowicki, *Colloids Surf., A: Physicochem. Eng. Aspects*, **194**, 159 (2001).
49. T.M. Birshtein and O.V. Borisov, *Vysokomol. Soedyneniya. Ser. A.*, **28**, 2265 (1986) (in Russian).
50. C.M. Marques and J.F. Joanny, *J. Phys.* (France), **49**, 1103 (1988).
51. D.K. Klimov and A.R. Khokhlov, *Polymer*, **33**, 2181 (1992).
52. V.K. La Mer, *Dis. Farad. Soc.*, **42**, 248 (1996).
53. A.A. Litmanovich, Yu.E. Kuzovlev and E.V. Polyakova, *Vysokomol. Soedyneniya. Ser. B*, **39**, 1527 (1997) (in Russian).
54. Yu.S. Lipatov, *Colloid Polym. Sci.*, **264**, 377 (1986).
55. Yu.S. Lipatov, V.V. Shilov, G.P. Kovernik and Yu.P. Gomza, *Dopovidi AN UkrSSR. Ser. B*, (1), 41 (1986) (in Ukrainian).
56. Yu.S. Lipatov, N.P. Gudima, A.E. Nesterov and T.S. Chramova, *Dopovidi AN UkrSSR. Ser. B*, (11), 44 (1989) (in Ukrainian).
57. L.V. Zherenkova, P.G. Khalatur and A.R. Khokhlov, *Colloid J.*, **59**, 634 (1997) (in Russian).
58. P.G. Khalatur, L.V. Zherenkova and A.R. Khokhlov, *Colloid J.*, **59**, 646 (1997) (in Russian).
59. P.G. Khalatur, L.V. Zherenkova and A.R. Khokhlov, *J. Phys. II* (France), **7**, 543 (1997).
60. L.V. Zherenkova, D.A. Mologin, P.G. Khalatur and A.R. Khokhlov, *Colloid Polym. Sci.*, **276**, 753 (1998).
61. A.S. Grodskiy, *Colloid J.*, **59**, 1082 (1989) (in Russian).
62. Yu.V. Shulepov, L.K. Koopal, Y. Lyklema and S.S. Dukhin, *Colloids Surf., A: Physicochem. Eng. Aspects*, **131**, 51 (1997).
63. Y. Mao, M.E. Cates and H.N.W. Lekkerkerker, *Physica A*, **222**, 10 (1995).
64. G.A. Vliegenthart and H.N.W. Lekkerkerker, *J. Chem. Phys.*, **111**, 4153 (1999).
65. R.P. Sear, *Phys. Rev. Lett.*, **82**, 4244 (1999).
66. A.Yu. Zubarev, L.Yu. Iskhakova, *Colloid J.*, **66**, 296 (2004).
67. P. Sollich, *J. Phys.: Condens. Matter*, **14**, R79 (2002).
68. V. Tohver, A. Chan, O. Sakurada and J.A. Lewis, *Langmuir*, **17**, 8414 (2001).
69. V. Tohver, J.E. Smay, M. Braem, P.V. Braun and J.A. Lewis, *Proc. Natl. Acad. Sci. U.S.A.*, **98**, 8950 (2001).
70. C.J. Martinez, J. Liu, S.K. Rhodes, E. Luijten, E.R. Weeks and J.A. Lewis, *Langmuir*, **21**, 9978 (2005).
71. W. Lee, A. Chan, M.A. Bevan, J.A. Lewis and P.V. Braun, *Langmuir*, **20**, 5262 (2004).
72. D.J. Fairhurst, Polydispersity in colloidal phase transitions, *PhD Thesis* (University of Edinburgh, UK, 1999).
73. T.B. Zheltonozhskaya, V.M. Kharlanov, B.V. Eremenko and I.A. Uskov, *Ukr. Khim. Zhurn.*, **59**, 987 (1993) (in Ukrainian).
74. N. Permyakova, T. Zheltonozhskaya, N. Strilchuk, N. Zagdanskaya, L. Momot and V. Syromyatnikov, *Macromol. Symp.*, **203**, 247 (2003).

75. B.V. Eremenko, N.P. Melnik, N.V. Kutsevol, L.N. Momot and T.B. Zheltonozhskaya, in *MinChem'92, The Fourth Symposium on Mining Chemistry* Symposium Proceedings (Kiev, Ukraine, 1992), p. 303.

76. R.K. Iler, *The Chemistry of Silica: Solubility, Polymerization, Colloid and Surface Properties and Biochemistry of Silica* (A Wiley-Interscience Publication, New York-Chichester-Brisbano-Toronto, 1982).

77. O.A. Kazakova, V.M. Gunko, E.F. Voronin, S.S. Silchenko and A.A. Chuiko, *Colloid J.*, **60**, 563 (1998).

78. T.B. Zheltonozhskaya, L.N. Podolyak, B.V. Eremenko and I.A. Uskov, *Vysokomol. Soedyneniya. Ser. A.*, **29**, 2487 (1987) (in Russian).

79. T.B. Zheltonozhskaya, B.V. Eremenko and I.A.Uskov, *Ukr. Khim. Zhurn.*, **47**, 948 (1981) (in Russian).

80. N.M. Permyakova, T.B. Zheltonozhskaya, B.V. Eremenko, N.E. Zagdanskaya, M.L. Malysheva and V.G. Syromyatnikov, in *CERECO'2003, The 4th International Conference on Carpathian Euroregion Ecology,* Conference Proceedings (Miskolc-Tapolca, Hungary, 2003), p. 228.

81. V.A. Kabanov and I.M. Papisov, *Vysokomol. Soedyneniya. Ser. A.*, **21**, 243 (1979) (in Russian).

82. I.C. Leyte and M. Mandel, *J. Polym. Sci., Part A: Polym. Chem.*, **2**, 1879, (1964).

83. B.V. Eremenko, I.A. Uskov and Zh.V. Chernenko, in *Physical Chemistry of Polymer Compositions* (Naukova Dumka, Kiev, 1974), p. 60 (in Russian).

84. B.V. Eremenko, B.E. Platonov, I.A. Uskov and I.N. Lyubchenko, *Colloid J.*, **36**, 240 (1974) (in Russian).

85. T.B. Zheltonozhskaya, B.V. Eremenko, Yu.A. Nuzhdina and I.A. Uskov, *Vysokomol. Soedyneniya. Ser. A.*, **29**, 265 (1987) (in Russian).

86. T.B. Zheltonozhskaya, V.B. Rudnitskaya, B.V. Eremenko and V.G. Syromyatnikov, *Colloid J.*, **58**, 319 (1996).

87. T. Sato, C.L. Sieglaff, *J. Appl. Polym. Sci.*, **25**, 1781 (1980).

88. I.L. Sharifullin, S.V. Krupin, V.P. Barabanov, *Colloid. J.*, **48**, 383 (1986) (in Russian).

89. B.V. Eremenko, M.L. Malysheva, O.D. Rusina, N.V. Kutsevol and T.B. Zheltonozhskaya, *Colloids Surf., A: Physicochem. Eng. Aspects*, **98**, 19 (1995).

90. O,V. Akopova, T.B. Zheltonozhskaya and B.V. Eremenko, *Colloid J.*, **56**, 655 (1994).

91. T.B. Zheltonozhskaya, O.O. Romankevich, N.V. Kutsevol, V.G. Syromyatnikov, in *CERECO'1997, The 2nd International Conference on Carpathian Euroregion Ecology,* Conference Proceedings (Miskolc-Lillafüred, Hungary, 1997), p. 149.

92. T. Zheltonozhskaya, *Macromol. Symp.*, **222**, 125 (2005).

93. O.V. Demchenko, T.B. Zheltonozhskaya, V.L. Budzinskaya, N.M. Permyakova, T.G. Ejova, V.G. Syromyatnikov and N.N. Osadchaya, in *CERECO'2000, The 3rd International Conf. on Carpathian Euroregion Ecology,* Conference Proceedings (Miskolc-Lillafüred, Hungary, 2000), p. 17.

94. O. Demchenko, T. Zheltonozhskaya, N. Kutsevol, V. Syromyatnikov, I. Rakovich and O. Romankevich, *Chem. Inz. Ekol.* **8**, 463 (2001).

95. V. Syromyatnikov, T. Zheltonozhskaya, J.-M. Guenet, I. Rakovich, O. Demchenko, N. Strilchuk and N. Permyakova, *Macromol. Symp.*, **166**, 237 (2001).
96. T. Zheltonozhskaya, O. Demchenko, J.-M. Guenet, V. Syromyatnikov, I. Rakovich and L. Kunitskaya, in *CERECO'2000, The 3ʳᵈ International Conf. on Carpathian Euroregion Ecology*, Conference Proceedings (Miskolc-Lillafüred, Hungary, 2000), p. 101.
97. S. Filipchenko, T. Zheltonozhskaya, O. Demchenko, T. Vitovetskaya, N. Kutsevol, L. Sadovnikov, O. Romankevich and V. Syromyatnikov, *Chem. Inz. Ekol.*, **9**, 1521 (2002).
98. T.B. Zheltonozhskaya, T.V. Vitovetskaya, O.V. Demchenko, J.-M. Guenet, S.A. Filipchenko, L.V. Sadovnikov and V.G. Syromyatnikov, in *CERECO'2003, The 4ᵗʰ International Conf. on Carpathian Euroregion Ecology*, Conference Proceedings (Miskolc-Tapolca, Hungary, 2003), p. 215.

# CHAPTER 9

# PHARMACEUTICAL APPLICATIONS OF INTERPOLYMER COMPLEXES

Vitaliy V. Khutoryanskiy

*Reading School of Pharmacy, University of Reading, Whiteknights,
PO Box 224, Reading RG6 6AD, United Kingdom
E-mail: v.khutoryanskiy@reading.ac.uk*

Polymers are widely used in pharmaceutical applications as suspending and emulsifying agents, flocculants, adhesives, packaging and coating materials, and as components of controlled and site-specific drug delivery systems.[1] Many pharmaceutical polymers are water-soluble or at least are able to swell in water, forming hydrogels. In mixtures, these polymers can often interact with each other and sometimes form interpolymer complexes (IPC). The complexation between polymers influences their structure and physicochemical properties such as solubility, viscosity in solutions, and ability to adhere to different substrates.

In this chapter, we will consider the application of hydrogen-bonded complexes for design of mucoadhesive dosage forms and tablets, formulation of lipophilic and hydrophilic drugs, encapsulation technologies, micro- and nano-particulate systems, hydrogels and *in situ* gelling systems. A review covering a similar topic has recently been published by Khutoryanskiy.[2]

## 1. Mucoadhesive dosage forms: tablets, films and microspheres

Mucus is a fully hydrated viscoelastic gel that covers the surface of the eye, nose, mouth, respiratory tract, cervix, and gastrointestinal tract.[3] The adhesive attachment of a material to mucus is called mucoadhesion and this phenomenon is widely used in the design of dosage forms.[4-6] Currently, mucoadhesive dosage forms are used for administration of drugs via buccal, ocular, nasal, oral, vaginal and rectal routes.

Mucus consists of mucins (up to 5 wt %), water (up to 95 wt %), inorganic salts (about 1 wt %), carbohydrates and lipids.[7,8] Mucins are macromolecules with a relative molecular weight range of 1 MDa–40 MDa consisting of a protein core with covalently attached carbohydrate side chains. Most of them carry a net negative charge due to the presence of sialic acids and ester sulphates at the terminus of some sugar units.

Some of the polymeric structural characteristics necessary for mucoadhesion can be summarised as follows: (1) strong hydrogen bonding groups, e.g., carboxyl, hydroxyl, amino- and sulphate groups, (2) strong anionic or cationic charges, (3) high molecular weight, (4) chain flexibility, (5) surface energy properties favouring spreading onto mucus.[9]

Poly(acrylic acid) (PAA) and its weakly cross-linked derivatives (Carbopol®, Polycarbophil® and Carbomer®) have been extensively used in design of dosage forms due to their excellent mucoadhesive performance.[10, 11] It is believed that these properties are due to the ability of the polymers to interact with mucins via hydrogen-bonding.[12,13] However, PAA and its derivatives have some limitations including their tendency to cause irritation of mucosal tissues,[14] high glass transition temperature in the dry state making the materials less elastic,[15] and high hydrophilicity leading to fast disintegration and poor mechanical properties in the swollen state. Blending of PAA with some non-ionic polymers has been demonstrated to be one of the ways to produce non-irritant drug delivery systems with optimal mucoadhesive characteristics.

When PAA interacts with non-ionic polymers in a blend, it may alter the mucoadhesive properties of a formulation significantly. Polymer blending may involve either blending of polymer powders or mixing polymers in solutions. In the later case, the interpolymer interactions may affect the properties of the resulting materials more significantly. Satoh and co-workers[16] have developed a series of bioadhesive tablets by compressing powder mixtures of Carbopol®934 with hydroxypropylcellulose (HPC) at different ratios. The mucoadhesive properties of these tablets were assessed by measuring the force required to detach a tablet from mouse peritoneal membrane. It was found that the mucoadhesive ability of the tablets is a function of Carbopol ® / HPC ratio, with the weakest adhesive force observed at 3:2 mixing ratio (**Fig. 1**). The authors related the observed trend to the possibility of interpolymer interaction between the polymers via hydrogen-bonding.

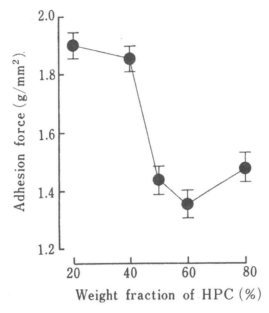

**Fig. 1.** Adhesion of Carbopol / HPC tablet to mice peritoneal membrane as a function of HPC content. Reproduced with permission from Chemical & Pharmaceutical Bulletin Vol. 37, No 5 (Ref. 16). Copyright 1989. Pharmaceutical Society of Japan.

This hypothesis was confirmed by investigating the complexation between Carbopol®934 and HPC in aqueous solutions by viscometric and turbidimetric techniques as well as the FTIR analysis of dry IPC precipitate. It was found that Carbopol®934 and HPC form hydrogen-bonded IPC at the mixing ratio 3:2 and the hydrophobicity of the resulting product was considered as a reason for inhibition in mucoadhesive properties. In a second paper, Satoh and co-workers[17] reported the disintegration and dissolution characteristics of tablets obtained by compressing IPC powder and Carbopol®934 / HPC physical mixtures. The IPC powder was obtained by mixing 1% Carbopol®934 and HPC solutions in weight ratio 2:3, precipitating the polycomplex formed and drying it under vacuum. In disintegration experiments, these tablets were immersed in a beaker with purified water and morphological changes were documented for 24 hours. The disintegration of tablets based on pure Carbopol®934 was slow and completed within 24 hours. The tablets based on pure HPC did not disintegrate, but gradually dissolved, exhibiting swelling and gelation properties. A rapid disintegration was observed for the tablets based on IPC at the initial stage of experiment, whereas the tablets based on polymer physical mixtures exhibited neither disintegration nor dissolution. The unusually fast disintegration of IPC was related to its hydrophobic nature, which weakens the inter-particle binding forces when water penetrates into the tablets. Brilliant blue dye was used as a model compound to mimic a water-soluble drug in tablet dissolution experiments. A rapid dissolution was observed within 1 hour when the tablet was prepared by compressing solid IPC. The slowest dissolution profile was obtained from the polymers physical mixture, which was explained by the complex formation between the polymers following water penetration into the tablet. From these observations, the use of a physical mixture of Carbopol®934 and HPC was found to be advantageous for controlling drug release. The same group has also demonstrated the use of Carbopol®934 / HPC mixtures for formulation of mucosal dosage forms of insulin[18] and lidocaine.[19]

Gupta *et al.*[20] studied the effect of interpolymer complexation of Carbopol-934 with HPC, poly(N-vinyl pyrrolidone) (PVP) and sodium salt of carboxymethylcellulose (NaCMC) on the bioadhesive strength and dissolution characteristics of buccal tablets. They established that the interpolymer complexes are formed under acidic conditions (pH < 4.5) and the complex formation is most pronounced in the case of Carbopol-934 mixtures with HPC and PVP, whereas NaCMC shows significantly lower complexation ability. The maximal IPC yield is observed at the weight ratios of 2:3, 1:1 and 4:1 for Carbopol-934 / HPC, Carbopol-934 / PVP and Carbopol-934 / NaCMC, respectively. The tablets were prepared by compressing powder mixtures of verapamil hydrochloride as a model drug, Carbopol-934 and either HPC or PVP. The dissolution study of tablets has shown that the drug release increases with the higher content of HPC or PVP in the mixtures and a lag time is observed for all the studied formulations. It is interesting to note that the maximal lag time was found for the tablets where the polymer weight ratio corresponds to the maximum of complexation. The authors explain the existence of a lag time by the formation of a three-dimensional network-like structure due to the complexation between the polymers. Hamster cheek pouch was used as a model mucosal tissue for assessment of the mucoadhesive properties of the tablets. The highest mucoadhesive ability was observed for the formulation containing 17%, 68% of HPC or PVP and 15% of verapamil hydrochloride. Unfortunately, the authors have not discussed this result properly.

The development of mucoadhesive tablets based on compressed powder blends of Carbomer and hydroxypropylmethylcellulose (HPMC) loaded with morphine sulphate[21] and carbamazepine[22] were reported. The formation of interpolymer complexes between Carbomer and HPMC in aqueous solutions was confirmed by turbidity and viscosity measurements as well as by infrared spectroscopy of dry precipitate. The investigation of adhesion of Carbomer/HPMC tablets towards bovine sublingual mucosa showed that mucoadhesive properties are practically unaffected by mucoadhesive interactions.

The complexation between poly(methacrylic acid) (PMAA) and starch under acidic conditions was used by Clausen and Bernkop-Schnurch[23] to design a compressible pH-responsive excipient for controlled drug delivery. The IPC precipitate was obtained by mixing aqueous solutions of PMAA and starch at pH 3.0, with subsequent lyophilisation and pulverisation using mortar. The tablets were prepared by direct compression of the IPC with the model drugs including amoxicillin, rifampicin and a model peptide (horseradish peroxidase). The disintegration studies of these tablets showed a completely different profile of PMAA-starch compositions compared to the tablets based on pure starch. The disintegration of the tablets based on PAA-starch complexes was found to be pH-dependent and the tablets tested at pH 1.2 were stable for several days; however, at higher pH (pH 5.0 and 7.0) the disintegration time decreased to $110 \pm 5$ and $40 \pm 8$ min, respectively.

Hao and co-workers[24] reported an attempt to evaluate the effect of complexation between polymers via hydrogen-bonding on the properties of bioadhesive formulations. They studied the complexation in the aqueous mixtures of poly(methyl vinyl ether-maleic acid) (Gantrez AN 169) and PVP aiming at the development of the composite films for controlled release of sodium diclofenac. The precipitation of IPC in the aqueous mixtures of these polymers was unfavorable for formulation of polymeric films, due to their opacity. Clear films were developed by adding 1.5 w/w % of N-methyl-2-pyrrolidone to the casting solutions, which helped to avoid the formation of insoluble IPC. The bioadhesive properties of the films were studied by measuring the force required to detach a polymeric film from silicone elastomer as a model of biological tissue. The composite films yielded larger detachment forces compared to the films based on individual polymers. The maximal adhesion force was observed for the formulation containing 50 wt % of PVP, which was related to the formation of IPC and increased cohesiveness between polymers.

The precipitation of IPC formed between PAA and poly(ethylene oxide) as well as some cellulose ethers such as HPC and methylcellulose (MC) also caused a problem in formulating clear and homogeneous mucoadhesive films.[25-28] A careful analysis of the film's structure and morphology as a function of the casting solution pH allowed identification of the conditions for preparation of homogeneous materials. It was demonstrated that the uniform mucoadhesive films can be prepared by mixing polymer solutions at pH above $pH_{crit1}$ (pH, below which insoluble IPC are formed) and below $pH_{crit2}$ (above which the hydrogen-bonding is completely disrupted). This selection of solution pH allows preparation of uniform polymeric blends, where miscibility is ensured by weak hydrogen-bonding between the component polymers. The effects of pH of casting solutions on the miscibility of polymers are discussed in detail in Chapter 1. The cross-linked films based on miscible blends of PAA with different non-ionic polymers (cellulose ethers and polymeric alcohols) may be prepared by their thermal treatment at 80-120 $^0$C for several hours.[29-32] Alternatively, ionising radiation may be used to cross-link the blends based on PAA and synthetic polymers such as poly(vinyl alcohol),[33] poly(methyl vinyl ether)[34] and poly(2-hydroxyethyl ether).[30] The blends containing polysaccharides cannot be cross-linked by radiation because of their degradation; however, the degradation process may be reversed by adding cross-linkers to the casting solutions.[27] Dubolazov *et al.*[27] have reported the mucoadhesive properties of PAA-HPC films towards porcine buccal mucosa. **Figure 2** shows the dependence of detachment/dissolution time of cross-linked and uncross-linked films retained on a surface of mucosal tissue.

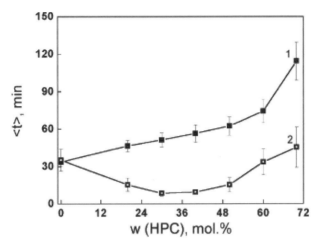

**Fig. 2.** Adhesion of soluble (1) and cross-linked PAA-HPC films (2) toward porcine buccal mucosa at pH 6.86 as a function of HPC content in the films. Reprinted with permission from Ref. 27. Copyright (2006) American Chemical Society.

The dissolution of soluble films adhered to mucosal tissue is observed within 30-110 min depending on the polymer's ratio. An increase in HPC content in the film results in slower dissolution, which is possibly due to poorer hydrophilic properties and higher rigidity of HPC macromolecules. The cross-linked films are retained on the surface of mucus for significantly shorter times compared to soluble polymer samples. This may be attributed to the reduced ability of macromolecules to diffuse into the mucus layer.

The effects of the preparation methods on the properties of films based on PAA and hydroxypropylmethylcellulose (HPMC) blends were studied by Wang and co-workers.[35] They used water and a water-ethanol mixture for casting of films and found that mixing the polymers in water results in formation of cloudy solutions, caused by interpolymer complexation. In contrast with the fully aqueous system, the use of water-ethanol mixture (1:1 by volume) leads to clear solutions. The films were dried at two different temperature regimes (60 and 110°C), which resulted in both cases in the formation of materials insoluble in water and phosphate buffer solution. The authors related it to the cross-linking reaction caused by esterification between carboxylic groups of PAA and hydroxyl groups of HPMC. The results of scanning electron microscopy investigation show that the morphology of the films cast from the water-ethanol mixture is smoother compared to the samples prepared by using water as a solvent. The films containing model drugs such as water-soluble propranolol hydrochloride and liphophilic ketoprofen were prepared by adding them after mixing two-component polymer solutions with subsequent drying. It was demonstrated that the thermal treatment of the films at 110°C does not affect the structure of the drugs. The analysis of the release kinetics from the films revealed different mechanisms for propranolol hydrochloride and ketoprofen. The water-soluble drug releases through diffusion, whereas the release of the liphophilic drug proceeds through erosion.

Template polymerisation of acrylic acid in the presence of poly(ethylene glycol),[36] poly(N-vinyl pyrrolidone)[37] and poly(vinyl alcohol)[38] can also be used to prepare mucoadhesive systems based on interpolymer complexes. The details about template polymerisation can be found in Chapter 6. After polymerisation, the IPCs were repeatedly washed with water to remove the unreacted monomers and polymeric films were prepared.[36-38] The adhesion of these films towards polypropylene plate as a model of biological tissue was studied by measuring the detachment force. In all three studies, it was demonstrated that polycomplexes have improved adhesive properties.

In a series of publications, Chun and co-workers[39,40] have demonstrated the possibility of preparing mucoadhesive microspheres using interpolymer complexation and solvent diffusion methods. The method of microspheres preparation involves a separate dissolving PAA and PVP in water/ethanol mixtures with sequential syringing of these solutions into corn oil containing sorbitan monooleate as a surfactant. Corn oil was chosen as the external phase because it is not miscible with water/ethanol mixtures, used as an internal phase and the IPCs remain insoluble. The mixtures were stirred at 500 rpm for 36 hours and the hardened microspheres were collected by filtration. The solidification of the IPC droplets happened as water and ethanol diffused out of the internal phase. The adhesion of the microspheres compressed into a tablet was studied using poly(propylene) plate as a substrate, mimicking biological tissue. The microspheres exhibited adhesive properties similar to the tablet based on pure Carbopol 971. The possibility of loading several drugs such as acetaminophen, clarithromycin and amoxicillin was also demonstrated by dissolving them in polymer solutions before syringing into corn oil. The drug release rate from the IPC was lower compared to the microspheres based on pure PVP.

Gel-forming erodible inserts for ocular delivery of ofloxacin were developed by Di Colo *et al.*[41] based on mixtures of PEO and Eudragit L100 (copolymer of methacrylic acid and methyl methacrylate). Immediately after the application in the lower conjuctival sac of the rabbit eyes, the inserts formed mucoadhesive gels, well tolerated by the animals; then the gels spread over the corneal surface and eroded. The insert erosion rate depended on the strength of interpolymer interactions between PEO and Eudragit L100.

Lele and Hoffman[42] have designed novel mucoadhesive drug delivery systems by combining indomethacine–polyethyleneglycol conjugate (IND-PEG) with poly(carboxylic acids) such as PAA and PMAA. The poly(carboxylic acids) were mixed with IND-PEG at pH 3.0, which resulted in precipitation of IPC formed via hydrogen-bonding between carboxylic groups of polyacids and ether groups of PEG. When the dry IPC particles were placed inside the dialysis membrane containing model tear fluid at pH 7.4 and 34°C, they swelled and dissociated to give a clear solution in 2–3 hours. Since IND was linked to PEG via labile anhydride bonds, they easily hydrolysed, releasing the free drug. More than 60% of the total drug was released from IND-PEG conjugate in 50 min compared to 80 min when the conjugate was in complexes with PAA. It was also demonstrated that the release rate can be controlled by changing the molecular weight of a polyacid. In a more recent study, Hoffman and co-workers[43] have used a similar approach for PEGylated protein (papain) that formed IPC with PAA as a viscous mucoadhesive gel at pH 3.0. These complexes dissociate at pH 7.4, which results in release of PEGylated papain. They demonstrated that when the molecular weight of

PEG is high enough (20–40 kDa), the release of PEGylated papain from the complex with PAA is significantly slower than the release of native papain physically-mixed with PAA. This did not occur when the papain was PEGylated with PEG of low molecular weight (5 kDa). This is related to the inability of short macromolecules to form stable IPCs with PAA. The authors believe that this approach can be used for developing novel protein delivery systems for nasal, ophthalmic and vaginal administration.

Thus, interpolymer complexation may be successfully used to design novel mucoadhesive systems in the form of tablets, films, microspheres and gels, although the mucoadhesive properties of these dosage forms may be seriously affected by the degree of interpolymer complexation.

## 2. Formulation of poorly water-soluble drugs

According to British Pharmacopoeia, a drug is considered to be "practically insoluble" if 1 g of such material requires more than 10,000 mL of solvent (e.g. water) to be solubilised, or alternatively, a material which has a solubility of less than 0.1 mg/mL in water.[44] The poor solubility of about 50% of the drugs in biological fluids is the main limitation in oral, parenteral, and transdermal administration.[45]

To overcome the problem of poorly water-soluble drugs and improve their bioavailability different strategies have been explored. These include their chemical modification (prodrug approach),[46] use of lipid-based formulations,[47] emulsions,[48] cyclodextrins,[49] polymeric micelles[50, 51] and solid dispersions.[52]

Solid dispersions are considered one of the most promising strategies for improving the oral bioavailability of poorly water-soluble drugs.[52] These are molecular mixtures of poorly water-soluble drugs in hydrophilic polymeric carriers, whose release profile is driven by properties of a polymeric matrix. In solid dispersions the crystallinity of drugs is reduced and smaller particle size and better wettability results in faster release and improved bioavailability.

In a series of publications Ozeki *et al.*[53-56] have demonstrated the possibility of using interpolymer complexes formed by Carbopol and poly(ethylene oxide) (PEO) for developing solid dispersions of phenacetin (PHE), an antipyretic drug. Powders of phenacetin, PEO and Carbopol were dissolved in water/ethanol mixture (1/1 v/v) and then solvents were evaporated, which resulted in solid dispersions. These were ground and dried at 50°C for 24 hours under reduced pressure. It was found that the release of PHE from solid dispersions can be controlled by varying the PEO/Carbopol ratio. The slowest release was observed at PEO/Carbopol ratio of 1/1, which corresponds to the IPC stoichiometry. The release of PHE from solid dispersion was also affected by solution pH with higher release rates observed under more basic conditions, which the authors related to dissociation of the IPC. It was also shown that the release of PHE from the solid dispersions can be controlled by varying the molecular weight of PEO and cross-linking degrees of Carbopol. Powder X-ray diffraction analysis and differential scanning calorimetry thermograms of these samples revealed that the crystallinity of PHE is markedly decreased in a solid dispersion, which was not observed in a physical mixture, when the powders of PEO, Carbopol and PHE were simply mixed. In a more recent study Ozeki and co-workers[57] have also demonstrated the possibility of formulating phenacetin using the IPC formed by Carbopol and methylcellulose.

Broman and co-workers[58] prepared solid dispersions of probucol using PVP, PEO, PAA as well as the combinations of PAA/PVP and PEO/PVP using solvent evaporation, compression moulding and co-melting methods. Probucol formed a single phase amorphous system with PVP when this binary system was evaporated from ethanol solutions, whereas drug crystallinity was not fully disrupted for the compression moulded sample. No disruption in drug crystallinity was observed in probucol mixtures with PAA obtained by any method. Some interaction was observed in probucol-PEO solid dispersions prepared by co-melting and compression moulding, however, it did not result in a complete disruption of the drug crystallinity. When the binary polymer systems PAA/PVP and PEO/PVP were used for preparation of *solid drug dispersions* by the compression moulding method it resulted in amorphous probucol. However, when the solvent method was used for preparing drug dispersions, probucol was mainly amorphous in PAA/PVP blend and formed crystalline form II polymorph in PEO/PVP. Although the authors did not discuss this observation in details, the difference between PAA/PVP and PEO/PVP observed in the presence of solvent may be related to stronger interpolymer interactions between PAA and PVP.

Kumar and co-workers[59] reported a new method involving in situ complexation between polyvinyl acetate phthalate (PVAP) and PVP for entrapment of ibuprofen, a bitter-tasting anti-inflammatory drug. PVAP-PVP granules with entrapped ibuprofen were prepared by two methods. The first method involved dissolving PVAP and PVP at 2:1 weight ratio in aqueous ammonium hydroxide solution and water respectively, with subsequent mixing, dissolving of the solid ibuprofen and addition of 0.1 N HCl to pH 1.0. It resulted in formation of precipitate, which was filtered, washed with water and dried under vacuum. In the second method PVAP and PVP were separately dissolved in ethanol and solutions were mixed at 2:1 weight ratio. It resulted in immediate precipitation of a gummy IPC. It was subsequently washed with hot ethanol and dried under vacuum. Six grammes of dry IPC and 4.0 g of ibuprofen powder were then suspended in 10 ml of water, to which aqueous ammonium hydroxide solution (28% v/v) was added until a clear solution was formed. The resulting solution was then treated with 0.1 N HCl leading to the formation of precipitate. This was filtered, washed with water and dried under vacuum. The ibuprofen content in the samples prepared by both methods varied within 90–98% of the theoretical amounts, indicating a near quantitative entrapment of the drug. The analysis of obtained solid drug dispersions by X-ray diffraction and differential scanning calorimetry revealed a partial reduction in crystallinity of ibuprofen.

## 3. Formulation of hydrophilic drugs

Interpolymer complexes can also be used for formulating highly water-soluble hydrochloride-type drugs due to the cationic nature of these compounds, which results in formation of electrostatic complexes with poly(carboxylic acids). The complexation between cationic drugs and poly(carboxylic acids) is often affected by the presence of non-ionic polymers in solutions, through the formation of triple polymeric complexes. The reduced solubility of these triple complexes in water allows retention of hydrophilic drugs on the administration site as well as precise control of their release.

Aksenova and co-workers[60] studied the interactions in aqueous solutions of a triple system involving PMAA, PEO and aminazine. It was found that an addition of small portions of aminazine solutions to PMAA-PEO interpolymer complex results in formation of PMAA-aminazine-PEO triple complex with reduced solubility. The further addition of aminazine to IPC dispersion leads to a complete dissolution of precipitate, displacement of PEO and formation of soluble drug-PMAA complex.

The interactions of local anesthetic lidocaine hydrochloride (Lid·HCl) with PAA, poly(2-hydroxyethyl vinyl ether) (PHEVE) and poly(methyl vinyl ether) (PMVE) in aqueous solutions have been reported in previous publications.[61,62] Formation of water-insoluble complexes is observed upon mixing 0.05 mol/L solutions of Lid·HCl and PAA, which are stabilised by electrostatic attraction forces. Non-ionic polymers PHEVE and PMVE acquire polyelectrolyte properties upon mixing with solutions of Lid·HCl, presumably due to macromolecule-drug association via hydrogen-bonding and hydrophobic effects. Mixing of 0.1 mol/L Lid·HCl, 0.2 mol/L PHEVE and 0.2 mol/L PAA solutions at the molar ratio 1:10:10 is accompanied by appearance of turbidity and formation of precipitate. After isolation of this precipitate by centrifugation, washing with distilled water and drying under vacuum, it was analysed by infrared spectroscopy. This revealed the presence of all three components PAA, PHEVE and the drug, confirming formation of a triple complex. Similar triple complex was detected in the mixture of PAA, PMVE and Lid·HCl. It is believed that within the triple complex the drug molecules can be bound to PAA via Coulombic interactions with carboxylate ions –COO$^-$ and hydrogen bonding with non-dissociated carboxylic groups – COOH. It may also be associated with the functional groups of the non-ionic polymer via hydrogen-bonding or hydrophobic effects. The structure of this complex is shown schematically in **Fig. 3**.

A combination of PAA with a second non-ionic polymer such as PVP, poly(vinyl alcohol) or Synperonic® F-127 in aqueous solutions was found to be useful for formulating pilocarpine hydrochloride, a drug for treatment of glaucoma.[63] The presence of a non-ionic polymer in the mixture leads to significant decrease of *in vitro* drug release and also to improvement in drug bioavailability.

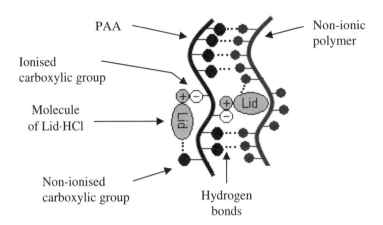

**Fig. 3.** Structure of PAA - Lid·HCl – non-ionic polymer complex. Reproduced from Ref. 61 with permission from Koninklijke Brill NV.

## 4. Encapsulation of biomacromolecules and microorganisms

Formation of insoluble IPC can be utilised for entrapment of biomacromolecules, colloidal particles and microorganisms through co-precipitation as shown in **Fig. 4**:

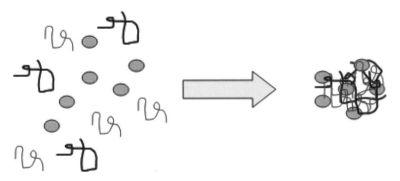

**Fig. 4.** Entrapment of macromolecules / microbial cells via interpolymer complexation. Cycles represent colloidal particles or living cells.

Kwon and co-workers[64] reported an entrapment of insulin through the complexation between PMAA and poly(ethyloxazoline) (PEOX) via hydrogen-bonding. For this purpose, 10 mg of zinc insulin were suspended in 10 mL of PMAA-PEOX mixture (pH 5.5) and 0.1 mol/L HCl was added to reduce pH below pH 5.0, which resulted in co-precipitation of IPC and insulin. The amount of the entrapped insulin was 0.5 ± 0.2 wt %. The precipitate was isolated, dried and compressed into disks. The disks were then soaked in 0.9% saline solution for three days to reach equilibrium swelling. During this pre-swelling period, the release of insulin was hardly detectable, however, when the samples were placed into the electric field, it resulted in up to 70% release.

Moolman *et al.*[65] have demonstrated the possibility of encapsulating probiotics using interpolymer complexes formed by poly(vinyl acetate-*co*-crotonic acid) and PVP. Probiotics are cultures of live microorganisms, which can improve microflora in the gastrointestinal tract. In this study, Moolman *et al.* blended dry PVP and poly(vinyl acetate-*co*-crotonic acid) powders with bacterial cultures of *B. longium* or *B. lactis* in a 1 L reaction chamber and saturated it with sterile pressurised carbon dioxide to a pressure of 300 bar at 40°C. Under these conditions, carbon dioxide forms a supercritical fluid, mediating the formation of interpolymer complex between the two polymers. The mixture was left to equilibrate for 2 hours and then sprayed through a 500 $\mu$m capillary. The survival of encapsulated bacteria in the gastrointestinal tract was estimated by exposure of the product to the simulated gastric juice for 2 hours and subsequently to simulated intestinal fluid for 8–24 hours. The encapsulated *B. longium* exhibited more than an order of magnitude improved survival compared to free bacteria control. In a similar manner, the same interpolymer complexation was also used for encapsulating indomethacin. The release of the drug was retarded under acidic conditions, while in more alkaline conditions 85% release was achieved within 24 hours. It is believed that this encapsulation methodology is of significant potential for use in food and pharmaceutical applications.

## 5. Nanoparticles and nanogels

Interpolymer complexation in solutions may lead to formation of colloidally-stable dispersions of nanoparticles and nanogels, in which size, surface charge and stability can be controlled by varying the nature of polymers, solution pH and temperature.

Recently, Khutoryanskaya and co-workers[66] demonstrated the formation of IPC nanoparticles in solution mixtures of PAA and methylcellulose (MC). The nanoparticles produced at pH 1.4 and 2.4 were spherical and dense structures ranged from 80 to 200 nm, whereas particles formed at pH 3.2 were 20–30 nm and were stabilized against aggregation by a network of uncomplexed macromolecules (**Fig. 5**).

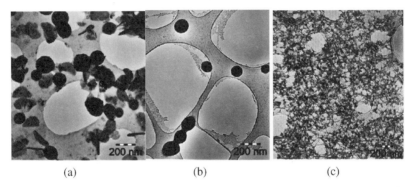

(a)                    (b)                    (c)

**Fig. 5.** Transmission electron microscopy image of interpolymer complexes obtained from 0.2 wt % solution mixtures of poly(acrylic acid) and methylcellulose (30:70 wt %) at pH 1.4 (a), 2.4 (b), and 3.2 (c). Reprinted with permission from Ref. 66. Copyright (2007) American Chemical Society.

Lu *et al.*[67] reported the preparation of nanogels by polymerising acrylic acid in the presence of HPC and N,N'-methylenebisacrylamide as a cross-linking agent. The size of the resulting nanogels was pH-dependent with smaller particles formed at pH 3.2 and their expansion upon pH increase to 7.4. This pH-dependence was related to the reversible interpolymer complexation between PAA and HPC within the nanogels. A similar strategy was also used by Sun and co-workers[68] for synthesis of nanoparticles based on PAA and dextran.

Hollow spheres were prepared by Dou *et al.*[69] using graft-copolymers of hydroxyethylcellulose-g-PAA, which were intramoleculary cross-linked by adding 2,2'-(ethylenedioxy)-bis(ethylamine) as a cross-linker in the presence of 1-(3-dimethylaminopropyl)-3-ethylcarbodiimide methiodide. The transitions between micellar structures and hollow spheres were detected in cross-linked graft-copolymers upon changes in solution pH.

Henke and co-workers[70] used ionising radiation for preparation of PAA and PVP based nanogels. They irradiated PAA-PVP complexes with pulses of fast electrons in dilute, deoxygenated solutions. This resulted in intramolecular cross-linking of macromolecules with IPC particles leading to the formation of nanogels. More details on this approach can be found in Chapter 10.

Sotiropoulou *et al.*[71] have reported the preparation of nano-particles based on complexes between PAA and poly(acrylic acid-*co*-2-acrylamido-2-methyl-1-propane sulfonic acid)-*graft*-poly(*N,N*-dimethylacrylamide) (P(AA-*co*-AMPSA)-*g*-PDMAM) at pH = 2.0. Dynamic and static light scattering measurements, in conjunction with small-angle neutron scattering experiments, suggest the formation of core-shell colloidal nanoparticles in dilute solution, comprised of an insoluble PAA/PDMAM core surrounded by an anionic P(AA-*co*-AMPSA) corona.

Insulin-loaded nanoparticles were prepared by Deutel and co-workers[72] through the formation of hydrogen-bonded IPC between PVP and PAA or PAA-cystein conjugates in the presence of insulin. The complex formation was triggered by lowering solution pH from 8.7 to 2.0 and the IPC suspension was then stirred for 24 hours at room temperature. When PAA-cystein conjugates were used this stirring resulted in nanoparticles cross-linking via disulphide bonds (-S-S-) formation due to oxidation of free thiol-groups by air. The oxidised PAA-cystein/PVP/insulin nanoparticles displayed a mean particle size of $255 \pm 25$ nm and a zeta-potential of $-11.2 \pm 1.0$ mV, whereas the size of nonoxidised particles was $185 \pm 40$ nm and zeta-potential $-18.3 \pm 2.4$ mV. When thiolated PAA nanoparticles were used instead of unmodified PAA, *in vivo* experiments with insulin-loaded nano-particulate suspensions demonstrated a 2.3-fold improvement in insulin bioavailability.

## 6. In-situ gelling systems

Aqueous solutions of polymers having the ability to form a gel in response to the changes in environmental conditions are termed in-situ gelling systems and have great potential in the design of novel dosage forms.[73-76] These systems have recently been considered to have great potential for ophthalmic, vaginal and parenteral drug delivery.

Changes in environmental temperature, pH or concentration of small molecules in solution may serve as a trigger for sol-gel transitions in some hydrophilic polymers. For example, Carbopol® forms a slightly viscous 0.5 v/w % solution in water but undergoes a rapid gelation upon increase in pH (**Fig. 6**).

**Fig. 6.** A 0.5 v/w % solution of Carbopol converts into gel upon slight increase in pH.

The ability of the IPC to form and dissociate upon changes in environmental factors offers an excellent opportunity to design in-situ gelling drug delivery systems. Haglund and co-workers[77] used the ability of PMAA-PEG complexes to precipitate in aqueous medium at low pH and dissolve in the presence of ethanol due to weakening of hydrogen bonds. An insoluble complex obtained by mixing 10% PMAA (15 kDa) with 20% PEG (18.5 kDa) under acidic pH was re-dissolved in 75% ethanol solution in water and used as in-situ gelling liquid. When this solution is injected into the site of administration, ethanol diffuses out quickly leading to gel formation. This gel can dissociate slowly at physiological pH, releasing a drug.

A number of non-ionic water-soluble polymers can also undergo physical gelation upon increase in temperature.[73] Typical pharmaceutical polymers of this type include Pluronics® and some cellulose ethers such as HPC, MC or HPMC. Although the gelation temperatures of HPC, MC and HPMC solutions are significantly higher than physiological temperatures (50–80°C), they can be reduced by mixing these polymers with poly(acrylic acid) or Carbopol®. It was demonstrated[78] that a combination of HPMC with PAA may reduce the phase transition temperature from 55–60°C to –7°C due to the interpolymer complexation. Kumar *et al.*,[79] Lin *et al.*,[80] Srividya *et al.*[81] and Qi *et al.*[82] have also demonstrated the use of Carbopol mixtures with HPMC, Pluronics® and Poloxamers® for developing in-situ gelling ophthalmic drug delivery systems. When PAA (Carbopol) is used in a combination with a temperature-responsive polymer, both pH- and temperature-changes can trigger in-situ gelation and interpolymer interaction between the components of the binary system plays an important role.

## 7. Hydrogels with interpolymer complexation

Hydrogels are three-dimensional cross-linked networks of hydrophilic polymers, which have the unique ability to swell in water. In the swollen state, they are soft and elastic materials, resembling biological tissues. The porous structure of their networks allows diffusion of both small and relatively large molecules through hydrogels. Depending on the chemical nature of hydrogels, they can be neutral, cationic or anionic. They can also be divided into two groups, depending on the nature of the cross-links: physically cross-linked and chemically cross-linked. The applications of hydrogels in biomedical and pharmaceutical sciences include soft contact lenses, drug delivery systems, and wound dressings.[83]

Many hydrogels exhibit so-called stimuli-sensitive behavior, i.e., they are able to change their swelling properties and undergo volume phase transitions in respond to environmental factors such as changes in temperature, pH, solvent quality, external electric and magnetic fields, and the presence of small molecules in solutions.[84,85]

When hydrogels are based on combinations of macromolecules able to form hydrogen-bonded interpolymer complexes, the specific interactions may govern the stimuli-responsive behavior of the whole network. For example, Nishi and Kotaka[86,87] prepared complex-forming hydrogels by polymerisation of acrylic acid within cross-linked PEO, resulting in interpenetrating network. These hydrogels exhibited reversible swelling / deswelling, permeability and also mechanochemical creep / recovery behavior upon varying pH and ionic strength of the medium. These changes were related to protonation / deprotonation of the PAA network, followed by reversible formation and dissociation of the IPC due to hydrogen-bonding between PEO and PAA networks within the IPN.

An interesting stimuli-responsive hydrogel system was reported by Ilmain *et al.*,[88] who utilised the ability of PAA and polyacrylamide (PAAM) to form hydrogen-bonded IPC. Hydrogels were synthesised by polymerising acrylic acid within a swollen polyacrylamide gel. Upon increase in temperature these hydrogels exhibited an abrupt expansion at 20°C, which was related to dissociation of interpolymer hydrogen bonds between PAA and PAAM. To confirm the involvement of hydrogen-bonding, the authors also studied the effect of urea on the swelling behavior of hydrogels. Urea is a well known agent for disrupting hydrogen bonds. In 1 mol/L urea solution the hydrogels did not exhibit any volume phase transition and remained swollen over the entire temperature range.

The complex-forming hydrogels based on poly(methacrylic acid) grafted with poly(ethylene glycol), developed by Peppas and co-workers,[89-91] have been considered as potential carriers for insulin via the oral route.[92,93] It is believed that these hydrogels could completely protect insulin in the gastric fluid. In the small intestine, the hydrogels inhibit protease activity due to deprivation of calcium ions from the structure of enzymes.

## 8. Conclusions

Formation of hydrogen-bonded complexes between poly(carboxylic acids) and nonionic polymers can offer many unique opportunities to pharmaceutical scientists. These include development of mucoadhesive dosage forms, reduction in crystallinity of poorly soluble drugs, manipulation with solubility of highly hydrophilic drugs, encapsulation of biomacromolecules and microorganisms, nanoparticles and nanogels, in-situ gelling systems and stimuli-responsive hydrogels.

## References

1. A. T. Florence and D. Attwood, *Physicochemical Principles of Pharmacy*, Pharmaceutical Press, London (2006).
2. V. V. Khutoryanskiy, *Int. J. Pharm.* **334**, 15 (2007).
3. X. Yang and J. R. Robinson, in Biorelated Polymers and Gels. Controlled Release and Applications in Biomedical Engineering (T. Okano., d.), Academic Press, Boston (1998).
4. D. Dodou, P. Breedveld, and P. A. Wieringa, *Eur. J. Pharm. Biopharm.* **60**, 1 (2005).
5. J. Smart, *Adv. Drug Delivery Reviews* **57**, 1556 (2005).
6. K. Edsman and H. Hagerstrom, *J. Pharm. Pharmacol.* **57**, 3 (2005).
7. C. Marriott and N. P. Gregory, in Bioadhesive Drug Delivery Systems (V. Lenaerts and R. Gurny, eds.), CRC Press (1990).
8. N.A. Peppas, Y. Huang, *Adv. Drug Delivery Reviews* **56**, 1675 (2004).
9. N. A. Peppas and P. A. Burim, *J. Control. Release* **2**, 257 (1985).
10. M. J. Tobyn, J. R. Johnson, and P. W. Dettmar, *Eur. J. Pharm. Biopharm.* **43**, 65 (1997).
11. S. A. Mortazavi, B. G. Carpenter, and J. D. Smart, *Int. J. Pharm.* **83**, 221 (1992).
12. H. Park and J. R. Robinson, *Pharm. Res.* **4**, 457 (1987).
13. M.M. Patel, J.D. Smart, T.G. Nevell, R.J. Ewen, P.J. Eaton, and J. Tsibouklis, *Biomacromolecules* **4**, 1184 (2003).
14. S. Geresh, G.Y. Gdalevsky, I. Gilboa, J. Voorspoels, J.P. Remon, and J. Kost, *J. Control. Release* **94**, 391 (2004).
15. A. H. Shojaei and X. Li, *J. Control. Release* **47**, 151 (1997).
16. K. Satoh, K. Takayama, Y. Machida, Y. Suzuki, M. Nakagaki, and T. Nagai, *Chem. Pharm. Bull.* **37**, 1366 (1989).
17. K. Satoh, K. Takayama, Y. Machida, Y. Suzuki, and T. Nagai, *Chem. Pharm. Bull.* **37**, 1642 (1989).
18. M. Ishida, Y. Machida, N. Nambu, and T. Nagai, *Chem. Pharm. Bull.* **29**, 810 (1982).
19. M. Ishida, N. Nambu, and T. Nagai, *Chem. Pharm. Bull.* **30**, 980 (1982).
20. A. Gupta, S. Garg, and R.K. Khar, *Drug Devel. Ind. Pharm.* **20**, 315 (1994).
21. S. Anlar, Y. Capan, O. Guven, A. Gogus, A. Dalkara, and A. A. Hincal, *Pharm.Res.* **11**, 231 (1994).
22. G. Ikinci, Y. Capan, S. Senel, E. Alaaddinoglu, T. Dalkara, and A.A. Hincal, *Die Pharmazie* **55**, 762 (2000).
23. A. E. Clausen and A. Bernkop-Schnurch, *J. Control. Release* **75**, 93 (2001).
24. J.S. Hao, L.W. Chan, Z.X. Shen, and P.W.S. Heng, *Pharm. Devel. Technol.* **9**, 379 (2004).
25. V.V. Khutoryanskiy, M.G. Cascone, L. Lazzeri, N. Barbani, Z.S. Nurkeeva, G.A. Mun, and A.V. Dubolazov, *Polym. Int.* **53**, 307 (2004).
26. V.V. Khutoryanskiy, A.V. Dubolazov, Z.S. Nurkeeva, and G.A. Mun, *Langmuir* **20**, 3785 (2004).

27. A.V. Dubolazov, Z.S. Nurkeeva, G.A. Mun, and V.V. Khutoryanskiy, *Biomacromolecules* **7**, 1637 (2006).
28. V.V. Khutoryanskiy and O.V. Khutoryanskaya, *British Pharmaceutical Conference*, Manchester, UK, A1 (2007).
29. Y.J. Bo, V.V. Khutoryanskiy, G.A. Mun, and Z.S. Nurkeeva, *Polym. Sci. Ser. A* **44**, 1094 (2002).
30. Z.S. Nurkeeva, G.A. Mun, V.V. Khutoryanskiy, A.B. Bitekenova, A.B. Dzhusupbekova, and K. Park, *J. Appl. Polym. Sci.* **90**, 137 (2003).
31. V.V. Khutoryanskiy, M.G. Cascone, L. Lazzeri, Z.S. Nurkeeva, G.A. Mun, and R.A. Mangazbaeva, *Polym.Int.* **52**, 62 (2003).
32. O. V. Khutoryanskaya, V. V. Khutoryanskiy, and R. A. Pethrick, *Macromol. Chem.Phys.* **206**, 1497 (2005).
33. Z.S. Nurkeeva, G.A. Mun, A.V. Dubolazov, and V.V. Khutoryanskiy, *Macromol. Biosci.* **5**, 424 (2005).
34. Z.S. Nurkeeva, G.A. Mun, V.V. Khutoryanskiy, and A.B. Dzhusupbekova, *Radiat. Phys. Chem.* **69**, 205 (2004).
35. L.-F. Wang, W.B. Chen, T.-Y. Chen, and S.C. Lu, *J. Biomat. Sci. Polym. Edn.* **14**, 27 (2003).
36. H.-K. Choi, O.-J. Kim, C.-K. Chung, and C.-S. Cho, *J. Appl. Polym. Sci.* **73**, 2749 (1999).
37. M.-K. Chun, C.-S. Cho, and H.-K. Choi, *J. Control. Release* **81**, 327 (2002).
38. J. M. Oh, C.-S. Cho, Y.-B. Lee, S.-C. Shin, and H.-K. Choi, Proceed. 30th Annual Meeting of Controlled Release Society, 155, Glasgow, UK (2003).
39. M.-K. Chun, C.-S. Cho, and H.-K. Choi, *Int. J. Pharm.* **288**, 295 (2005).
40. M.-K. Chun, H. Sah, and H.-K. Choi, *Int. J. Pharm.* **297**, 172 (2005).
41. G. Di Colo, S. Burgalassi, P. Chetoni, M.P. Fiaschi, Y. Zambito, and M.F. Saettone, *Int. J. Pharm.* **215**, 101 (2001).
42. B.S. Lele and A.S. Hoffman, *J. Control. Release* **69**, 237 (2000).
43. Y. Hayashi, J. Milton Harris, and A. S. Hoffman, *React. Func. Polym.* **67**, 1330 (2007).
44. British Pharmacopoeia, Volume 1, London (2005)
45. D.A. Chiappetta and A. Sosnik, *Eur. J. Pharm. Biopharm.* **66**, 303 (2007).
46. V.J. Stella and K. W. Nti-Addae, *Adv. Drug Delivery Reviews* **59**, 677 (2007).
47. A.J. Humberstone and W.N. Charman, *Adv. Drug Delivery Reviews* **25**, 103 (1997).
48. P. P. Constantinides, A. Tustian, and D. R. Kessler, *Adv. Drug Delivery Reviews* **56**, 1243 (2004).
49. M.E. Brewster and T. Loftsson, *Adv. Drug Delivery Reviews* **59**, 645 (2007).
50. G. Gaucher, M.-H. Dufresne, V. P. Sant, N. Kang, D. Maysinger, and J.-C. Leroux, *J. Control. Release* **109**, 169 (2005).
51. A.N. Lukyanov and V.P. Torchilin, *Adv. Drug Delivery Reviews* **56**, 1273 (2004).
52. T. Vasconcelos, B. Sarmento, and P. Costa, *Drug Discovery Today* **12**, 1068 (2007).
53. T. Ozeki, H. Yuasa, and Y. Kanaya, *Int. J. Pharm.* **171**, 123 (1998).
54. T. Ozeki, H. Yuasa, and Y. Kanaya, *Int. J. Pharm.* **165**, 239 (1998).
55. T. Ozeki, H. Yuasa, and Y. Kanaya, *J. Control. Release* **58**, 87 (1999).
56. T. Ozeki, H. Yuasa, and Y. Kanaya, *J. Control. Release* **63**, 287 (2000).

57. T. Ozeki, H. Yuasa, and H. Okada, *AAPS Pharm. Sci. Technol.* **6**, E231 (2005).
58. E. Broman, C. Khoo, and L. S. Taylor, *Int. J. Pharm.* **222**, 139 (2001).
59. V. Kumar, T. Yang, and Y. Yang, *Pharm. Devel. Technol.* **6**, 71 (2001).
60. N.I. Aksenova, V.A. Kemenova, A.V. Kharenko, A.B. Zezin, G.B. Bravova, and V.A. Kabanov, *Vysokomolek. Soed. Ser. A.* **40**, 403 (1998).
61. Z.S. Nurkeeva, G.A. Mun, V.V. Khutoryanskiy, A.B. Bitekenova, and A.B. Dzhusupbekova, *J. Biomat. Sci. Polym. Edn.* **13**, 759–768 (2002).
62. Z.S. Nurkeeva, V.V. Khutoryanskiy, G.A. Mun, and A.B. Bitekenova, *Polym. Sci. Ser. B.* **45**, 365–369 (2003).
63. M. Oechsner and S. Keipert, Proceed. 2nd World Meeting APGI/APV, 967, Paris (1998).
64. I.C. Kwon, Y.H. Bae, and S.W. Kim, *Nature* **354**, 291 (1991).
65. F.S. Moolman, P.W. Labuschagne, M.S. Thantsa, T.L. van der Merwe, H. Rolfes, and T.E. Cloete, *South African J. Sci.* **102**, 349 (2006).
66. O.V. Khutoryanskaya, A.C. Williams, and V.V. Khutoryanskiy, *Macromolecules* **40**, 7707 (2007).
67. X. Lu, Z. Hu, and J. Schwartz, *Macromolecules* **35**, 9164 (2002).
68. H. Dou, M. Tang, and K. Sun, *Macromol. Chem. Phys.* **206**, 2177 (2005).
69. H. Dou, M. Jiang, H. Peng, D. Chen, and Y. Hong, *Angew. Chem. Int. Ed.* **42**, 1516 – 1519 (2003).
70. A. Henke, S. Kadlubowski, P. Ulanski, J.M. Rosiak, and K.-F. Arndt, *Nucl. Instr. Meth. Phys. Res. B.* **236**, 391 (2005).
71. M. Sotiropoulou, F. Bossard, E. Balnois, J. Oberdisse, G. Staikos, *Langmuir* **23**, 11252-11258 (2007).
72. B. Deutel, M. Greindl, M. Thaurer, and A. Bernkop-Schnurch, *Biomacromolecules*, **9**, 287 (2008).
73. B. Jeong, S.W. Kim, and Y.H. Bae, *Adv. Drug Delivery Reviews* **54**, 37 (2002).
74. C.B. Packhaeuser, J. Schnieders, C.G. Oster, and T. Kissel, *Eur. J. Pharm. Biopharm.* **58**, 445 (2004).
75. B.K. Nanjawade, F.V. Manvi, and A.S. Manjappa, *J. Control. Release* **122**, 119 (2007).
76. L. Klouda and A.G. Mikos, *Eur. J. Pharm. Biopharm.* **68**, 34 (2008).
77. B.O. Haglund, R. Joshi, and K.J. Himmelstain, *J. Control. Release* **41**, 229 (1996).
78. R.A. Mangazbaeva, G.A. Mun, Z.S. Nurkeeva, and V.V. Khutoryanskiy, *Polym. Int.* **55**, 668 (2006).
79. S. Kumar and K.J. Himmelstain, *J. Pharm. Sci.* **84**, 344 (1995).
80. H.-R. Lin and K. C. Sung, *J. Control.Release* **69**, 379 (2000).
81. B. Srividya, R.M. Cardoza, and P.D. Amin, *J. Control .Release* **73**, 205 (2001).
82. H. Qi, W. Chen, C. Huang, L. Li, C. Chen, W. Li, and C. Wu, *Int. J. Pharm.* **337**, 178 (2007).
83. N.A. Peppas (ed.), *Hydrogels in Medicine and Pharmacy.* CRC Press (1987).
84. K. Dusek (ed.), *Responsive Gels: Volume Transitions I.* Advances in Polymer Science. Springer-Verlag Berlin and Heidelberg, GmbH & Co. K, 1993.
85. K. Dusek (ed.), *Responsive Gels: Volume Transitions II.* Advances in Polymer Science. Springer-Verlag Berlin and Heidelberg, GmbH & Co. (1993).

86. S. Nishi and T. Kotaka, *Macromolecules* **18**, 1519 (1985).
87. S. Nishi and T. Kotaka, *Macromolecules* **19**, 978 (1986).
88. F. Ilmain, T. Tanaka, and E. Kokufuta, *Nature* **349**, 400 (1991).
89. C.L. Bell and N.A. Peppas, *J. Control. Release* **37**, 277 (1995).
90. C.L. Bell and N.A. Peppas, *Biomaterials* **17**, 1203 (1996).
91. A.M. Lowman and N.A. Peppas, *Macromolecules* **30**, 4959 (1997).
92. T. Yamagata, M. Morishita, N.J. Kavimandan, K. Nakamura, Y. Fukuoka, K. Takayama, and N.A. Peppas, *J. Control. Release* **112**, 343 (2006).
93. M. Morishita, T. Goto, K. Nakamura, A.M. Lowman, K. Takayama, and N.A. Peppas, *J. Control. Release* **110**, 587 (2006).

# CHAPTER 10

# RADIATION CROSS-LINKED HYDROGEN-BONDED INTERPOLYMER COMPLEXES

Artur Henke, Piotr Ulański, Janusz M. Rosiak

*Institute of Applied Radiation Chemistry, Technical University of Łódź*
*Wróblewskiego 15, 93-590 Łódź, Poland*

Hydrogen-bonded interpolymer complexes are in the area of interest due to their potential applications, especially in the field of biomedical science. One of the obstacles limiting the potential broad and versatile use of materials based on weak physical interactions is the lack of their stability when subjected to changes of external conditions like pH, ionic strength or temperature. In the literature, one can find a growing number of reports on studies aimed at development of new ways of stabilizing the structure of H-bonded complexes. In this chapter, we briefly review approaches that can be used to reach this goal. We mainly concentrate on radiation-induced synthesis and stabilization of materials based on interpolymer complexes. We also give examples of chemical methods applied for synthesis of systems consisting of polymers that can mutually interact through hydrogen-bonding.

## 1. Stability of hydrogen-bonding interpolymer complexes

Interactions based on hydrogen bonds take place between atoms that possess proton donor and proton acceptor character.[1,2] If they occur between macromolecules in a solution, formation of polymolecular structures, like spheres, micelles or vesicles, is observed.[2] An important class of such structures is comprised of hydrogen-bonded interpolymer complexes (IPC), which result from spontaneous association of complementary macromolecules in solutions through multiple hydrogen bonds. The most frequently described hydrogen-bonded systems are these composed of a polyelectrolyte (e.g. poly(acrylic acid)) and a proton accepting polymer like poly(ethylene oxide) or poly(N-vinylpyrrolidone).[3-5] The stability of interpolymer complex can be understood in two different ways. It can refer to the intramolecular stability resulting from direct connection of the interacting macromolecules through hydrogen-bonding. This kind of stabilization is responsible for the integrity of the IPC's structure. One can also distinguish interparticle- or, in other words, colloidal stability of IPCs which prevents them from flocculation. The latter subject is discussed in some details in one of the following paragraphs. Here, we concentrate on the primary phenomenon which is especially important from the point of view of basic understanding of complex behaviour.

Stable polyanion-based hydrogen-bonded interpolymer complexes can exist only below certain solution pH, i.e., when most of the carboxylic groups are protonated and thus macromolecules are capable of forming uninterrupted sequence of H-bonds along a chain segment, consisting of a certain minimum number of monomer units.[6-10] In consequence, process of interpolymer association is reversible and one can disintegrate complex to individual chains by addition of a base. The pH value at which a detectable onset of complexation or rather aggregation takes place (which is somewhat imprecise and dependent on the observation method) and which reflects formation of stable IPCs is often called critical pH (usually denoted as $pH_{crit.}$).[3,11,12]

Hydrogen-bonding is, however, not the only factor responsible for the stability of the interpolymer complex. There may be also contribution from the hydrophobic effect which is especially important in aqueous solution. In the complexation process, it results from direct hydrogen-bonding of the pendant groups undergoing dehydration to ensure the closest possible contact between proton acceptor and donor groups.[3,13,14] In consequence, polymer-solvent interactions within complexing areas are disrupted and polymeric backbone assembly occurs as a result of low polarity of formed microenvironment. This rearrangement is of thermodynamic nature and balances the entropy loss caused by forced ordering of water molecules. At solution conditions that promote spontaneous associations of polymers, both types of interactions act cooperatively.

Studies on the formation of H-bonded interpolymer complexes in aqueous solutions of synthetic and natural polymers are aimed not only at physicochemical description and better understanding of this phenomenon, but also at exploring its potential practical significance.[12,15,16] Hydrogen bond is directional and in consequence interactions between polymers can result in formation of well-defined structures, e.g. ladder-like or spherical.[2] Intermolecular hydrogen-bonding is not restricted to homopolymers. Non-covalent interactions of amphiphilic block copolymers with complementary macromolecules can result in their assembling in aqueous solution into micelles with hydrogen-bonded core and shell.[5,17-19] Very interesting structures can be also observed in organic solvents for water insoluble polymers.[20-22] There are also known complexes formed by polyacids and temperature-sensitive polymers like poly(N-isopropylacrylamide) (PNIPAM), poly(N-vinylcaprolactam) (PVCL), hydroxypropylcellulose (HPC) or poly(vinyl methyl ether) (PVME).[23-26]

Described structures of hydrogen-bonding interpolymer complexes can be interesting from the point of view of their applications in the field of drug delivery. Due to cooperative hydrogen-bonding and hydrophobic domain formation, interpolymer complexes can be used as promising carriers of hydrophobic substances, for example, drugs that are poorly soluble in water.[16,27-29] However, one of the obstacles on the way to the broad and versatile use of IPC-based materials is their lack of stability when subjected to changes of solution conditions. In fact, many products for biomedical purposes, e.g., formulations for oral drug delivery, nanoparticles for long-term blood circulation, etc., should be characterized by long-term stability and good mechanical properties, preventing them from disintegration upon encountering shear effects, fluctuations of pH, ionic strength and temperature. This is not the case for most materials based on relatively weak physical interactions. As a consequence, significant number of investigations performed in the last two decades on interpolymer complexes was directed towards elaborating permanent IPC systems with maintained hydrogen-bonding interactions. Scientific interest in this area is focused on complexes that are formed readily in aqueous solution, i.e. between hydrophilic or amphiphilic polymers. As weak polyelectrolyte is one of the constituents, synthesized material is usually pH-sensitive, what provides a possibility of controlled and reversible response to environmental changes. Permanent IPCs of these kinds should retain the ability of pH-induced H-bond formation and the resulting pH-dependence of their properties (hydrophilic-hydrophobic balance, pore size in case of IPC-based hydrogels, etc.), but the whole system should not disintegrate at conditions where hydrogen-bond interactions are weak or non-existent.

In the following paragraph, we briefly summarize chemical and UV-induced synthetic approaches dedicated to the formation of permanent systems containing polymers capable of hydrogen-bonding. In Section 3.1 we describe an alternative method of radiation-initiated cross-linking of complex-forming polymers and application of this approach for synthesizing nanogels based on interpolymer complex of poly(N-vinylpyrrolidone) and poly(acrylic acid).

## 2. Chemical- and UV-induced permanent stabilization of hydrogen-bonded interpolymer complex systems

Covalent cross-linking of a system consisting of polymers (and/or monomers) leads to the formation of gels. Introduction of additional "transverse" bonds between segments of the macromolecules causes that the formed network is insoluble but possesses ability to swell in appropriate solvent. A gel consisting of hydrophilic and/or amphiphilic macromolecules and thus capable of swelling in aqueous medium is termed a hydrogel. Hydrogels combine properties of solutions, due to their high capacity of swelling, and solids, because of binding all polymeric constituents into a network of a stable shape and structure.[30-32] Depending on the size of the formed network, one can divide them into macroscopic gels, microgels and nanogels.[33,34] Because of the porous character of hydrogels and presence of water in the pores, transport of molecules smaller than the network mesh can be easily provided, what makes hydrogels very good candidates for several applications including drug delivery.[30] What is more, permanent networks can be further functionalized with specific groups. Such targeting strategy can be very useful for specific purposes, for example, in recognition of antibodies and other biologically active substances.[35]

One can distinguish the following possible substrate systems from which synthesis of covalently stabilized materials based on interpolymer complex-forming systems can be started:

a)  a system where interpolymer complex is actually formed,
b)  a system consisting of polymers capable of hydrogen-bonding interactions, but not forming interpolymer complex at the chosen treatment conditions,
c)  a mixture of monomers or monomer and polymer.

The first two cases are distinguished as separated points due to a difference in the physicochemical properties of the starting materials. In general, when no interactions take place between polymers, changes of physical parameters as a function of mutual polymer composition follow the additivity law. On the other hand, for hydrogen-bonded interpolymer complexes, one can observe negative or positive deviation from simple weight average of investigated quantities.[36,37]

For a system consisting of monomers, the whole process is usually defined as cross-linking polymerization.[30] As it was mentioned earlier, polymer association through hydrogen bonding takes place only when a complexable sequence exists along the polyelectrolyte chain. This sets the specific requirements for synthesis routines. For example, copolymerization of two or more monomers should result in a copolymer with appropriate length of blocks to ensure stable hydrogen-bonding between complementary components. In consequence, using classical polymerization in a mixture or common solution of two monomers, one should make sure that the chosen monomers considerably differ in reactivity or the mutual monomer composition is carefully selected in order to avoid formation of a chain with monomers connected in a random manner (statistical copolymer).[10]

### 2.1. Synthesis of hydrogels

Hydrogels based on hydrogen-bonded complexing components can be synthesized by a variety of methods. Among the most frequently used, free-radical solution polymerization of monomers in the presence of initiator and cross-linking agent is of great importance.[30] Polymerization can be induced by redox initiators (e.g. mixture of ammonium persulfate and N,N,N',N'-tetramethylethylenediamine), or by initiators giving rise to transient active species upon their thermal- or UV-induced decomposition. Cross-linking reaction is achieved by addition of substance that possesses at least two functional cross-linkable groups (e.g. vinyl moieties) which can be covalently incorporated into a growing chain.[30] The most often used cross-linking agents are: ethylene glycol dimethacrylate (EGDMA) and its derivatives or N,N'-methylenebisacrylamide (BIS).[30] Networks of different chemical structures can be obtained.[38] Hydrogels based on copolymerized constituents are one of the examples. Using this approach hydrogels based on poly(N-isopropylacrylamide-co-(meth)acrylic acid), poly(N-vinylpyrrolidone-co-(meth)acrylic acid), poly(methacrylic acid-co-methacryloxyethyl glucoside) and ternary poly(N-vinylpyrrolidone-co-acrylic acid)-poly(ethylene glycol) were prepared.[39-50] Such gels are particularly interesting because of their ability to respond to changes of solution pH and ionic strength by pronounced changes in their swelling ability. In the first example mentioned above, beside pH-sensitivity, hydrogel responses to temperature changes as PNIPAM is a polymer characterized by the lower critical solution temperature (LCST) at around 32°C.[51,52] Above this temperature, polymer undergoes coil-to-globule transition which is caused by weakening of polymer-solvent interactions and in consequence by lower solubility of macromolecules. This can further result in precipitation of the polymer from solution. When PNIPAM is present in a gel structure, a volume phase transition of the gel is observed close to LCST, which manifests itself as a pronounced de-swelling. The PNIPAM-based ionic copolymer hydrogel is one of the most frequently studied systems undergoing volume phase transition.[53-55] Hydrogen bonds formed between carboxyl of MAA and amide group of NIPAM units are strongly stabilized by hydrophobic effect. With increasing amount of (meth)acrylic acid units in a network and increasing ionization degree, the critical temperature of PNIPAM solubility is usually shifted to higher value and the phase change is not as abrupt as in case of hydrogel based on the homopolymer. At very high concentration of charge-bearing component in a network and pH significantly above apparent $pK_a$, temperature-induced transition of NIPAM blocks is totally suppressed. Detailed investigations on the swelling process of P(NIPAM-co-MAA) hydrogels were performed by Diez-Pena *et al.*[45,46,56-59]

Peppas group prepared poly(methacrylic acid-graft-ethylene glycol) hydrogels by UV-vis-induced cross-linking polymerization.[60,61] Poly(ethylene glycol) was introduced into reaction mixture in monomethacrylated form.[62,63] The swelling properties of this type of a hydrogel depend on PEG chain length, mutual ratio of monomeric units within network and environmental conditions.[64] At low pH, the mesh size of the network is strongly reduced by hydrogen bond formation between carboxyl groups of PMAA and oxygen of PEG. The strongest gel collapse was observed for equimolar participation of PEG and PMAA monomeric units in a hydrogel and the longer was the chain of a graft the stronger deswelling was observed due to more effective complexation between PEG and PMAA. Moreover, it was reported that the process of gel contraction due to hydrogen bonds formation is faster than expansion of the network under neutral conditions. The diffusion of solutes like vitamins or proteins was decreased when longer PEG grafts were introduced into the network.[62,63] P(MAA-g-PEG) hydrogels were extensively tested towards biomedical application.[65-67] Hydrogel based on poly(methacrylic acid) and poly(ethylene glycol) di- and triacrylates were also prepared in bulk to yield highly cross-linked networks.[68-70]

Free-radical polymerization can also be applied for preparation of two-component, complex hydrogels in the form of interpenetrating network (IPN). For these purposes, usually so-called sequential method is used in which the initial gel is synthesized, followed by polymerization of the second component within the primary hydrogel. The polymerization stage of the second component can occur via template or matrix polymerization step as monomer can form hydrogen bonds with the existing polymer and upon initialization of polymerization process the forming chain grows along the matrix.[71] IPNs composed of poly(N-isopropylacrylamide), polyacrylamide, poly(N,N-dimethylacrylamide) or poly(vinyl alcohol) (PVA) and weak polyacids (PAA or PMAA) were synthesized.[15,72-79] IPN based on poly(vinyl alcohol) can be prepared using glutaraldehyde as a PVA cross-linking agent or by successive freezing and thawing processes resulting in formation of highly crystalline domains within PVA chains.[76,80-83] Thus, the latter interpenetrating network consists of chemically and physically cross-linked parts. For reasons described earlier, formation of interpolymer hydrogen bonds within IPN is more effective than in case of copolymer hydrogels, as confirmed by nuclear magnetic resonance.[56]

Complementary pair of polymers can also form layer-by-layer interpolymer complex on flat surfaces when deposited in alternate manner from their solutions.[84-86] Formed film can be cross-linked by applying a method based on reaction of polyacid carboxylic groups with a specially modified non-ionic macromolecule possessing amine moieties. Chemical bonds can be introduced by condensation reaction in process supported by carbodiimide compounds. In such way, stable films of poly(N-vinylpyrrolidone) with poly(methacrylic acid) were prepared by group of Sukhishvili.[87]

Hydrogels composed of complex-forming constituents can be also prepared by direct stabilization with ultraviolet light. Such a synthesis may be based not only on cross-linking polymerization of monomers, but also on cross-linking of macromolecules. The latter approach, where the starting material is a polymer solution, has some advantages in comparison with cross-linking polymerization process as one can avoid using monomers and cross-linking agents, which are usually toxic and difficult to remove from the final product. Using this method, one can prepare a blend from complementary constituents and expose the material, prepared for example in a form of a foil, to the direct action of UV light with wavelength shorter than 300 nm. As a consequence of ionization process, formation of radicals on polymeric chains takes place what finally leads to cross-linking of both components. One can also carry out UV-induced cross-linking of polymers in a solution upon addition of photoinitiator, e.g., hydrogen peroxide. Irradiation of such sample with UV light causes formation of hydroxyl radicals by $H_2O_2$ decomposition. These reactive species react further with polymeric chains and form macroradicals that undergo intermolecular re-combination resulting in a formation of cross-links. Both approaches were used in preparation of hydrogels based on poly(N-vinylpyrrolidone) and poly(acrylic acid).[88,89] These hydrogels can be used as simple chemical detectors when containing immobilized active substances. For example, glucose oxidase can be immobilized within a PVP-PAA matrix, and such system may act as a simple detector of glucose. As a result of gluconic acid formation, solution pH decreases and local phase separation due to hydrogen-bonding occurs, causing shrinkage and appearance of opacity within a hydrogel.[89]

### 2.2. Synthesis of nano/microgels

One of the drawbacks of using macroscopic stimuli-sensitive gels is their long response time to changes of environmental conditions. Often immobilization of biologically active substances within a gel goes through a step of network swelling in appropriate amount of fluid. Especially for highly cross-linked gels, it can take several hours until the solvent diffuses to the inner parts of the network and equilibrium state is reached. It seems that nano- and microgels can help to solve these problems due to their small sizes and in consequence shorter response time, as the latter is correlated with sample dimensions according to $time = length^2/D$ relationship, where D is the diffusion coefficient of a solute.[90]

Majority of methods utilized for polymeric nano- and microparticle synthesis require dilute conditions in order to prevent formation of bulk material. From the point of view of reaction mechanism, methods applied for microscopic hydrogel synthesis are the same as these for bulk gels. However, in order to avoid macrogelation synthesis of nano- and microgels is often performed in emulsion or microemulsion to separate monomer or polymer domains and prevent their mutual contact.[33,34,91]

Besides cross-linking polymerization in microemulsion, nanoparticles of hydrogel-type, containing hydrogen-bonding components, can be synthesized by means of surfactant-free emulsion polymerization (SFEP).[33,34,91,92] For instance, in some circumstances, electrostatic stabilization can be achieved in the absence of surfactants, as a consequence of incorporation of ionic initiator in the structure of forming chains. Both synthetic approaches are among the most common methods dedicated to microscopic network formation. It is very often used for preparation of microgels based on poly(N-isopropylacrylamide).[93]

Synthesis is performed at elevated temperature (around 60°C) in order to ensure decomposition of polymerization initiator. However, as polymerization of monomers proceeds, formed PNIPAM chains collapse when they reach certain molecular weight. Microgel is formed as a consequence of simultaneous polymerization, aggregation and cross-linking processes. Surfactant-free approach with distinct precipitation character of the polymerization process was also employed in preparation of PNIPAM microgel with anionic comonomers. Several works were dedicated to synthesis of poly(N-isopropylacrylamide-co-(meth)acrylic acid) hydrogel microparticles.[94,95] Peppas group applied microemulsion and surfactant-free emulsion method to the UV-induced synthesis of micro- and nanoparticles based on poly(methacrylic acid)-graft-poly(ethylene glycol) (PMAA-g-PEG). As in case of macroscopic hydrogels, poly(ethylene glycol) monoderivative formed network with PMAA.[96-98]

Free-radical cross-linking polymerization can proceed as template polymerization if matrix molecules are added to the reaction mixture.[71] As it was already mentioned, in such case propagating chain grows along complementary polymer. Such chain, with cross-linking agent incorporated in its structure (e.g. N,N'-methylenebisacrylamide), can then form a network. This technique was successfully applied in preparation of hydroxypropylcellulose-poly(acrylic acid) complex. In consequence, matrix macromolecule becomes locked within a microgel shell which swells or de-swells upon changes of solution pH.[99] Sub-micron P(NIPAM-co-AA) hydrogels can be also prepared by two-step approach. The first stage includes formation of one-component microgel particles. Synthesized particles can be then immersed in aqueous solution of the second monomer and the above-described surfactant-free emulsion cross-linking polymerization can be used with appropriate initiator and cross-linker. Similarly to the latter case, the second step proceeds as template polymerization because the introduced microgel acts as a matrix for growing chain due to the possibility of hydrogen bond formation between both components. The resulted structure is micron-size, pH-sensitive interpenetrating network composed of two complementary polymers. In concentrated solution, at pH higher than 5 and temperature higher than the volume phase transition of PNIPAM block, P(NIPAM-co-AA) microgels form a physical gel. Matrix polymerization is a crucial stage that decides about formation of interpenetrating network. IPN is not formed if sodium acrylate is the starting material in the system.[100]

Chemical stabilization can be also realized within self-assembled structures formed by hydrogen-bonding interpolymer complexes in dilute aqueous solution. It seems that this method can be a very effective way to prepare particles starting from macromolecules rather than mixture of monomers. Dou *et al.* used copolymer composed of hydroxylethylcellulose (HEC) grafted with poly(acrylic acid) (PAA). Interpolymer complexation was observed in water upon decreasing pH below 4.[101] Aggregation of HEC-g-PAA IPCs led to micelle-like structures with no signs of precipitation due to sterical stabilization of particles by non-complexed HEC segments. The hydrodynamic size of resulted aggregates could be regulated by changes in the number of PAA grafts on HEC unit. Formed micelles were "frozen" in acidic conditions by chemically-induced cross-linking of aggregates' core using 2,2'-ethylenedioxybisethylamine. Upon increasing solution pH aggregates transformed to hollow-like structures due to cross-linking of core's periphery and disintegration of hydrogen bonds between pendant groups at pH higher than complexation value. The process of micelle-to-hollow-like structure transition was totally reversible. The degree of core cross-linking depends on reaction time and the amount of cross-linking agent. If synthesis lasted more than 24 hours and the amount of cross-linker was significantly increased, the whole volume of the core underwent cross-linking and particles of typical microgel structure were obtained.

Hydrogen-bonding interpolymer complexes can be also deposited on spherical templates like silica or polystyrene particles. If the template is removed in acidic conditions where hydrogen-bonding interpolymer complexation is ensured, formation of polymeric capsules is observed.[86,102] The radius of a capsule depends on the size of used template and can range from nano- to micrometers, while a thickness of a wall depends on the number of deposited layers. Similarly to films deposited on flat surface, the structure of a capsule can be stabilized by carbodiimide chemistry. By this approach poly(N-vinylpyrrolidone) and poly(methacrylic acid) were successfully cross-linked to form stable capsules.[87,103-106]

## 3. Radiation-induced synthesis of materials based on hydrogen-bonded interpolymer complex systems

As it was mentioned, direct stabilization of macromolecules associated in a form of IPC can be very convenient method to produce pH-sensitive materials.[89,101] Such procedure has some advantages in comparison with cross-linking polymerization method where monomer is a starting material. Besides the limitations of the latter which were discussed above, one should also consider that due to incomplete conversion of monomers there is a need to perform purifying procedures in order to remove un-reacted substance (monomer removal is of particular importance if the product is intended for biomedical purposes). However, chemically-induced cross-linking of macromolecules may in some cases lead to similar problems, due to the presence of a cross-linking agent and initiator, which should also be separated from the final materials.

An alternative approach towards permanent stabilization of complex-forming systems can be realized by inducing cross-linking of polymers in aqueous solution with high-energy radiation, e.g., γ-rays or fast electrons.[107] In comparison with chemical cross-linking procedures, this method does not require any additives, i.e., it is performed in cross-linker-free and initiator-free systems. In consequence, no further purifying steps are necessary, which reduces time and costs of the synthesis. What is also important, especially in production of materials for biomedical application, in some cases the synthesis can be combined in one process with sterilization of the final product.

In the following paragraph, we summarize processes that take place in aqueous polymer solutions under the action of ionizing radiation. In the further sections, we present examples of application of radiation technique for cross-linking of various polymeric complex-forming systems.

### 3.1. *Radiation-induced processes in aqueous solution of polymers*

When a multi-component system in subjected to ionizing radiation, the amount of energy absorbed by each component is proportional to its weight fraction.[108,109] For synthesis of micro- or macroscopic gels, dilute and semi-dilute solutions are usually used, respectively. Thus, the main part of radiation energy is absorbed by the solvent, while the direct influence of radiation on macromolecules can usually be neglected. In consequence, interaction of high-energy radiation with polymers in aqueous solution occurs mainly through an indirect effect. The interaction of γ-rays or fast electrons with water leads in the first order to ionization of water molecules (reaction 1).[109]

$$H_2O \xrightarrow{\text{ionizing radiation}} H_2O^{+\bullet} + e^- \tag{1}$$

A part of molecules undergoes also excitation which may be followed by dissociation into radicals (reaction 2).

$$H_2O^* \rightarrow {}^{\bullet}OH + H^{\bullet} \tag{2}$$

Process of ionization takes place within $10^{-16}$ seconds. In the time range of $10^{-13}$–$10^{-11}$ seconds, products of water ionization undergo reactions that lead to formation of the hydroxyl radicals (reactions 2 and 3) and thermalization of an electron, i.e., its hydration (reaction 4).[109]

$$H_2O^{+\bullet} + H_2O \rightarrow {}^{\bullet}OH + H_3O^+ \qquad (10^{-13} \text{ sec}) \tag{3}$$

$$e^- + nH_2O \rightarrow e^-_{aq} \qquad (10^{-11} \text{ sec}) \tag{4}$$

The species produced in the reactions 1–4 are called primary products of water radiolysis. The areas of accumulated products, formed as a result of energy absorption step, are called spurs. For times up to $10^{-7}$ seconds, cascade of reactions of these primary products starts to occur within and outside spurs. Upon diffusion of reactive species from a spur to the bulk solution reactions with the solutes start. Efficiency of these reactions, which proceed in competition to self-recombination of the primary products and some side processes, depends on concentration of the solutes and their reactivity towards the primary products, which may be quantified in the form of appropriate rate constants. In oxygen-free conditions the main primary products with a life-time in the range of milliseconds that show very high reactivity towards most organic solutes are: hydrated electrons ($e^-_{aq}$), OH radicals ($^{\bullet}OH$) and hydrogen atoms ($H^{\bullet}$).[109] The amount of species produced upon absorption of a unit of energy can be represented by radiation-chemical yield G, expressed in mol/J. For example, in aqueous solution saturated with argon radiation yield of hydroxyl radical formation is equal to $2.8 \times 10^{-7}$ mol/J, hydrogen atom – $0.6 \times 10^{-7}$ mol/J, $e_{aq}^{-}$ – $2.7 \times 10^{-7}$ mol/J.[109] These radiation-chemical yields are constant in the range of pH = 3–11. The most reactive species towards macromolecules are hydroxyl radicals, which typically abstract hydrogen atoms and leave behind carbon-centered polymer radicals. The reaction of $^{\bullet}$OH with polymers is usually diffusion-driven and thus the selectivity in the reaction is very low. In this way, the reactivity is transferred from the primary radiolysis products to polymer chains.

Hydrated electron is the strongest known reducing agent that reacts mainly with aromatic groups and systems of coupled double bonds of carbon-carbon type.[109] Its reactivity towards aliphatic polymers is lower than in case of hydroxyl radicals. However, it shows very strong reactivity towards carbonyl group what should be taken into account due to the fact that many hydrophilic polymers contain such group (e.g. poly(acrylic acid), poly(N-vinylpyrrolidone)).[109] Usually, hydrated electrons can be converted into additional hydroxyl radicals upon solution saturation with nitrous oxide (reaction 5):

$$e_{aq}^{-} + N_2O \rightarrow {}^{\bullet}OH + N_2 + OH^{-} \qquad (5)$$

Polymer radicals, formed as a result of hydrogen abstraction from macromolecules, can undergo several one- and two-radical reactions. Detailed reaction mechanism depends to some extent on the chemical structure of radical-bearing chains, but the general reaction types are similar for most of the studied systems (**Scheme 1**).

One-radical reactions include degradation of the main polymeric chain and hydrogen atom transfer, which may occur in intra- or intermolecular manner. The former reaction leads to a decrease in average molecular weight of macromolecules, and thus is a process which, if possible, should be suppressed in typical gel synthesis. The latter reaction is usually of minor practical importance, since it influences neither the polymer structure nor the number of polymer radicals, however radical structure and localization may be changed.

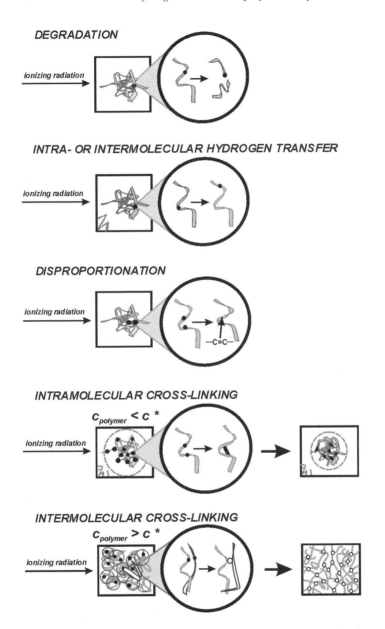

**Scheme 1.** One- and two-radical reactions in irradiated aqueous solution of polymers. Black dots represent mid-chain radicals, white dots and black lines – cross-links between chains and polymeric segments respectively and dashed circles – the size of the single polymeric coil.

In the group of two-radical reactions one can distinguish disproportionation or recombination. Disproportionation, similarly to hydrogen atom transfer, can occur within the same macromolecule or between separate chains. It results in formation of double bonds. While this reaction does not itself contribute to gel formation, the double bond formed may enhance the ability of a polymer to form cross-links.

The most important reaction in the process of gel formation is, of course, radical recombination. When two radicals are generated on separate chains, they can recombine to form a covalent bond which permanently links both macromolecules. Radical recombination occurs preferentially in the intermolecular way if one uses polymer concentration higher than overlap concentration and applies low dose rates of radiation, so that, in the stationary state of the process, on average less than one radical is present at a time on each macromolecule. Irradiation of such system leads finally to macroscopic gelation, i.e., formation of a "wall-to-wall" gel. Radiation-induced cross-linking has been successfully applied for synthesis of macroscopic hydrogels in aqueous solution of several hydrophilic polymers including poly(N-vinylpyrrolidone), poly(acrylic acid), poly(ethylene oxide) or poly(vinyl methyl ether).[107,110] In practice, this method is the basis of implemented, industrial-scale production technology of hydrogel wound dressings.[111]

On the contrary, high dose rates and low concentration of a polymer in solution favor simultaneous formation of many radicals on each polymer chain and, as a result, intramolecular recombination predominates.[33,112,113] Recombination of radicals within a single chain leads to cross-linking of adjacent polymeric segments and in consequence internally cross-linked macromolecules are formed, of similar size but different structure and properties when compared to free polymer coils. Such nanostructures bear structural resemblance to macroscopic hydrogels, and thus are called nanogels.[33] Process of nanogel formation is accompanied by a decrease in the radius of gyration due to contraction of a chain by formation of covalent bonds between segments, while, at least in theory, average molecular weight in purely intramolecular cross-linking remains unchanged. In practice, some changes in average molecular weight are usually observed as a result of contribution of intermolecular recombination and, more often than not, also of degradation. Pulse irradiation of hydrophilic polymers in dilute solution, so-called preparative pulse radiolysis, was successfully utilized for synthesizing nanogels of non-ionic hydrophilic polymers like poly(vinyl alcohol), poly(N-vinylpyrrolidone), poly(ethylene oxide) or poly(vinyl methyl ether).[113-116] Nanogels of simple polyelectrolyte, poly(acrylic acid), were also obtained.[113,115] A modified version of this process, involving two consecutive irradiation steps, can be applied for synthesis of gel particles (in the submicron range) of independently controlled molecular weight and dimensions. This so-called two-step routine consists of continuous irradiation in semi-dilute solution when mainly intermolecular recombination occurs and subsequent pulse irradiation of dilute solution in order to promote intraparticle cross-linking and size reduction of particles. Such routine has been successfully tested on poly(N-vinylpyrrolidone) and more recently on poly(ethylene oxide).[117] It was additionally shown that PAA nanogels formed interpolymer complexes with non-ionic polymers.[118]

Preparative pulse radiolysis can be applied for cross-linking complex-forming systems, as has been shown on the example of synthesizing two-component nanogels based on poly(N-vinylpyrrolidone) and poly(acrylic acid).[119] This issue will be discussed in some detail in section 3.4.3. Before that, in the following section, selected literature data on other radiation-based techniques used for stabilization of hydrogen-bonding interpolymer complexes will be presented.

### 3.2. Radiation-induced cross-linking of hydrogen-bonding interpolymer complexes in a blend

Permanent stabilization of a complex-forming polymer system can be realized in a composition that already contains formed complex or in a mixture of complementary polymers at conditions where no hydrogen-bonding interactions take place. As mentioned before, these two cases represent substrates with different properties. The differences result from specific interactions within IPC or their absence in case of a non-complexed physical mixture of polymers.

For radiation-induced stabilization of interpolymer complex systems, the starting material for the synthesis was often prepared in the form of a film. One of the problems in the preparation of films based on interpolymer complexes is the lack of homogeneity when cast polymer solution is prepared below critical pH. Due to high concentration usually used in such studies, complexed systems undergo phase separation. One of the ways to avoid this problem is preparation of a blend at high dissociation degrees of used polyacid. However, as it was shown by Khutoryanskiy *et al.* one should carefully investigate the phase behaviour as a function of pH for given pair of polymers.[120,122] Upon γ-irradiation macromolecules underwent cross-linking. Irradiated films, when immersed in water, swelled and formed hydrogel. However, one should have in mind that radiation cross-linking of polymers in solid state usually requires the use of relatively high doses.[108] For example, in order to obtain 80% of a gel fraction in poly(vinyl alcohol)-poly(acrylic acid) film, prepared from 0.1 M solution at pH < 4 and equimolar composition, one should use dose as high as *ca.* 60 kGy. The reasons for that are the following. Radiation interacts with solid-state polymers directly, and the yield of radical formation is lower than in aqueous solutions. Moreover, in order to generate, for instance, 1 cross-link per chain, much higher dose (energy per kg of the system) is needed in the solid state (high number of chains per kg) than in the solution (much lower number of chains per kg). Secondly, polymer chains in films possess restricted segmental mobility. In consequence, recombination of formed radicals does not progress as effectively as in a solution. Another problem arises from significant participation of degradation reaction during radiation treatment of polymers in the solid state. Scission of the main polymeric chain is additionally accelerated by a presence of oxygen, which often cannot be totally removed from a film. All these facts cause an increase of gelation dose. Nevertheless, this technique was successfully applied to prepare radiation cross-linked poly(vinyl alcohol)-poly(acrylic acid), poly(2-hydroxyethyl vinyl ether)-poly(acrylic acid) and poly(vinyl methyl ether)-poly(acrylic acid) films.[120-122] More recently, Dubolazov *et al.* prepared hydroxypropylcellulose-poly(acrylic acid) (HPC-PAA) mucoadhesive films cross-linked by γ-radiation in the presence of a cross-linker, N,N'-methylenebisacrylamide (BIS).[123] The addition of BIS was dictated by a fact that in pure film HPC underwent degradation and it was not possible to obtain a cross-linked material. In general, polysaccharides are strongly prone towards chain scission upon irradiation due to cleavage of glycosidic bond. Hydrogels from cellulose derivatives can be synthesized only under high polymer concentration (samples usually in paste-like form) and high dose rate irradiation.[124-126]

### 3.3. *Radiation-induced cross-linking of hydrogen-bonded interpolymer complexes and monomer systems in aqueous solutions*

Hydrogel with complex-forming components can be also prepared by radiation-induced cross-linking of complementary polymers in a solution. Recently, poly(N-vinylpyrrolidone)-poly(acrylic acid) hydrogels were prepared in our group by electron-beam irradiation of their aqueous solutions.[127]

A characteristic feature of such IPC-forming systems in concentrated solution is the process of phase separation which is manifested by appearance of turbidity and thus complexation can be conveniently studied with optical techniques. For example, on the basis of UV-vis measurements one can easily estimate aggregation pH due to very sharp transition (**Fig. 1**).

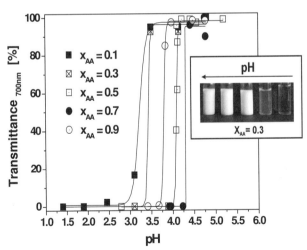

**Fig. 1.** Transmittance of PVP-PAA solutions at different $x_{AA}$ as a function of pH. Transmittance measured at 700 nm. Weight–average molecular weight of polymers: PVP – 450 kDa, PAA – 50 kDa. Total concentration of polymers in solution – 1 mol dm$^{-3}$. Inset: phase behaviour of PVP-PAA solution at $x_{AA} = 0.3$ and different pH. The arrow indicates decreasing solution pH.[127]

The gelation dose $D_g$ of irradiated system can be estimated on the basis of Charlesby-Rosiak equation[128]:

$$s + \sqrt{s} = \frac{p_0}{q_0} + \left(2 - \frac{p_0}{q_0}\right) \cdot \left(\frac{D_V + D_g}{D_V + D}\right) \qquad (6)$$

where s is the content of a sol fraction ($0 < s < 1$), $p_0$ is so-called degradation density, i.e., the average number of main chain scissions per monomer unit and per unit dose, $q_0$ – cross-linking density, i.e., the average number of cross-linked units per monomer unit and per unit dose, $D_g$ – gelation dose and $D_V$ – virtual dose, i.e., dose necessary to transform the actual polymer sample into a sample characterized by the most probable molecular weight distribution $M_w/M_n = 2$. From a gravimetric analysis of synthesized hydrogel, so-called sol-gel analysis, one can obtain the amount of a sol and a gel fraction in a sample irradiated with certain dose D. In **Fig. 2**, an example of sol-gel data plotted on the basis of Eq. (6) are presented for PVP-PAA system irradiated at different $x_{AA}$.

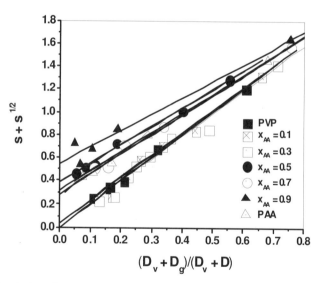

**Fig. 2.** Sol-gel data for argon-saturated aqueous solutions of PVP-PAA prepared at different $x_{AA}$ and irradiated with different doses. Weight–average molecular weight of polymers: PVP – 450 kDa, PAA – 50 kDa. Total concentration of polymers in solution – 1 mol dm$^{-3}$. [127]

In **Table 1**, radiation parameters obtained on the basis of Eq. (6) are shown for investigated compositions.

**Table 1.** Results of cross-linking parameters obtained for PVP-PAA solutions at different $x_{AA}$.[127]

| $x_{AA}$ | Gelation dose, $D_g$ [kGy] | $p_0/q_0$ |
|---|---|---|
| 0.0 | 2.80 | 0.00 |
| 0.1 | 3.26 | 0.00 |
| 0.3 | 3.44 | 0.00 |
| 0.5 | 3.87 | 0.38 |
| 0.7 | 4.02 | 0.31 |
| 0.9 | 4.56 | 0.55 |
| 1.0 | 5.03 | 0.27 |

It was observed that for PVP-PAA system, gelation dose increased almost linearly with increasing the amount of poly(acrylic acid) in the solutions. This fact is related to the polyelectrolyte character of this polymer – irradiation of PAA at dissociation degree $\alpha > 0$ is accompanied with restricted intermolecular recombination of radicals due to repulsion of charges localized on a chain. In consequence radicals present on a macromolecule can live up to hours at relatively high pH in aqueous PAA solutions.[129] Up to $x_{AA} = 0.3$, formation of hydrogel proceeded according to pure cross-linking reaction, as expected for solutions with dominant amount of PVP. On the other hand, at higher PAA fractions chain scission was clearly visible in the system as increasing ratio of degradation to cross-linking events.

These hydrogels showed an abrupt response to solution pH. In pH range below $pH_{complex}$, swelling degrees decreased significantly due to hydrogen-bonding complexation between polymeric segments. This process was accompanied by appearance of turbidity due to local phase separation of complexed segments of PVP and PAA within hydrogel.[127]

Materials based on hydrogen-bonding interpolymer complexes can also be synthesized by radiation-induced copolymerization of monomers or monomer-polymer mixture in aqueous solution. By the former routine hydrogels based on poly(N-vinylpyrrolidone) and different weak polyacids, like poly(acrylic acid) or poly(itaconic acid), were obtained.[130-133] As already mentioned in section 3.1, in radiation-induced process there is no need of using any initiators or cross-linking agents due to formation of highly reactive radicals from water radiolysis stage. In consequence, simultaneously with chain formation cross-linking process takes place and as a result hydrogel is formed. Similarly to the processes described in section dedicated to chemical- and UV-induced hydrogel synthesis, in case of monomer-polymer mixture polymerization of monomer can occur according to template mechanism where the polymer serves as a matrix for a new growing chain.

### 3.4. *Radiation-induced synthesis of two-component nanogels based on poly(N-vinylpyrrolidone)-poly(acrylic acid) interpolymer complexes*

Physicochemical aspects of interpolymer complex formation are usually investigated in dilute solution. Concentrations below critical hydrodynamic range are also suitable for synthesis of polymeric nanogels. However, there is still lack of detailed investigations on some subjects related to hydrogen-bonding complexation process, e.g. the influence of solution conditions and polymer molecular weight on aggregation of PVP-PAA interpolymer complexes. As shown in case of polyelectrolyte complexes, the sample preparation procedure can additionally affect some properties of the material like structure of resulted particles. Till now, there is no evidence that the way of polymer mixing can influence processes in hydrogen-bonding interpolymer complex systems.

#### 3.4.1. *Formation of poly(N-vinylpyrrolidone)-poly(acrylic acid) interpolymer complexes and their aggregates*

Interpolymer complexes of poly(N-vinylpyrrolidone) and weak polyacids are known as one of the most stable hydrogen-bonding IPCs.[134] Complexation phenomenon in poly(N-vinylpyrrolidone)-polyacids system was studied in some details by variety of methods, including potentiometric and conductometric titration, calorimetry, viscometry, nuclear magnetic resonance, fluorescence and infrared spectroscopy or to some extent also by the light scattering method.[6-8,88,135-143]

It was shown that complexation as well as aggregation can be influenced by temperature, addition of low molecular-weight salts which change the efficiency of hydrogen-bonding and accelerate flocculation process.[143-146] One of the methods that allow detecting interactions between macromolecules is laser light scattering technique (LLS).[5] In order to strengthen observable effect of interpolymer complexation in solution one can choose for example a system in which polymers differ significantly in refractive index increment or there is a big difference in size, i.e., high-molecular weight polymer and oligomer. Recent investigation of Henke *et al.* showed that formation of hydrogen-bonding IPC is highly sensitive to the way of sample preparation.[147] There are two simple ways of setting appropriate value of solution pH: in the first case two polymer solutions can be mixed after setting their pH to desired value. However, one should have in mind that upon polymer mixing pH increases due to hydrogen bond formation. On the other hand, one can firstly mix both solutions, followed by regulation of solution pH. In the latter case, however, strong local fluctuations in concentration of inorganic acid are produced, even upon strong stirring. In consequence complexation may occur in less controllable manner (pronounced aggregation processes). The resulted product differed in the colloidal stability and the size of formed IPCs (**Fig. 3**).

Moreover, in light scattering experiments appearance of complexation region in the investigated systems was reflected by a decrease in the hydrodynamic radius and radius of gyration of high-molecular weight constituent due to its contraction upon hydrogen bond formation with complementary chain of lower molecular weights.[147] Such behavior was also confirmed in the literature by other techniques like viscometry.[5] When pH is decreased significantly, complexation is followed by aggregation process which, on the contrary, causes an increase of the size. Observed minima in dimensions of formed complexes can be treated as an indication of complexation region. At the highest molecular weight of PAA (50 kDa, denoted further in the text as PAA50) tested by Henke *et al.*, the stoichiometry corresponding to the highest pH of complexation (i.e., to the strongest tendency towards complex formation) is close to equimolar. Surprisingly, for the shortest

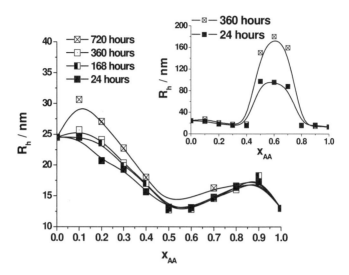

**Fig. 3.** Changes of hydrodynamic radius for PVP-PAA system at pH = 3.4 prepared according to two different procedures. Main graph – samples obtained by mixing polymer solutions after setting their pH to 3.4, inset – samples prepared through mixing followed by addition of acid to set the pH to 3.4. Molecular weights of polymers: PVP – 800 kDa, PAA – 50 kDa. Total concentration of polymers in solution – 0.01 mol dm$^{-3}$.[147]

poly(acrylic acid), with weight-average molecular weigt of 0.9 kDa (PAA1), the maximum occurred at $x_{AA}$ as high as 0.75–0.80. This fact reflected different ability of short and long polyacid chains to stabilize interpolymer complexes containing high-molecular weight PVP. For PAA1 the length of its chain is comparable with calculated by Iliopoulos and Audebert minimal sequence of protonated carboxyl groups that supports formation of stable complexes with PVP.[10] In consequence, dissociation of any group should result in destabilization of PVP-PAA1 complex and its disintegration. Going further, one can not exclude that only a fraction of all available short chains was capable of forming IPCs at any pH. This effect can be mainly a result of polydispersity of the sample, which was not very high ($M_w/M_n$ = 1.15), however, at such chain length even that degree of polydispersity can influence complexation process. In fact, our potentiometric titration measurements revealed that complexation degree at pH = 3.4 (much lower than the complexation value) and $x_{AA}$ = 0.7–0.8 was only 20%–25%. This was further confirmed by gravimetrical analysis of PVP-PAA1 complexes/aggregates. Our results have also shown that short PAA chains are more or less uniformly distributed among all PVP chains. In consequence, e.g., at equimolar composition efficient complexation can be observed only when most of the polyacid macromolecules are protonated. If one distributes now very large number of PAA1 chains among small fraction of PVP chains the probability of finding complementary macromolecules that fulfill complexation condition is significantly higher. On the contrary, longer PAA chains possess many complexation sequences and optimal hydrogen-bonding can be observed at almost equimolar composition. Both cases can be schematically presented as in **Scheme 2**.

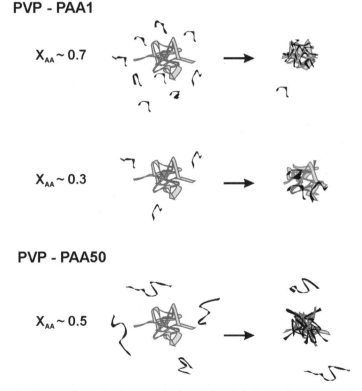

**Scheme 2.** Proposed mechanism of hydrogen-bonded interpolymer complexation between high molecular weight poly(N-vinylpyrrolidone) (gray chain) and poly(acrylic acid) (black chains) of low and high molecular weight and at different mutual ratios of polymers in a solution.

Complexation is followed by aggregation process. From kinetic point of view the aggregation can occur according to diffusion-limited cluster aggregation mechanism (DLCA) when repulsion barrier is diminished and every collision leads to the particle merging, or by reaction-limited cluster aggregation (RLCA) (also known as potential) when the barrier is high enough to cause repulsion between particles thus not every collision is effective.[148] In the latter case, the system forms a stable aggregate only after particles find optimal spatial arrangement. As reported by Hemker and Frank, for PEO-PAA IPCs at pH value near the complexation threshold reaction-limited aggregation was observed.[149] This is related to existence of repulsive barrier probably due to the presence of non-complexed segments within a complex, including ionized polyacid groups. On the other hand, when pH was decreased significantly below the complexation point, aggregation turned into diffusive. The values of fractal dimension which correlates aggregate mass with its size ($M = R^{df}$) were obtained in that work on the basis of mathematical model allowing to describe the experimental data on hydrodynamic size distribution with aggregation kinetics equation. In DLCA $d_f = 1.7–1.8$ while in RLCA fractal dimension changed form 2.1 to 2.2. While for DLCA regime, the scaling was very good, there was a large uncertainty of $d_f$ value obtained for RLCA regime.[149]

For the above-mentioned PVP-PAA systems investigated by Henke *et al.*, some studies of their aggregation behavior have been undertaken as well.[147] Of all PVP-PAA systems investigated by us the highest aggregation dynamics was observed for PVP-PAA1 IPCs at a composition that differed significantly from equimolar ($x_{AA} \sim 0.7–0.8$). Similarly to findings made by Hemker and Frank, flocculation mechanism was pH dependent. A decrease of solution pH to 3.4 caused that aggregation in PVP-PAA1 system was slow and proceeded according to the RLCA mechanism. At pH = 3.2, clear transition between RLCA and DLCA was marked while starting from pH 3.0 "pure" diffusion-controlled process took place. . The mechanism of aggregation was also confirmed by means of estimated fractal dimension. While in DLCA, the scaling was very good with theory ($d_f = 1.7$), at pH = 3.4, $d_f$ was somewhat higher and reached value 2.8. This suggested formation of very compact aggregate structures. Similar fractal dimension was also observed recently for PNIPAM aggregates.[150]

### 3.4.2. *Structure of poly(N-vinylpyrrolidone)-poly(acrylic acid) interpolymer complexes and their aggregates*

From the point of view of nanogel synthesis, it is very important to know the structure of formed complex. As was described in earlier section, complexation strongly depends on solution pH and molecular weight of used polyacid. These factors influenced complexation parameters such as pH value of complex formation or stoichiometry of the complex. In consequence the architecture of PVP-PAA IPCs, should depend on the mentioned factors as well. Moreover, there is no systematic study on the influence of polymer molecular weight on resulted structure of hydrogen-bonding interpolymer complexes. It was observed by Small Angle Neutron Scattering that for PEO-PAA IPCs, the structure changed from ladder-type near complexation threshold to compact upon stronger decrease in solution pH.[151,152]

In our investigations, the structures of IPCs and their aggregates for all investigated PVP-PAA systems were studied by means of combined static and dynamic light scattering measurements.[147] For oligo(AA)-based IPCs the polymers formed compact complexes at the complexation threshold – hydrodynamic size and radius of gyration significantly decreased before complexes started to aggregate. The $R_g/R_h$ ratio, which reflects molecular architecture, was close to hard sphere model. One can imagine such complex particle as schematically presented in **Fig. 4A**. IPCs based on the other oligoacid sample, PAA with $M_w = 5$ kDa (PAA5), were more compact. This was confirmed by fluorescence probe solubilization experiments in which stronger intensity of emitted fluorescence and shift of maximum wavelength were observed for PVP-PAA5 complexes than for PAA1-based system. In case of PVP-PAA50 IPCs, it was observed that at complexation pH the particles adopted extended structure probably similar to ladder-type ($R_g/R_h > 2$). It is known from theoretical considerations on hydrogen-bonding complexation that ladder-type structure is one of the most probable at least at the complexation threshold. This is also in agreement with our

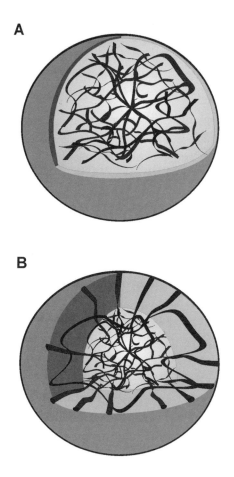

**Fig. 4.** Proposed structures of PVP-PAA interpolymer complexes (A) hard sphere, (B) core-shell.

findings, as a step of hydrogen bonding must follow according to zipping mechanism. In other case the effect of minimum chain length should not be observed. However, one can not detect directly this process in case of interpolymer complexes built by short PAA chains due to fast complex compaction, a process supported by hydrophobic effect. Secondly, it is very unlikely that each short PAA chain is bound to a PVP macromolecule close to another PAA chain – a situation that can potentially cause rigidity of the resulted complex. This can take place only at high fraction of polyacid in the mixture. However, one should take into account that even in such case matrix macromolecule can contain uncomplexed sequences. On the other hand, long PAA chains possess many adjacent complexing sequences on one chain and stretching effect should be very strong. Further decrease of solution pH caused compaction of PVP-PAA50 complexes and formation of core-shell particles (**Fig. 4B**). In such case, inner part of the particle contains complexed segments while shell consists mainly of uncomplexed parts of both chains. It should be also emphasized that in all investigated cases the IPCs' structure strongly depended on molar fraction of poly(acrylic acid) units in solution.

The structure of aggregates formed by PVP-PAA interpolymer complexes, as already mentioned in the following paragraph, varied depending on the mechanism of cluster formation. However, one could calculate on the basis of obtained fractal dimensions that compaction of RLCA-aggregate structure was restricted by IPCs packing and in consequence these aggregates could adopt only *ca.* 65% of packing efficiency characteristic for hard spheres despite small difference in $d_f$ values of these systems ($d_f$ for spherical objects like polystyrene latex is equal to 3).[153]

### 3.4.3. *Radiation-induced stabilization of poly(N-vinylpyrrolidone)-poly(acrylic acid) interpolymer complexes in dilute aqueous solution*

Based on the data regarding the architecture of investigated system, one can carefully choose conditions for stabilization of the complex structure. In the preliminary experiments, Henke *et al.* made an attempt to "freeze" the structure of aggregates prepared in PVP-PAA1 system at pH = 3.4 and three different complex compositions. It was decided to irradiate samples at pH = 3.4 due to almost spherical shape of formed aggregates.[119] The results are summarized in **Fig. 5**.

It was observed that an increase of absorbed dose caused a decrease in radius of gyration of PVP-PAA-1 aggregates due to intramolecular cross-linking process. Upon irradiation, inter-complex covalent bonds formation was observed as well. The latter process was expressed

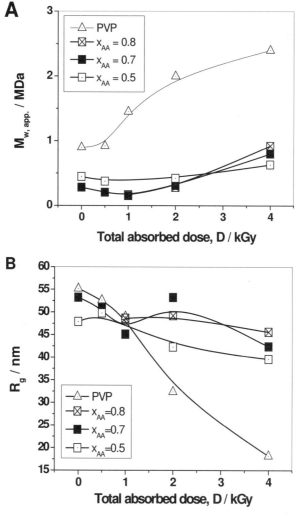

**Fig. 5.** Changes of apparent weight average molecular weight (A) and radius of gyration (B) for aqueous solutions of PVP-PAA interpolymer complexes at pH = 3.4 for selected $x_{AA}$ as a function of total absorbed dose. The weight-average molecular weights of polymers: PVP – 800 kDa, PAA – 0.9 kDa. For the calculation of $M_w$ according to Zimm algorithm in each case the same value of (dn/dc) = 0.185 ml/g was taken into account. Total concentration of polymers in solution – 0.01 mol dm$^{-3}$.[119] Reprinted from "Radiation-induced cross-linking polyvinylpyrrolidone-poly(acrylic acid) complexes", 236, Henke Artur, Kadlubowski Slawomir, Ulanski Piotr, Rosiak Janusz M. and Arndt Karl-Friedrich, *Nuclear Instruments and Methods in Physics Research* Section B: Beam Interactions with Materials and Atoms, 391-399, (2005) with the permission of Elsevier.

as an increase of weight-average molecular weight. Light-scattering measurements were performed at pH = 10 with addition of 0.5 M NaClO$_4$ as in such condition all polyacid chains are ionized and do not form complexes with PVP. Should no covalent bonds be formed between PAA and PVP chains (complexed during irradiation at pH 3.4), the R$_g$ and M$_w$ results at pH 10 (no complexation) would correspond to those for unirradiated samples.

Reported decrease of aggregate dimension was rather weak in comparison with nanogel-forming one-component systems, e.g. aqueous solutions of PVP or PAA. One should, however, bear in mind that the resulted structure of a complex is compact. These particles form further aggregates with fractal dimension 2.8 and thus they also reveal some packing efficiency representing certain optimum arrangement of particles within a cluster. On the other hand, in case of polymeric coil most of the inner volume is occupied by solvent. In consequence, due to high compactness of PVP-PAA system, the decrease of its size upon irradiation is slight, while upon coil irradiation stronger contraction occurs. Molecular weight changes suggested that the possibility of fixing PVP-PAA1 aggregate structure was rather limited in the applied range of irradiation doses. The slight decrease of M$_w$ in the initial stage of irradiation process can be attributed to degradation reaction that can be present in a system containing polyelectrolyte-centered radicals. While in case of PVP, degradation is of minor importance at applied dose range and solution conditions, PAA is highly prone towards chain scission especially at relatively high pH.[129,154,155]

In consequence, one should choose a complex system with restricted participation of aggregation process. In further experiments, PVP-PAA5 IPC was exposed to the action of pulsed electron beam from linear accelerator 24 hours after polymer mixing at equimolar composition at pH = 3.4 in N$_2$O-saturated aqueous solution.[156] The addition of nitrous oxide, as already mentioned, doubles the yield of hydroxyl radical formation and thus increases the number of potential cross-link points. Dynamic and static light scattering measurements were used to characterize changes of macromolecular parameters upon irradiation. In **Fig. 6**, changes of scattered light intensities, hydrodynamic radius and radius of gyration are presented as a function of total absorbed dose.

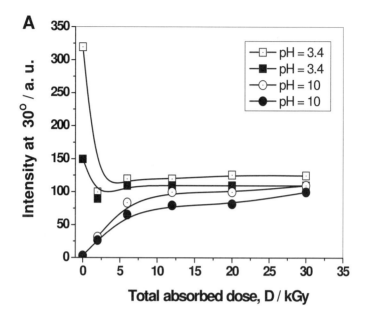

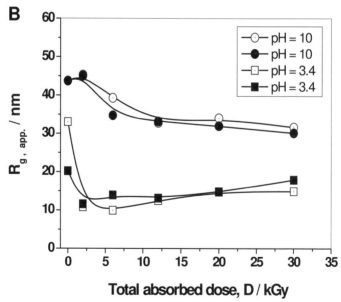

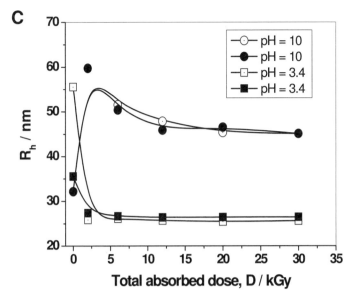

**Fig. 6.** Changes of scattered light intensity (A), radius of gyration (B) and hydrodynamic radius (C) as a function of total absorbed dose for PVP-PAA system at equimolar composition. The weight-average molecular weights of polymers: PVP – 800 kDa, PAA – 5 kDa. Measurements were performed at pH = 3.4 (synthesis) and 10. Irradiation performed 24 hours after polymer mixing. Total concentration of polymers in solution – 0.01 mol dm$^{-3}$. Empty and full symbols represent changes reported one day and one week after irradiation, respectively. Hydrodynamic radius was obtained from extrapolation of angular dependency of $R_h$.[156]

Intensity of scattered light recorded at pH = 3.4 was constant in time except for non-irradiated samples. It indicated that for D = 0 kGy aggregation of IPC takes place in the system, however at a limited rate in comparison with PVP-PAA1 IPCs. On the other hand, irradiation results in enhanced colloidal stability in PVP-PAA5. At pH = 10, where no hydrogen-bonding was present, intensity increased with irradiation dose suggesting that the number of PAA chains covalently connected with PVP molecule increased (while the refractive index increment did not change significantly). Intensity changes vs. dose reached a plateau starting from *ca.* 15 kGy and are comparable with these observed for particles at pH = 3.4. $R_h$ and $R_g$ values decreased with total absorbed dose due to dominance of intramolecular cross-linking reaction within a complex. The initial increase of hydrodynamic radius at pH = 10 results from not very high intramolecular recombination efficiency of macroradicals at applied dose range 2–5 kGy. In consequence, loosely bound PAA segments can diffuse from the center of a molecule due to repulsion of charges localized on PAA chains, causing an increase in $R_h$. To prove that permanently stabilized two-component nanogel was obtained, the swelling properties of the synthesized material were tested. For this purpose, PVP-PAA5 IPCs after dialysis were analyzed at different pH conditions. The purification step was undertaken in order to remove non-cross-linked polymer. The results of hydrodynamic diameter changes are presented in **Fig. 7**.

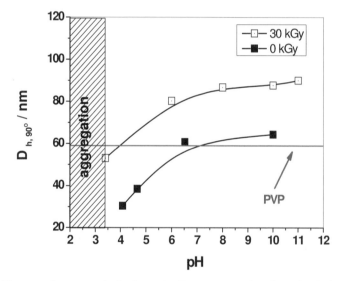

**Fig. 7.** Changes of apparent hydrodynamic diameter as a function of solution pH for PVP-PAA interpolymer complexes and nanogels. The weight-average molecular weights of polymers: PVP – 800 kDa, PAA – 5 kDa. Sample irradiated with 30 kGy was dialyzed against water (membrane molecular cut-off 10 kDa). Total concentration of polymers in solution – 0.01 mol dm$^{-3}$. Measurements were performed at scattering angle of 90° immediately after setting pH.[156]

The ionization of polyacid within interpolymer complex results in a slight increase in size over $R_h$ value observed for pure PVP chain in aqueous solution. The small increase observed at pH = 10 is a consequence of polyelectrolyte effect, i.e. upon dissociation carboxyl groups repeal each other what cause slight swelling of PVP chain due to incomplete disentanglement of both polymers. On the other hand, nanogel swelled to a size being about 30% larger in comparison with the hydrodynamic size of PVP molecule. The degree of swelling is restricted by formed covalent bonds between PVP segments. All presented facts suggest that both polymers are covalently linked within nanogel.

The structure of formed particles at pH = 3.4 and in the swollen state at pH = 10 is shown in **Fig. 8**.

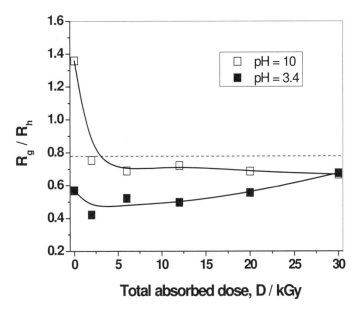

**Fig. 8.** $R_g/R_h$ ratio as a function of total absorbed dose for PVP-PAA nanogels at pH = 3.4 and 10. The weight-average molecular weights of polymers: PVP – 800 kDa, PAA – 5 kDa. Total concentration of polymers in solution – 0.01 mol dm$^{-3}$. Measurements were performed 24 hours after preparative pulse radiolysis. The dotted line represents $R_g/R_h$ value for hard sphere model.[156]

While the shape factor for system studied at pH = 3.4 falls in the range typical for a core-shell particle ($R_g/R_h$ in the range of 0.3-0.6), the swollen nanogels adopt structure similar to that of a hard sphere. It seems to indicate that recombination of polymer radicals took place in the whole core's volume of each particle and there is a high probability that the shell underwent at least partial cross-linking as well because hydroxyl radicals are characterized by low selectivity in reaction with most solutes. As discussed earlier, irradiated complexes showed colloidal stability at pH = 3.4 in contrast to unirradiated sample. It is a consequence of stiffening of the shell structure and when particles are close, they must repel each other due to limited possibility of polymeric chain penetration and compression (high osmotic pressure). However, one can not exclude that during irradiation some functional groups changed their chemical structure including formation of new ionic group and are responsible for additional colloidal stabilization of PVP-PAA nanogels.

Doses in the order of 30 kGy are relatively high for irradiation of dilute aqueous solution of polymers. Taking into account radiation-chemical yield of hydroxyl radical formation in $N_2O$-saturated aqueous solution, at 30 kGy and polymer concentration of 0.01 M we produce, at least in the theory, nearly twice as much hydroxyl radicals as the number of monomeric units present in the system. It means that every polymeric unit should react with •OH at least once. Of course, one should bear in mind that in reality the number of radicals that reach polymeric chain is smaller in comparison with theoretically calculated number, due to OH self-combination. Nevertheless, one should investigate the possible impact of irradiation dose on the chemical structure of polymeric chain. One of the results of irradiation action on polymeric chain may be elimination of pendant groups from polymeric chain or the change in their structure. The results from UV-vis and infrared spectroscopy showed that in fact there were some changes in the polymer structure upon irradiation. For example, in UV-vis spectra formation of double bonds was manifested by significant increase of absorbance at $\lambda$ <500 nm. What is more, the number of hydrogen bonds decreases as reported on the basis FT-IR results. Potentiometric titration revealed further that the real decrease of carboxyl groups in the mixture was *ca.* 10%. Taking this into account, in order to reduce these side effects one should limit the magnitude of irradiation dose. As shown in **Fig. 4A**, intensity of scattered light, which can be treated as a measure of a degree of PAA cross-linking with PVP chain, reaches constant value near 15 kGy, where radiation-induced changes in polymer structure are still relatively small.

## 4. Concluding remarks

Hydrogen-bonding interpolymer complexes attract growing interest, especially in application-oriented studies. In our investigations on interpolymer complexation, we presented some new data and ideas, especially on aggregation process and the structure of these supramolecular particles. One of the disadvantages of hydrogen-bonding interpolymer complexes, i.e., their lack of stability to the changes of external conditions, can be eliminated by fixing these structures with covalent bonds formed between the IPC components. Besides a variety of chemical or UV-induced methods for IPCs stabilization, radiation-induced approach seems to be a promising alternative.

## Acknowledgments

This work was supported by Polish Ministry of Science and Higher Education (grants 3 T09A 161 29, 3 T08E 078 29 and 138/E-370/SPB/IAEA/KN/DWM77/2005-2007), International Atomic Energy Agency (project POL/6/007) and NATO (grant PST.MEM.CLG 980622).

## References

1. Y. He, B. Zhu and Y. Inoue, *Prog. Polym. Sci.*, **29**, 1021 (2004).
2. J. M. Lehn, in *Supramolecular Polymers*, Ed. A. Ciferri, (CRC Press, Boca Raton, 2005), p. 3.
3. E. Tsuchida, and K. Abe, *Adv. Polym. Sci.*, **45**, 1 (1982).
4. E. Bekturov and L. A. Bimedina, *Adv. Polym. Sci.*, **41**, 100 (1981).
5. M. Jiang, M. Li, M. Xiang and H. Zhou, *Adv. Polym. Sci.*, **146**, 122 (1999).
6. Y. Osada, *J. Polym. Sci., Polym. Chem. Ed.*, **17**, 3485 (1979).
7. I. Iliopoulos and R. Audebert, *Eur. Polym. J.*, **24**, 171 (1988).
8. I. Iliopoulos, J. L. Halary and R. Audebert, *J. Polym. Sci., Part A: Polym. Chem.*, **26**, 275 (1988).
9. I. Iliopoulos and R. Audebert, *J. Polym. Sci., Part B: Polym. Phys.*, **26**, 2093 (1988).
10. I. Iliopoulos and R. Audebert, *Macromolecules*, **24**, 2566 (1991).
11. T. Ikawa, K. Abe, K. Honda and E. Tsuchida, *J. Polym. Sci. Pol. Chem. Ed.*, **13**, 1505 (1975).
12. Z. S. Nurkeeva, V. V. Khutoryanskiy and G. A. Mun, *Macromol. Biosci.*, **3**, 283 (2003).
13. D. J. Hemker, V. Garza and C. W. Frank, *Macromolecules*, **23**, 4411 (1990).
14. M. Koussathana, G. Staikos and P. Lianos, *Macromolecules*, **30**, 7798 (1997).

15. S. Nishi and T. Kotaka, *Macromolecules*, **19**, 978 (1986).
16. V. V. Khutoryanskiy, *Int. J. Pharm.*, **334**, 15 (2007).
17. X. Yuan, M. Jiang, H. Zhao, M. Wang, Y. Zhao and C. Wu, *Langmuir*, **17**, 6122 (2001).
18. S. C. Lee, K. J. Kim, Y. K. Jeong, J. H. Chang and J. Choi, *Macromolecules*, **38**, 9291 (2005).
19. S. C. Lee and H. J. Lee, *Langmuir*, **23**, 488 (2007).
20. M. Xiang, M. Jiang, C. Wu and Y. Zhang, *Macromolecules*, **30**, 5339 (1997).
21. S. Liu, L. Zhu, H. Zhao, M. Jiang and C. Wu, *Langmuir*, **16**, 3712 (2000).
22. W. P Gao, Y. Bai, E. Q. Chen, Z. C. Li, B. Y. Han, W. T. Yang and Q. F. Zhou, *Macromolecules*, **39**, 4894 (2006).
23. K. Karayanni and G. Staikos, *Eur. Polym. J.*, **36**, 2645 (2000).
24. I. M. Okhapkin, I. R. Nasimova, E. E. Makhaeva and A. R. Khokhlov, *Macromolecules*, **36**, 8130 (2003).
25. G.-G. Bumbu, C. Vasile, J. Eckelt and B. A. Wolf, *Macromol. Chem. Phys.*, **205**, 1869 (2004).
26. T. V. Burova, N. V. Grinberg, V. Y. Grinberg, E. V. Kalinina, V. I. Lozinsky, A. R. Khokhlov, V. O. Aseyev, S. Holappa and H. Tenhu, *Macromolecules*, **38**, 1292 (2005).
27. M. K. Chun, S. C. Cho and H. K. Choi, *J. Controlled Release*, **81**, 327 (2002).
28. M. K. Chun, S. C. Cho and H. K. Choi, *Int. J. Pharm.*, **288**, 295 (2005).
29. Y. Tian, P. Ravi, L. Bromberg, T. A. Hatton and K. C. Tam, *Langmuir*, **23**, 2638 (2007).
30. N. A. Peppas in *Hydrogels in Medicine and Pharmacy: Fundamentals* (CRC Press, Boca Raton, 1987).
31. W. E. Hennink and C. F. Van Nostrum, *Adv. Drug Delivery Rev.*, **54**, 13 (2002).
32. J. M. Rosiak and F. Yoshii, *Nucl. Instr. Meth. B*, **151**, 56 (1999).
33. P. Ulanski and J. M. Rosiak, in *Encyclopedia of Nanoscience and Nanotechnology*, Ed. S. Nalwa, (2004) p. 845.
34. S. Nayak and L. A. Lyon, *Angew. Chem. Int. Ed.*, **44**, 7686 (2005).
35. M. E. Byrne, K. Park and N. A. Peppas, *Adv. Drug Delivery. Rev.*, **54**, 149 (2002).
36. M. T. Garaya, C. Alava, and M. Rodriquez, *Polymer*, **41**, 5799 (2000).
37. C. Lau, and J. Mi, *Polymer*, **43**, 823 (2004).
38. F. Ikkai, T. Suzuki, T. Karino, and M. Shibayama, *Macromolecules*, **40**, 1140 (2007).
39. C. S. Brazel and N. A. Peppas, *Macromolecules*, **28**, 8016 (1995).
40. M. Shibayama, Y. Fujikawa and S. Nomura, *Macromolecules*, **29**, 6535 (1996).
41. P. Ravichandran, K. L. Shantha and K. Panduranga Rao, *Int. J. Pharm.*, **154**, 89 (1997).
42. A. M. Mathur, K. F. Hammonds, J. Klier and A. B. Scranton, *J. Controlled Release*, **54**, 177 (1998).
43. R. A. Stile, W. R. Burghardt and K. E. Healy, *Macromolecules*, **32**, 7370 (1999).
44. M. K. Yoo, Y. K. Sung, Y. M. Lee and C. S. Cho, *Polymer*, **41**, 5713 (2000).
45. E. Diez-Peña, I. Quijada-Garrido and J. M. Barrales-Rienda, *Macromolecules*, **35**, 8882 (2002).

46. E. Diez-Peña, I. Quijada-Garrido and J. M. BarralesRienda, *Polymer*, **43**, 4341 (2002).
47. D. M. Devine and C. L. Higginbotham, *Polymer*, **44**, 7851 (2003).
48. D. M. Devine and C. L. Higginbotham, *Eur. Polym. J.*, **41**, 1272 (2005).
49. D. M. Devine, S. M. Devery, J. G. Lyons, L. M. Geever, J. E. Kennedy and C. L. Higginbotham, *Int. J. Pharm.*, **326**, 50 (2006).
50. S. K. Bajpai and S. Dubey, *React. Funct. Polym.*, **62**, 93 (2005).
51. M. Heskins and J. E. Guillet, *J. Macromol. Sci., Chem. Ed.*, **A2**, 1441 (1968).
52. J. Zhang, L. Y. Chu, Y. K. Li and Y. M. Lee, *Polymer*, **48**, 1718 (2007).
53. M. Shibayama and T. Tanaka, *Adv. Polym. Sci.*, **109**, 1 (1993).
54. E. Kokufuta, B. Wang, R. Yoshida, A. R. Khokhlov and M. Hirata, *Macromolecules*, **31**, 6878 (1998).
55. E. Kokufuta, (2001) in *Physical Chemistry of Polyelectrolytes*, Ed. T. Radeva (Marcel Dekker Inc., New York, 2001) p. 591.
56. E. Diez-Peña, I. Quijada-Garrido, J. M. Barrales-Rienda, M. Wilhelm and H. W. Spiess, *Macromol. Chem. Phys.*, **203**, 491 (2002).
57. E. Diez-Peña, I. Quijada-Garrido, J. M. Barrales-Rienda, I. Schnell and H. W. Spiess, *Macromol. Chem. Phys.*, **205**, 438 (2004).
58. E. Diez-Peña, I. Quijada-Garrido, P. Frutos and J. M. Barrales-Rienda, *Macromolecules*, **35**, 2667 (2002)
59. E. Diez-Peña, I. Quijada-Garrido, P. Frutos and J. M. Barrales-Rienda, *Polym. Int.*, **52**, 956 (2003).
60. J. Klier, A. B. Scranton and N. A. Peppas, *Macromolecules*, **23**, 4944 (1990).
61. A. M. Lowman and N. A. Peppas, *Macromolecules*, **30**, 4959 (1997).
62. C. L. Bell and N. A. Peppas, *J. Controlled Release*, **39**, 201 (1996).
63. C. L. Bell and N. A. Peppas, *Biomaterials*, **17**, 1203 (1996).
64. N. A. Peppas and J. Klier, *Biomaterials*, **16**, 203 (1991).
65. B. Kim and N. A. Peppas, *Int. J. Pharm.*, **266**, 29 (2003).
66. B. Kim and N. A. Peppas, *Biomed. Microdevices*, **5**, 333 (2003).
67. K. Nakamura, R. J. Murray, J. I. Joseph, N. A. Peppas, M. Morishita and A. M. Lowman, *J. Controlled Release*, **95**, 589 (2004).
68. R. A. Scott and N. A. Peppas, *Macromolecules*, **32**, 6139 (1999).
69. R. A. Scott and N. A. Peppas, *Biomaterials*, **20**, 1371 (1999).
70. R. A. Scott and N. A. Peppas, *Macromolecules*, **32**, 6149 (1999).
71. S. Połowiński, *Prog. Polym. Sci.*, **27**, 537 (2002).
72. S. Nishi and T. Kotaka, *Macromolecules*, **18**, 1519 (1985).
73. T. Aoki, M. Kawashima, H. Katono, K. Sanui, N. Ogata, T. Okano and Y. Sakurai, *Macromolecules*, **27**, 947 (1994).
74. Y. M. Lee, S. H. Kim and C. S. Cho, *J. Appl. Polym. Sci.*, **62**, 301 (1996).
75. J. Byun, Y. M. Lee and C. S. Cho, *J. Appl. Polym. Sci.*, **61**, 697 (1996).
76. N. A. Peppas and S. L. Wright, *Eur. J. Pharm. Biopharm.*, **46**, 15 (1998).
77. J. Zhang and N. A. Peppas, *Macromolecules*, **33**, 102 (2000).
78. J. Zhang and N. A. Peppas, *J. Appl. Polym. Sci.*, **82**, 1077 (2001).
79. J. Zhang and N. A. Peppas, *J. Biomater. Sci., Polym. Ed.*, **13**, 511 (2002).
80. A. S. Hickey and N. A. Peppas, *J. Membr. Sci.*, **107**, 229 (1995).

81.  A. S. Hickey and N. A. Peppas, *Polymer*, **38**, 5931 (1997).
83.  R. Hernández, E. Pérez, C. Mijangos and D. López. *Polymer*, **46**, 7066 (2005).
84.  W. B. Stockton and M. F. Rubner, *Macromolecules*, **30**, 2717 (1997).
85.  S. A. Sukhishvili and S. Granick, *J. Am. Chem. Soc.*, **122**, 9550 (2000).
86.  S. A. Sukhishvili, *Curr. Opin. Colloid Interface Sci.*, **10**, 37 (2005).
87.  V. Kozlovskaya, E. Kharlampieva, M. L. Mansfield and S. A. Sukhishvili, *Chem. Mater.*, **18**, 328 (2006).
88.  H. Kaczmarek, A. Szalla and A. Kaminska, *Polymer*, **42**, 6057 (2001).
89.  S. Kadlubowski, A. Henke, P. Ulanski, J. M. Rosiak, L. Bromberg. and L. A. Hatton, *Polymer*, **48**, 4974 (2007).
90.  K. Kratz, T. Hellweg, T. Eimer, *Polymer*, **42**, 6631 (2001).
91.  W. Funke, O. Okay and B. Joos-Müller, *Adv. Polym. Sci.*, **136**, 139 (1998).
92.  B. R. Saunders and B. Vincent, *Adv. Colloid Interface Sci.*, **99**, 1 (1999).
93.  R. Pelton, *Adv. Colloid Interface Sci.*, **85**,. 1 (2000).
94.  K. Kratz, T. Hellweg and W. Eimer, *Colloid Surf. A*, **170**, 137 (2000).
95.  C. D. Jones and L. A. Lyon, *Macromolecules*, **33**, 8301 (2000).
96.  R. K. Drummond, J. Klier, J. A Alameda and N. A. Peppas, *Macromolecules*, **22**, 3816 (1989).
97.  C. Donini, D. N. Robinson, P. Colombo, F. Giordano and N. A. Peppas, *Int. J. Pharm.*, **245**, 83 (2002).
98.  D. N. Robinson and N. A. Peppas, *Macromolecules*, **35**, 3668 (2002).
99.  X. Lu, Z. Hu and J. Schwartz, *Macromolecules*, **35**, 9164 (2002).
100. X. Xia and Z. Hu, *Langmuir*, **20**, 2094 (2004).
101. H. Dou, M. Jiang, H. Peng, D. Chen and Y. Hong, *Angew. Chem. Int. Ed.*, **42**, 1516 (2003).
102. S. Yang, Y. Zhang, G. Yuan, X. Zhang and J. Xu, *Macromolecules*, **37**, 10059 (2004).
103. V. Kozlovskaya, S. Ok, A. Sousa, M. Libera and S. A. Sukhishvili, *Macromolecules*, **36**, 8590 (2003).
104. R. N. Elsner, V. Kozlovskaya, S. A. Sukhishvili and A. Fery, *Soft Matter*, **2**, 966 (2006).
105. V. Kozlovskaya and S. A. Sukhishvili, *Macromolecules*, **39**, 6191 (2006).
106. V. Kozlovskaya and S. A. Sukhishvili, *Macromolecules*, **39**, 5569 (2006).
107. J. M. Rosiak and P. Ulanski, *Radiat. Phys. Chem.*, **55**, 139 (1999).
108. A. Charlesby in *Atomic Radiation and Polymers* (Pergamon Press, Oxford, 1960).
109. C. von Sonntag in *Free-Radical-Induced DNA Damage and Its Repair: a Chemical Perspective* (Springer, Berlin, 2006).
110. I. Janik, E. Kasprzak, A. Al.-Zier. and J. M. Rosiak, *Nucl. Instrum. Meth. B*, **208**, 374, (2003).
111. J. M. Rosiak, P. Ulanski, L. A. Pajewski, F. Yoshii and K. Makuuchi, *Radiat. Phys. Chem.*, **46**, 161 (1995).
112. P. Ulanski and J. M. Rosiak, *Nucl. Instr. Meth. B*, **151**, 356 (1999).
113. S. Kadlubowski, J. Grobelny, W. Olejniczak, M. Cichomski and P. Ulanski, *Macromolecules*, **36**, 2484 (2003).
114. P. Ulanski, I. Janik and J. M. Rosiak. *Radiat. Phys. Chem.*, **52**, 289 (1998).

115. S. Kadlubowski, P. Ulanski and J. M. Rosiak, *Radiat. Phys. Chem.*, **63**, 533 (2002).
116. T. Schmidt, I. Janik, S. Kadlubowski, P. Ulanski, J. M. Rosiak, R. Reichelt and K.-F. Arndt, *Polymer*, **46**, 9908 (2005).
117. S. Kadłubowski, in *Radiation Synthesis of Polymeric Nanogels*, Ph.D. thesis, Technical University of Łódź, Poland (in Polish) (2004).
118. Z. S. Nurkeeva, V. V. Khutoryanskiy, G. A. Mun, A. B. Bitekenova, S. Kadlubowski, Y. A. Shilina, P. Ulanski and J. M. Rosiak, *Colloid. Surface. A*, **236**, 141 (2004).
119. A. Henke, S. Kadlubowski, P. Ulanski, J. M. Rosiak and K.-F. Arndt, *Nucl. Instr. Meth. B*, **236**, 391 (2005).
120. Z. S Nurkeeva, G. A. Mun, V. V. Khutoryanskiy and A. B. Dzhusupbekova, *Radiat. Phys. Chem.*, **69**, 205 (2004).
121. Z. S Nurkeeva, G. A. Mun, V. V. Khutoryanskiy, A. B. Bitekenova, A. B. Dzhusupbekova and K. Park, *J. Appl. Polym. Sci.*, **90**, 137 (2003).
122. Z. S Nurkeeva, G. A. Mun, A. V. Dubolazov and V. V. Khutoryanskiy, *Macromol. Biosci.*, **5**, 424 (2005).
123. A. V. Dubolazov, Z. S. Nurkeeva, G. A. Mun and V. V. Khutoryanskiy, *Biomacromolecules*, **7**, 1637 (2006).
124. R. A.Wach, H. Mitomo, N. Nagasawa and F. Yoshii, *Nucl. Instr. Meth.. B*, **211**, 533 (2003).
125. R. A.Wach, H. Mitomo, N. Nagasawa and F. Yoshii, *Radiat. Phys. Chem.*, **68**, 771 (2003).
126. R. A.Wach, H. Mitomo, F. Yoshii and T. Kume, *Macromol. Mater. Eng.*, **287**, 285 (2002).
127. A. Adamczyk in *Synthesis and Properties of Hydrogels Containing Glucose Oxidase*, Master Thesis, Technical University of Łódź, Poland (in Polish) (2006).
128. J. Olejniczak, J. M. Rosiak, A. Charlesby, *Radiat. Phys. Chem.*, **37**, 499 (1991).
129. P. Ulanski, E. Bothe, K. Hildenbrand, J. M. Rosiak and C. von Sonntag, *Radiat. Phys. Chem.*, **46**, 909 (1995).
130. M. Şen, O. Kantoğlu and O. Güven, *Polymer*, **40**, 913 (1999).
131. C. Ozürëk, T. Çaykara, O. Kantoğlu and O. Güven, *J. Polym. Sci. Part B: Polym. Phys.*, **38**, 3309 (2000).
132. A. El-Hag Ali, H. A. Shawky, H. A. Abd El Rehim and E. A. Hegazy, *Eur. Polym. J.*, **39**, 2337 (2003).
133. H. A. Abd El-Rehim, E. A. Hegazy, F. H. Khalil and N. A. Hamed, *Nucl. Instr. Meth. B*, **254**, 105 (2007).
134. V. V. Khutoryanskiy, G. A. Mun, Z. S. Nurkeeva, A. V. Dubolazov, *Polym. Int.*, **53**. 1382 (2004).
135. H. Ohno, K. Abe and E. Tsuchida, *Macromol. Chem.*, **179**, 755 (1978).
136. D. W. Koetsier, G. Challa and Y. Y. Tan, *Polymer*, **22**, 1709 (1981).
137. H-L. Chen and H. Morawetz, *Eur. Polym. J.*, **19**, 923 (1983).
138. I. Iliopoulos and R. Audebert, *Polym. Bull.*, **13**, 171 (1985).
139. S. L.Maunu, J. Kinnunen, K. Soljamo and F. Sudholm, *Polymer*, **34**, 1141 (1993).
140. A. Leiva, L. Gargallo and D. Radic, *Polym. Int.*, **34**, 393 (1994).
141. A. Usaitis, L. S. Maunu and H. Tenhu, *Eur. Polym. J.*, **33**, 219 (1997).

142. W. Argüelles-Monal, A. Pérez-Gramatges, C. Peniche-Covas, J. Desbrieres and M. Rinaudo, *Eur. Polym. J.*, **34**, 809 (1998).

143. V. V. Khutoryanskiy, Z. S. Nurkeeva, G. A. Mun, A. B. Bitekenova, A. V. Dubolazov and Z. Sh. Esirkegenova, *Eur. Phys. J. E*, **10**, 65 (2003).

144. V. A. Prevysh, B.-C. Wang and R. J. Spontak, *Colloid Polym. Sci.*, **274**, 532 (1996).

145. V. V. Khutoryanskiy, Z. S. Nurkeeva, G. A. Mun and A. V. Dubolazov, *J. Appl. Polym. Sci.*, **93**, 1946 (2004).

146. H. Ohno, H. Matsuda and E. Tsuchida, *Makromol. Chem.*, **182**, 2267 (1981).

147. A. Henke, M. Wolszczak, V. Boyko, T. Schmidt, S. Kadłubowski, P. Ulański, K.-F. Arndt and J. M. Rosiak, submitted.

148. J. Goodwin in *Colloids and Interfaces with Surfactants and Polymers* (West Sussex, Willey, 2004) p. 127.

149. D. J. Hemker and C. W. Frank, *Macromolecules*, **23**, 4404 (1990).

150. V. Aseyev, S. Hietala, A. Laukkanen, M. Nuopponen, O. Confortini, F. E. Du Prez and H. Tenhu, *Polymer*, **46**, 7118 (2005).

151. M. Zeghal and L. Auvray, *Europhys. Lett.*, **45**, 482 (1999).

152. M. Zeghal and L. Auvray, M. Jebbari and A. Gharbi, *Macromol. Symp.*, **222**, 149 (2005).

153. S. Tang, *Colloid Surf. A*, **157**, 185 (1999).

154. P. Ulanski and J. M. Rosiak, *J. Radioanal. Nucl. Chem.*, **186**, 315 (1994).

155. C. von Sonntag, E. Bothe, P. Ulanski and A. Adhikary, *Radiat. Phys. Chem.*, **55**, 599 (1999).

156. A. Henke, P. Ulański and J. M. Rosiak, unpublished results.

# CHAPTER 11

# PREVENTION OF SOIL MIGRATION AND CAPTURING OF RADIONUCLIDES BY INTERPOLYMER COMPLEXES

Sarkyt E. Kudaibergenov[1*], Larisa A. Bimendina[1], Marziya G. Yashkarova[2]

[1]*Institute of Polymer Materials and Technology, Panfilov Str. 52/104
050004, Almaty, Republic of Kazakhstan*
*[*]E-mail: ipmt-kau@usa.net*
[2]*Semipalatinsk State Shakarim University, Faculty of Chemistry and Chemical
Expertise, Glinka Str. 20A, Semipalatinsk, Republic of Kazakhstan*
*E-mail: us@semgu.kz*

Soil contamination or pollution due to human activities is a global problem. One of the major impacts is topsoil contamination and related health problems. Major concerns are the long time needed to regenerate contaminated soil and the considerable investment required for remedial measures. Contamination with radioactivity has mainly taken place as a result of nuclear weapons tests, improper radioactive waste disposal and the Chernobyl accident.[1] The latter has caused a wide scale contamination of Belarus, Ukraine and Russia rural and urban lands with radionuclides. Population is still receiving over-permissible doses of radiation. The analysis carried out has shown that the soil is a prime source of the population external and internal exposure.

In 1998, the failure of a large mine tailing dam at Aznalcóllar (Seville) released about 4 $Mm^3$ of trace element-contaminated sludge into the Guadiamar River.[2] The resulting flood inundated 55 $km^2$ of the basin southward towards the Doñana National Park.[3] The affected soils, mostly under agricultural production, were burdened with high concentrations of As, Cd, Cu, Pb, Tl and Zn.[4] After the accident, an emergency cleanup removed sludge and contaminated topsoil, which were transported and deposited in the nearby opencast mine. Despite this remediation, the underlying soils still contained elevated amounts of trace elements.[5] Organic matter and Ca-rich amendments were added with the aim of immobilizing trace elements and improving soil fertility.

It is not possible at present to make a more comprehensive assessment of progress in the management of contaminated land in the EU, because the available information is far from complete.[6] Consequently, all relevant "hot spots" may not have been identified. Soil problems are influenced by the diversity, distribution and specific vulnerability of soils across Europe. They also depend on geology, topography and climate and on the distribution of the driving forces. Better integration of soil protection into the sectoral policies and better harmonization of information across Europe are needed to move to more sustainable use of soil resources and promotion of sustainable models of its use. In particular, soil contamination from diffuse inputs and local sources can result in the damage of several soil functions and the contamination of surface water and groundwater.

Wind erosion is responsible for secondary suspension of contaminated soil particles in the air and further spreading of contamination. For instance, the dust particles with diameter of 0.05–0.1 mm are dropped within a couple of kilometers of the erosion site, while particles of about 0.005–0.01 mm diameter can travel hundreds and thousands of kilometers. It is well-known that Semipalatinsk Test Site (STS) has been contaminated by radionuclides during 40 years of atmospheric, aboveground and underground intensive nuclear tests.[7] Surveys of residual radioactivity in the soil at ten STS areas showed that a great number of the long-lived radionuclides Cesium-137 and Strontium-90 are concentrated in the depth of a soil layer 0–8 cm.[8,9] Yamamoto *et al.*[10] present evidence that the residual radioactivity within the STS is tightly bound to the soil as a result of extreme heating and melting of the soils during the tests. According to Ref. 7, the maximal amount of radionuclides is accumulated on the fine soil particles having 0.1–1.0 mm size. The sites contaminated by "Cesium spots" with surface density of $24 \cdot 10^3$–$240 \cdot 10^3$ Bq·m$^{-2}$ occupy approximately 65 km$^2$, 42 km$^2$ and 28 km$^2$, respectively.

The so-called "Koshkar-Ata Water Basin", located 3 km from Aktau city and 7 to 8 km from the Caspian Sea, contains the waste of uranium mining industry.[11] The soil in coastal zone is mainly polluted by the following elements, most of which being highly toxic heavy metals: Cd, Hg, Cu, Ni, Zn, Co, Mn, Pb, Fe, Cr, and Mo. Due to the permanent evaporation of water, the coastal zone of "Koshkar-Ata Water Basin" dries up and increases every year. Today, the dried area of "Koshkar-Ata Water Basin" is more than 60 km². Along the former shoreline, salt and dust have accumulated due to evaporation. As a result of strong winds, salt and small dispersed dust containing harmful substances are being picked up and transported and then deposited over thousands of square kilometers of cultivated land. In summer time, the thin dust with average diameter 0.1–0.5 mm constantly migrates from the dried part of "Koshkar-Ata Water Basin" to Aktau city, as a result of wind erosion (16–17 m·sec⁻¹).[11] Therefore, the application of innovative materials and technologies that prevent the aerial migration of contaminated soils is important to protect the population, flora and fauna from harmful effects.

To our knowledge, the technological aspects of application of interpolymer complexes for remediating topsoil contaminated by radionuclides and polluted by heavy metal ions have not been developed yet. A method for soil decontamination from radionuclides was proposed, consisting in removal of the upper soil layer, followed by its decontamination.[12,13] This method includes extraction of a fine-dispersed soil fraction and treatment with water or a chemical reagent solution. Earlier in the frame of International Science and Technological Center (ISTC) Program, the project "Development of electrokinetic and chemical methods for rehabilitation of soil and ground water contaminated by radionuclides and heavy metals" was carried out.[14] The goal of this basic research was to develop an understanding of the soil-contamination kinetics, and ultimately to develop the so-called complexants, that will solubilize the "target" contaminant metals. The cleanup of radionuclides from soil and groundwater by electrokinetic remediation technology was suggested.

The fate of trace metals in soils is one of the main research areas of the Soil Protection Group at the Institute of Terrestrial Ecology of the Swiss Federal Institute of Technology, ETH, Switzerland.[15] Work has been done spreading from a regional scale and the application of geostatistics to metal distribution and transfer in soil profiles, to the molecular scale, e.g., the impact of metals on soil microbial activity. In order to reduce hazards from metal polluted soils, different remediation approaches have been studied such as metal immobilization by natural and modified minerals, phytoremediation and soil washing using chelating agents.

This review considers the application aspects of interpolymer complexes for aggregation and wind erosion of topsoils contaminated with radionuclides and/or heavy metals.

## 1. Using of polymers and composites for soil aggregation

One of the promising methods of soil structuring is its treatment with polymer solutions.[16-24] Polyacrylamide (PAAm), polyacrylic acid (PAA) and hydrolyzed polyacrylonitrile (HPAN) have been widely used for these purposes. In agrotechnology, partially hydrolyzed polyacrylamides have also been used. Commercially available composites, that exhibit a structuring effect at low doses and for a long time, are ammonium and calcium lignosulphonates. Natural polymers such as cellulose, brown coal, peat and their derivatives exhibit structuring action with respect to soil particles.

The interaction of HPAN, PAAm and polymer composites with soil particles has a complex character. It proceeds through polymer adsorption on the soil surface with formation of mono- and polymolecular layers.[16] The adsorption proceeds through the formation of specific bonds, for example, hydrogen bonds. Besides, the polymers are able to coagulate under the influence of the cations of the soil solutions. In this case, van der Waals interactions can take place in the aggregation process. It was determined by electron microscopy[16] that at low concentrations of polymers bridges between soil particles are formed, clamping soil particles into bigger aggregates. The formation of such bridges is more effective at optimal polymer doses, corresponding to their surplus, that does not react with the surface of the soil particles, and fills the free pores of the soil. It was found[16] that the structuring action of PAAm and HPAN is preserved during 3 years. The effectiveness of the oily acrylic acid, as structuring agent, becomes stronger with the increase in the ionization degree.[23] Polymers, containing different functional groups, are the most efficient structuring agents. The presence of free salts in the soil, particularly of $Ca^{2+}$ ions, decreases the soil structuring efficiency of the carboxyl- and carboxylate containing polymers, probably due to formation of insoluble calcium salts.

The structuring properties of the polymers considerably depend on the soil pH.[25] The non-hydrolyzed polyacrylamide is effective at pH 2.8–9.0, while the efficiency of the hydrolyzed polyacrylamide is limited at pH between 4.5–9.0. At pH < 4.0, it is necessary to combine the application of the hydrolyzed PAAm with the liming of the soil.

For structuring sand soils, the combination of mineral and organic surfactants has been used.[26] The role of surfactants for developing artificial structure in a soil-lime-ashes-water mixture is their ability to interact with both anions and cations with formation of new products.

The copolymers of maleic acid/acrylamide, vinylpyrrolidone/ acrylamide, polyoxyethylene and polyvinylalcohol have also been used as soil structuring agents.[27-29] The main conclusion was that the interaction of these polymers with the solid soil phase proceeds *via* both adsorption and adhesion.[30,31]

## 2. Interpolymer complexes as soil structuring agents

There is limited information on using interpolymer complexes for aggregation of soil particles.[16] Interpolymer complexes (IPC) are products of specific cooperative interactions of chemically and structurally complementary macromolecules.[32-35] There are several types of interpolymer complexes, depending on the nature of the reacting complementary macromolecules and interactions causing the complex formation.[36,37] They are the donor-acceptor IPC (Inter-PC), stabilized by a cooperative system of hydrogen bonds; interpolyelectrolyte complexes (IPEC), which are the products of interaction between oppositely charged macromolecular chains; intrapolymer complexes (Intra-PC), which are formed in macromolecular chains of graft copolymers, consisting of complementary main and grafted side chains. Schemes 1-3 represent the different types of IPC.

Appropriate and commercially available polymers to produce IPEC are the poly(acrylic acid) (PAA), the poly(methacrylic acid) (PMAA), the sodium salt of poly(styrenesulfonate) (PSSNa), the poly(ethyleneimine) (PEI), the sodium salt of carboxymethylcellulose (CMCNa), the poly(allylamine) derivatives, particularly the poly(*N,N*-dimethyldiallylammonium chloride) (PDMDAAC), etc. Proton-donating polymers such as poly(acrylic acid), poly(methacrylic acid), and proton-accepting polymers such as poly(N-vinylpyrrolidone) (PVP), poly(ethylene glycol) (PEG), poly(vinyl alcohol) (PVA), poly(acrylamide) (PAAm) are widely used to form inter-PC.

In the series of the developed Intra-PC, there are the graft copolymers of polyacrylamide to poly(vinyl alcohol) (PVA-*g*-PAAm), with different number and molecular weight (length) of grafts, and block-copolymers based on PAAm and poly(ethylene oxide) (PEO-*b*-PAAm). The intrapolymer complexes, in contrast to the majority of the interpolyelectrolyte and interpolymer complexes, are not destroyed into the individual components, during the changes in pH and concentration of the low-molecular-weight electrolytes in solution.[38] These heterogeneous macromolecular systems belong to binders of a new generation.

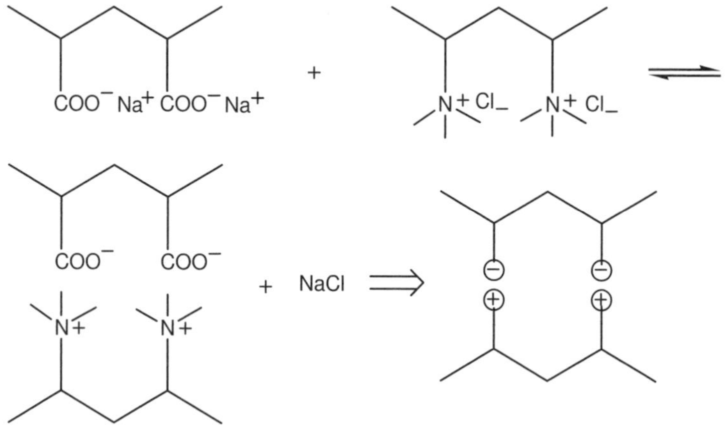

**Scheme 1**. Formation of interpolyelectrolyte complexes (IPEC) stabilized by ionic contacts.

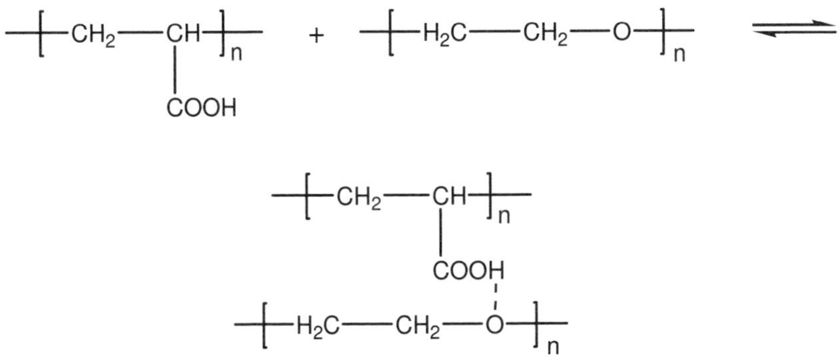

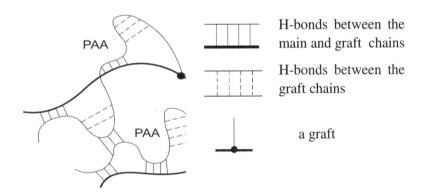

**Scheme 2.** Formation of interpolymer complexes (Inter-PC) between PAA and PEO stabilized by hydrogen bonds.

**Scheme 3.** Formation hydrogen-bonded intrapolymer complexes (Intra-PC) based on nonionic graft copolymers of polyacrylamide and poly(vinyl alcohol) (PVA-*g*-PAA).

Michaels and co-workers,[32] studying the complexes of strong polyelectrolytes, proposed that the IPEC structure is of an ideal "zip-up" type. However, later the investigators[39-42] shown that IPEC structure contains simultaneously the bounded parts (hydrophobic fragments) and defects or loops (hydrophilic fragments).

Earlier, IPECs were successfully applied for sedimentation and cementation of radioactive dust during the Chernobyl disaster, to prevent spreading of the contaminated topsoil.[43] Later, these polycomplexes were found to be effective dust and wind erosion suppressing agents, contributing to vegetation.[44-51] The high efficiency of IPEC soil binders is directly attributed to their specific structure and dynamics. They are amphiphilic macromolecular compounds containing both relatively hydrophobic and hydrophilic positively and negatively charged sites. These dissimilar sites are able to spontaneously and rapidly exchange their location within water swollen IPECs without their dissociation into the constituent polyions. The specific features of the IPEC structure provide a unique opportunity for IPECs to interact with surfaces of different nature and hence with different colloid and dispersed particles. These compounds are very effective and universal flocculants and binders for a variety of dispersions, in particular for soils and grounds, to prevent them from water and wind erosions and to localize contaminations, in particular to prevent spreading of the radionuclide contamination. The technological procedure of the preparation and utilization of IPEC binders is quite simple. It includes (i) preparation of dilute aqueous polyelectrolyte mixtures at high concentrations of low-molecular-weight salts (for instance, mineral fertilizers), when ionic interactions between oppositely charged polyions are completely suppressed, (ii) introduction of these mixtures into disperse systems, usually by spraying polyelectrolyte mixtures on soil or ground surfaces, and (iii) washing of the disperse systems with water to remove salts; the latter is accomplished through the rain or another natural moisture and favors the formation of IPECs because of a decrease in salt concentrations. The effective binders based on IPECs are environment-tally friendly polymeric systems. The soil-polymer crust, formed as a result of the soil treatments with IPECs, is favorable for seed sprouting and plant growth.

Inorganic materials in the topsoil have slightly negatively charged particles (**Fig. 1**). These charges result from the pH-dependent dissociation of the silanol (Si-OH) groups on the mineral surfaces.

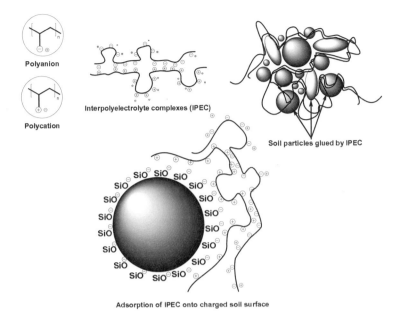

**Fig. 1.** Mechanism of protective coating formation.

The positively charged "loops" and/or "dangles" of IPECs interact with the negatively charged silanol groups on the mineral surfaces through ionic bonding. The IPEC treatment produces a soil-polymer protective coating 3-6 mm thick which tightly attaches fine grains to each other by aggregation into small particles, and by coagulation of soil colloids which are firmly bonded to each other by the polymer chains (**Fig. 2**). Some IPECs, based on polyacrylic acid/poly-2- or 4-vinylpyridine, polyacrylic acid/poly-N,N-dimethyldiallylammonium chloride, polyacrylic acid/polyethyleneimine were used for soil structuring and radionuclide sorption.[52-54]

Granulometric analysis of soils treated by IPEC solutions reveals that a considerable increase in large aggregated soil particles takes place.[50,51] The adsorption of [144]cerium on 0.25–0.5 mm size particles was investigated. The results showed that the distribution coefficient of [144]cerium between liquid and solid phases in the soil-soil solution system hardly depends on the presence of structuring agents in solution. The laboratory experiments show that the application of IPEC produces a polymeric coating resistant to strain-stress from water or wind (**Fig. 3**).

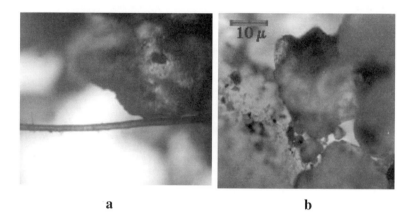

a                                        b

**Fig. 2.** (a) Thin elongate (above 20 mm) bridges, when the particles of different size are separated by a considerable distance (from 1 to 2 mm). (b) Attachment of small-sized particles (5-20 mm) to larger ones (over 700 mm). It occurs during evaporation of the polymeric solution covering the particles. The fine particles are picked up by the solution from interspaces and attached/attracted by large particles. Reprinted from IAEA-TECDOC-1403, S.V. Mikheykin, The long-term stabilization of uranium mill tailings, Pages 265-279, Copyright (2004) with permission from IAEA.

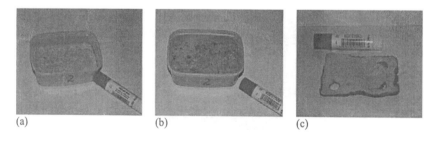

(a)                          (b)                          (c)

**Fig. 3.** Photos of IPEC-treated soil samples. Reprinted from IAEA-TECDOC-1403, S.V. Mikheykin, The long-term stabilization of uranium mill tailings, Pages 265-279, Copyright (2004) with permission from IAEA.

Lignin-based interpolymer complexes were developed by Shulga *et al.*[55,56] for application as soil conditioners. In particular, a new silicon-containing adhesive (Si-Ad) was developed to withstand the wind erosion.[57] The Si-Ad consists of interpolymer complexes of lignosulphonate and heterosiloxanolate oligomer, stabilized by salt a nd hydrogen bonds. Si-Ad is able to cement sand particles and form a layer with a thickness of 4–14 mm and a penetration resistance of 0.49–2.90 MPa. The recommendations proposed for the application of Si-Ad are summarized in **Table 1**.

**Table 1**. Recommendations for the application of Si-Ad polycomplexes. Reprinted from *J. Agric. Engng Res.*, Vol. 62, G. Telysheva, G. Shulga, Silicon containing polycomplexes for protection against wind erosion of sandy soil, Pages 221–227, Copyright (1995) with permission from Silsoe Research Institute.

| Purpose of application | Si content in structure, % | Application, $L \cdot m^{-2}$ | Concentration of solution, $g \cdot L^{-1}$ |
|---|---|---|---|
| Against wind erosion | 0.5–0.8 | 3–4 | 100–200 |
| Mulching of soil | 0.8–1.0 | 2–3 | 300–400 |

New amino-containing soil conditioners based on lignosulphonate (LS), a by-product of the papermaking industry, have been developed by the authors[58] to create an artificial soil macrostructure and to reinforce surface layers, withstanding wind and water erosion on light sandy soil. Commercial products, such as a polymer amine (PA) having molecular mass of 50,000, and its oligomeric analogue (OA) with a molecular mass of 200–300, were used for the formation of polycomplexes with LS. The mass ratio of LS to PA (OA) in the IPEC compositions was close to stoichiometric. The consecutive spraying of solutions of LS, and an amine-containing component with pre-determined concentrations, results in the formation of IPEC in the surface upper layers, which acts as glue by cementing the small particles of sandy soil and forms a reinforced surface layer in the air-dry state. It was established that the surface layers, reinforced by a double IPEC with the participation of an oligomer amine, are characterized by poorer mechanical properties and stability against erosion than the layers formed by an IPC containing a polymer amine (**Table 2**). It is clear that the high molecular mass of PA favors the increase in the number of the salt bonds and the topological entanglement with the oppositely charged molecules of LS in the IPEC chemical structure. This shifts the hydrophilic-hydrophobic balance of

IPECs towards hydrophobicity and causes the strengthening in the cohesion properties of the polymer matrix and an increase in its adhesion towards the soil particles.

**Table 2**. Thickness of soil surface layer stabilized with IPECs. Reprinted from *J. Agric. Engng. Res.*, Vol. 78(3), G. Shulga, F. Rekner, J. Varslavan, Lignin-based interpolymer complexes as a novel adhesive for protection against erosion of sandy soil, Pages 309–316, Copyright (1995) with permission from Silsoe Research Institute.

| Concentration of IPEC components, $g \cdot L^{-1}$ | Layer thickness, mm | | |
|---|---|---|---|
| | Total dosage of solutions applied, $g \cdot L^{-1}$ | | |
| | 1.5 | 2.0 | 3.0 |
| 50 LS + 10 PA | 1.5±0.2 | 2.2±0.1 | 3.5±0.4 |
| 100 LS + 15 PA | 2.9±0.3 | 5.3±0.4 | 6.1±0.2 |
| 200LS + 20 PA | 5.7±0.4 | 8.4±0.2 | 10.2±0.2 |
| 50 LS + 10 OA | 1.2±0.3 | 1.8±0.2 | 3.3±0.3 |
| 100 LS + 15 OA | 2.9±0.2 | 4.1±0.2 | 5.8±0.3 |
| 200 LS + 20 OA | 5.2±0.4 | 7.8±0.3 | 8.9±0.5 |

The resistance of IPECs to UV irradiation was also studied. The results of climatic ageing show that, after 15 and 150 hours of UV irradiation, practically all surface layers preserve both their exterior and mechanical properties. However, after a 400-hour exposure, the penetration resistance of the LS layers decreases by 20% on average. Local coloration and hair cracks appear on their surface. The binary IPEC layers do not change their exterior after a 400-hour exposure, and the change in their penetration resistance does not exceed 10% from the initial value.

The authors[59-64] developed new macromolecular systems based on graft and block copolymers, so called intrapolymer complexes, which are effective binders with respect to: (i) mono- and polydispersed colloid (silica) particles of a size from some nm to some $\mu$m; (ii) many organic substances of natural (humic acids) and anthropogenic (phenols) origin and iii) ions of the stable and radioactive metal isotopes. Intra-PCs based on polyacrylamide grafted to polyvinylalcohol (PVA-*g*-PAAm) or block-copolymers based on polyethylene oxide and polyacrylamide (PEO-*b*-PAAm) are efficient binders with respect to colloid particles of different size. In particular, they can strongly be absorbed on the large silica particles (r = 70–110 $\mu$m) and form a very thin adsorption layer ($\Omega$ = 100–200 nm) that strengthens the colloid particles aggregation. They can easily bind the smallest silica sol particles (r = 10–30 nm), as well as humic and fulvic acids, with formation of stable polymer-colloid complexes. The macromolecular system based on Intra-PC and called "Unicomfloc"[60] is able to extract up to 63% [137]Cs and 29% [90]Sr from radioactive polluted water (30–150 Bk·dm[-3]).[62]

The Institute of Polymer Materials and Technology (Almaty, Kazakhstan), the Center of Chemical Ecological Problems at Semipalatinsk Shakarim State University (Semipalatinsk city, Kazakhstan) and the Institute of Radiation Safety and Ecology (Kurchatov city, Semipalatinsk region) carried out model and real field experiments on application of polycomplexes for structuring STS soil particles, in order to prevent wind and water erosion of radionuclides.[65-74] Both water soluble polymers – polyacrylic acid (PAA), polyethylene glycol (PEG), poly-N,N-dimethyldiallylammonium chloride (PDMDAAC) – and interpolymer complexes (Inter-PC) based on PAA-PEG and IPEC based on PAA-PDMDAAC were used as soil structuring agents. **Table 3** summarizes the Inter-PC applied for topsoil aggregation and extraction of radioactive [90]Sr.

**Table 3**. Inter-PC and IPEC used for soil structuring

| Interpolymer complex | Composition | Type of complex | [$\eta$], dL·g$^{-1}$ |
|---|---|---|---|
| PAA – PEG | 1:1 | Inter-PC | 0.10 |
| PAA - PVP | 1:1 | Inter-PC | 0.10 |
| PAA - PEI | 1:1 | IPEC | 0.10 |
| PAA-PDMDAACl | 1:1 | IPEC | 0.08 |

The treatment of soil particles was carried out by two methods: the first method (I) is the uniform spraying of a polymer solution; the second one (II) is the pouring of a polymer solution on the soil surface. The results obtained are summarized as follows:

- The optimal concentrations of polymer solutions are arranged between $10^{-4}$–$10^{-2}$ mol·L$^{-1}$. The best results correspond to the concentration of $10^{-2}$ mol·L$^{-1}$.

- Pouring on soil surface (method II) is most preferable in comparison with spraying of polymer solution (method I).

- Both individual polymers (PAA, PEG, PDMDAAC) and PAA-PEG, PAA-PDMDAAC interpolymer complexes exhibit the structuring effect. However, the best results were observed for the Inter-PC solution. The treatment of soil with PAA-PEG solution with a concentration of $10^{-2}$ mol·L$^{-1}$ increases the amount of particles with a diameter > 0.25 mm from 38.13 (control) to 89.62% by method I and up to 91.60% by method II. Fractions with 5 mm diameter are considerably increased. For low concentrations of PAA-PEG solutions, the aggregating effect is lower.

- The treatment of soil with Inter-PC solution is the most efficient. The successive treatment of the soil first with PAA solution, then with PEG solution and vice versa gives the worst results.

- Deviation from the equimolar composition of Inter-PC, for instance, for [PAA]:[PEG] = 1:2 and [PAA]:[PEG] = 2:1 mol/mol leads to the worst results.

- The results of soil treatment with Inter-PC solutions having different pH (8.0, 4.5, and 3.0) show that acidification of the Inter-PC solutions causes a strengthening of the structuring effect, due to suppressing the dissociation of PAA and formation of more stable Inter-PC.

- For the PAA-PDMDAAC system the best results were obtained for nonstoichiometric [PAA]:[PDMDAAC] = 3:1 complexes.

Thus, the best structuring effect has been obtained for the equimolar [PAA]:[PEG] = 1:1 complex solution, with a concentration $10^{-2}$ mol·L$^{-1}$ and pH = 3.0, when the Inter-PC solution is poured on the soil surface. The capillary moisture capacity of forming aggregates is given in **Table 4**. The moisture capacity of the soil increases with increase in polymer concentration. The moisture capacity of the untreated soil is 20.65%, while the same parameter of soil treated with 0.1% complex solution is equal to 47.13%. It means that the structured soil particles effectively preserve the moisture and prevent its evaporation.

**Table 4.** Capillary moisture capacity of treated and untreated by InterPC soils. Reprinted from *J. Appl. Polym. Sci.*, Vol. 87, L.K. Orazzhanova, M.G. Yashkarova, L.A. Bimendina, S.E. Kudaibergenov, Binary and ternary polymer – strontium complexes and capture of radioactive strontium-90 from the polluted soil of Semipalatinsk Nuclear Test Site, Pages 759-764, Copyright (2003), with permission from John Wiley.

| Variant | Capillary moisture capacity |
|---|---|
| Untreated soil | 20.65 |
| Soil + 0.05% Inter-PC | 28.42 |
| Soil + 0.1% Inter-PC | 47.13 |

At concentration of [Inter-PC] = 0.05%, the velocity of moisture movement for treated soils increases in comparison with an untreated one. The further increase in Inter-PC concentration up to 0.1% leads to increase in the capillary rising of water, then a delay is observed. The delay of the capillary water movement is probably attributed to break of the continuity of the capillaries on the particles surface with formation of inter-aggregates and large voids that do not exhibit capillary properties. The soil treated with a 0.1% complex solution, shows the highest water stability, and the quantity of water stable aggregates having a diameter > 0.25 mm is 46.1% in comparison with the control experiments (5.8%). The soil structuring efficiency of the equimolar Inter-PCs, and water-soluble polymers at the same concentration, increases in the following order: PAA:PEG = 1:1 (pH 3.0) > PAA:PEG = 1:1 (pH 4.5) > PAA:PEG = 1:1 (pH 6.5) > PAA:PEG = 1:1 (pH 8.0) > PAA >PEG.

## 3. Ternary polymer-metal complexes and capturing of $^{90}$Sr from radioactively polluted topsoils

It is well known[75-79] that individual cationic, anionic and nonionic polymers, as well as interpolymer complexes, can form binary and ternary polymer-metal ion complexes. In particular, IPECs can be used not only to prevent the transfer of disperse particles, but they are also able to accumulate heavy metal ions and radionuclides with formation of ternary polymer-metal complexes.[80,81] For example, mixing an IPEC formed by poly(acrylate anion) (PAA) and poly(ethyleneimine cation) (PEI) with copper, cobalt, nickel or iron salts, produces a ternary polymer-metal complex, as schematically shown below (**Scheme 4**):

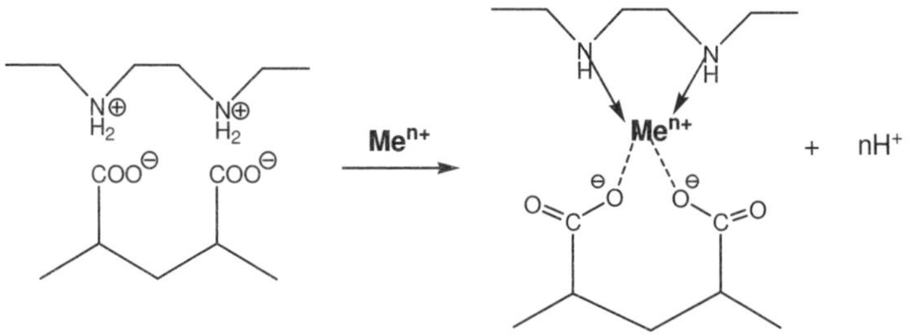

**Scheme 4.** Schematically representation of formation of ternary polymer-metal complexes.

The ternary polymer-metal complexes are stabilized by cooperative salt and coordination bonds between metal ions and both the polymer components. The quantitative extraction of transition metal ions in a wide range of their concentrations in a solution as low as $10^{-5}$ mol·L$^{-1}$ can be performed with IPECs.

The introduction of polyvalent metal salts, such as FeCl$_3$ and Cu(NO$_3$)$_2$, into the IPEC results in the formation of ternary polymer-metal complexes, whose structuring and stabilizing effect on sandy soil is more pronounced than for binary IPECs.[58] This reveals itself in a decrease in the soil losses by 20–25% at a wind speed of 25 m·s$^{-1}$, a drop in the soil losses of 30–35% and a reduction in the water surface runoff by 40–45% under the simulation of water erosion conditions. The application of 100–200 g·L$^{-1}$ solutions of LS, containing 2.5–5.0 g·L$^{-1}$ polyvalent metal salts, and 15-20 g·L$^{-1}$ solutions of PA or OA may be supposed to be effective for temporary protection of the eroded soil, while the seeds germination and plants growth, as well as the establishment of the roots, proceed.

The exterior and mechanical properties of the surface layers formed with a ternary IPEC have not changed, even after a 400-hour exposure to UV irradiation. The high photoresistance of ternary IPEC layers can be explained by the blocking of the main centers of photochemical transformations of lignin, i.e. of the phenolic hydroxyl groups with polyvalent metal ions.

The complex formation of PAA and PEG with Sr$^{2+}$-ions, as well as the formation of ternary polymer-Sr$^{2+}$-polymer complexes, has been investigated.[82,83] The titration curves of PAA, in the presence of

$Sr^{2+}$, are shifted to the acidic region, that is related to replacement of PAA protons by strontium ions, due to formation of binary polymer-metal complexes. The average coordination numbers of the metal ions and the stability constants of the polymer-metal complexes, calculated for the molar ratios $[PAA]:[Sr^{2+}] = 1:1$ and 6:1 are equal to 1 and $K = 1.25 \cdot 10^{-4}$ and $K = 5.62 \cdot 10^{-5}$, respectively. The influence of the nature of mixed solvents (water-methanol, water-ethanol, water-DMF), as well as of the temperature on the stability of the $[PAA]:[Sr^{2+}] = 1:1$ complexes, has been investigated. The common feature of the polymer-metal complexes is the gradual unfolding of the complex particles structure with increase in organic solvent content in the mixed solvents and with rising temperature. It was found that titration of $[PAA]:[PEG] = 1:1$ complexes by $Sr^{2+}$ ions and $[PAA]:[Sr^{2+}] = 1:1$ complexes by PEG leads to monotonous changes in conductivity and pH values. At the same time, a considerable deviation from the additive properties is observed when the $PEG-Sr^{2+}$ pair is titrated by a PAA solution, that is attributed to the formation of a ternary polymer-metal complex with composition of $n = [PEG-Sr^{2+}]/[PAA]=2:1$. Since PAA, PEG and PAA-PEG pairs form polymer-metal complexes with $Sr^{2+}$ ions, the capture of soil radionuclides by means of binding into the ternary polymer-metal complexes was patented.[84] The quantitative analysis of strontium-90 in treated by polymer solutions and an untreated soil was carried out by AAS. As is seen from Table 5, the $Sr^{90}$ content increases

**Table 5.** Content of strontium-90 in soils treated with individual polymer solutions and Inter-PC. Reprinted from *J. Appl. Polym. Sci.*, Vol. 87, L.K. Orazzhanova, M.G. Yashkarova, L.A. Bimendina, S.E. Kudaibergenov, Binary and ternary polymer – strontium complexes and capture of radioactive strontium-90 from the polluted soil of Semipalatinsk Nuclear Test Site, Pages 759–764, Copyright (2003), with permission from John Wiley.

| Variant | $C_{Polymer}$, $mol \cdot L^{-1}$ | $C_{Sr}$, $\mu g \cdot L^{-1}$ |
|---|---|---|
| Soil (control) | — | 2.31 |
| PAA | $10^{-2}$ | 4.13 |
| | $10^{-4}$ | 4.45 |
| PEG | $10^{-2}$ | 4.49 |
| | $10^{-4}$ | 3.69 |
| PAA-PEG=1:1 | $10^{-2}$ | 9.34 |

twice in soils treated with PAA and PEG in comparison with the untreated soil, due to formation of PAA-$Sr^{2+}$ and PEG-$Sr^{2+}$ complexes. However, the content of $Sr^{90}$ is much higher in soils treated with the [PAA]:[PEG] = 1:1 complex.

The strontium content increases up to 9.34 $\mu g \cdot L^{-1}$, in comparison with the control experiments (2.31 $\mu g \cdot L^{-1}$), that may be related to formation of ternary PAA-$Sr^{2+}$-PEG polymer-metal complexes. These results clearly show the potential applicability of polycomplexes as effective soil structuring composites and also as radioactive elements capturing agents.

## 4. Conclusion

The aggregation of soil particles with water-soluble polymers, polymer composites and various types of interpolymer complexes was considered. The main attention was paid to the prevention of wind migration of thin dust and capturing of radionuclides from the radioactively contaminated topsoils. It was found that the interpolymer complexes significantly suppress the wind erosion of topsoil and accumulate radionuclides *via* formation of ternary polymer-metal complexes. Moreover, interpolymer complexes are resistant to UV irradiation and favorable for seed sprouting and plant growth.

## References

1. UNEP *National report on the state of the environment in the Russian Federation*, (1998) http://ceeri.ecoinfo.ru/state_report_98/eng/introduction.htm
2. A. Garralón, P. Gómez, M.J. Turrero, M. Sánchez, A.M. Melón, *The Science of the Total Environment*, 242, 27 (1999).
3. J.O. Grimalt, M. Ferrer, E. Macpherson, *The Science of the Total Environment*, 242, 3 (1999).
4. F. Cabrera, L. Clemente, E. Díaz Barrientos, R. López, J.M. Murillo, *The Science of the Total Environment*, 242, 117 (1999).
5. F. Moreno, F. Cabrera, J.E. Fernández, I. Girón, *Boletín Geológico y Minero*, 112, 178 (2001).
6. N.B. Denisov, R.A. Mnatsakanian, A.V. Semichaevsky, *Environmental reporting in Central and Eastern Europe: a review of selected publications and frameworks*, UNEP and Central European University (1997).
7. S. Balmukhanov, G. Raissova, T. Balmukhanov, *Three generations of Semipalatinsk affected to radiation*, Almaty: Sakshy, 2002.

8.  F. Steinhauzler, M. Gasterberg, A.K. Humber, M. Spano, R. Ranaldi, I. Stronati, A. *Testa, Assessment of radioactive doze due to nuclear tests for residents in areas adjoined to the Semipalatinsk Test Site*, "Polygon", 115 p. (2001).

9.  T.M. Carlsen, L.E. Peterson, B.A. Ulsh, C.A. Werner, K.L. Purvis, A.C.Sharber, *US DOE Report*, June 2001 (UCRL-JC-143920) (2001).

10. M. Yamamoto, T. Tsukatani, T. Katayama, *Health Physics*, 71, 142 (1996).

11. K.K. Kadyrzhanov, K. Kuterbekov, in *Radiation safety problems in the Caspian Region*, NATO Science Series, 41, 69 (2005).

12. Yu.P. Davydov, I.G. Toropov, I.V. Rolevich, V.V. Toropova, T.V. Vasilevskaya, *Official bulletin of the Republic of Belarus*, 2(13), 141 (1997).

13. S.V. Mikheykin, P.P. Poluektov, A.S. Chebotarev, V.P. Simonov, E.A. Lagusin, A.B. Zezin, A.I. Stepanenko, P. Kalb, J. Heiser, *Int. Conf. Waste Management '05*, Tucson, AZ, USA (2005).

14. Yu. B. Kondratenkov, D. D. Pobedimskaya, L. V. Konstantinov, *ISTC Project #0016* (2001).

15. M. Berli, M. J. Kirby, S. M. Springman, R. Schulin, *Soil & Tillage Research* 73, 57 (2003).

16. L.A. Bimendina, M.G. Yashkarova, S.E. Kudaibergenov, E.A. Bekturov, *Polymer complexes: preparation, properties and application*, Semipalatinsk: Semipalatinsk State University, 2003.

17. N.A. Kachinskii, A.I. Mosolova, L.Kh. Toimurazova, *Using of polymers for structuring and melioration of soils*, Moscow: Pochvovedenie, 1967.

18. A.B. Gussak, *Influence of humic acids and polymer agents on physical properties of soil*, Tashkent: Fan, 1961.

19. K.S. Akhmetov, E.A. Aripov, *Interaction of water-soluble polymers with disperse systems*, Tashkent: Fan, 1969.

20. P.V. Vershinin, *Soil structure and conditions of its formation*, Moscow: Pochvovedenie, 1961.

21. V.P. Batyuk, *Application of polymers and surfactants in soils*, Moscow: Pochvovedenie, 1978.

22. L.B. Smolina, V.L. Volkova, *Influence of artificial structuring agents of humine acids on aggregation of particles*, Tashkent: Fan, 1961.

23. A.I. Mosolova, *Influence of polymers on the structure of soils and crop capacity of cultures*, *Pochvovedenie*, 9, 54 (1970).

24. S.S.Khamraev, Kh.Yu. Artykbaeva, S. Tuimetova, *Pochvovedenie*, 9, 129 (1982).

25. I.B. Revut, I.A. Romanov, *Pochvovedenie*, 1, 60 (1966).

26. Sh.A. Kuldasheva, A.A. Agkhodzhaev, *Izv.Chem J.*, 5, 73 (2000).

27. A.I. Sharipova, A.A. Asanov, T.S. Sirlibaev, *Izv. Chim .J.*, 1, 24 (1999).

28. S.M. Epstein, V.Ph. Utkaeva, A.I. Mosolova, *Pochvovedenie*, 1, 44 (1987).

29. S.M. Epstein, O.A. Agafonov, *Agrochemistry*, 6, 106 (1977).

30. S.M. Epstein, I.B. Revut, *Abstr. X Intern. Conf.*, Moscow: Nauka, 1, 181 (1974).

31. S.M. Epstein, Physical-chemical investigation of interaction of synthetic structuring agents with solid phase of model soil aggregates, *PhD Thesis*, Moscow, 1981.

32. H.J. Bixler, A.S. Michaels, *Encyclopedia of Polymer Science and Technology*, 10, 765 (1969).

33. A.B. Zezin, V.A. Kabanov, *Uspekhi Khimii (Rus. Rev. Chem.)*, 56, 1447 (1982).
34. E. Tsuchida, K. Abe, *Adv. Polym. Sci.*, 45, 1 (1982).
35. E.A. Bekturov, L.A. Bimendina, *Adv. Polym. Sci.*, 41, 99 (1981).
36. B. Nowack, J. Koetz, , A.B. Zezin, T.B. Zheltonozhskaya, N. Sokolov, L.A. Bimendina, M.G. Yashkarova, S.E. Kudaibergenov, *Materials. of 2nd Intern. Conf. "Semipalatinsk Test Site. Radiation heritage and non-proliferation problems"*, Semipalatinsk, 1, 220 (2005).
37. Zh.K. Ibraeva, *PhD Thesis*, Almaty, 2005).
38. 36a. V. V. Khutoryanskiy, A.V. Dubolazov, G. A. Mun, *Hydrogen-Bonded Interpolymer Complexes: Formation, Structure and Applications* (V.V. Khutoryanskiy, G. Staikos Eds). Ch.1, 2007.
39. E.A. Bekturov, L.A. Bimendina, *Interpolymer Complexes*, Alma-Ata: Nauka, 1977.
40. V.A. Kabanov, I.M. Papisov, *Vysokomol. Soedin.*, A21, 243 (1979).
41. B. Philipp, H. Dautzenberg, I.J. Linov, J. Koetz, W. Dawydoff, *Progr. Polym. Sci.*, 14, 91 (1989).
42. V.A. Kabanov, Uspekhi Khimii (Rus. Rev. Chem.) 74, 5 (2005).
43. S.V. Mikheykin, *The long term stabilization of uranium mill tailings*, IAEA-TECDOC-1403, 265 (2004).
44. R.I. Kalyuzhnaya, A.B. Zezin, V.B. Rogacheva, V.A. Kabanov, *Abstr. Conf. "Colloid-chemical problems of ecology"*, Minsk, 1990.
45. V.A. Kabanov, A.B. Zezin, V.A. Kasaikin, A.A. Yaroslavov, D.A. Topchiev *Uspekhi Khimii (Rus. Rev. Chem)* 60, 595 (1991).
46. A.B. Zezin, V.B. Rogacheva, S.V. Mikheykin, V.A. Kabanov, *Proceed. Intern. Monitoring Conf. "Development of rehabilitation methodology in environment of Semipalatinsk region polluted by nuclear test"*, Semipalatinsk, 2002.
47. A.B. Zezin, V.B. Rogacheva, S.V. Mikheykin, V.A. Kabanov, *Proceed. 7th Intern. Conf. "Radioactive Waste Management and Environmental Remediation"*, Japan, Nagoya, 1999.
48. A.B. Zezin, *Vysokomol.Soedin.*, A14, 1971 (1999).
49. V.A. Kabanov, A.B. Zezin, V.B. Rogacheva, Z.Kh. Gulyaeva, N.A. Ryabtseva, S.P. Valueva, S.V. Mikheykin, A.N.Alexeev, A.Yu Smirnov, L.V. Pronina *Patent RF.* RU 2142 492 (1998).
50. K.P. Kehrer, S.M. Atlas, V.A. Kabanov, A.B. Zezin, V.B. Rogacheva, *US Patent, #* 6,716,312 B2 (2004).
51. K.P. Kehrer, S.M. Atlas, V.A. Kabanov, A.B. Zezin, V.B. Rogacheva, *US Patent,* #6,755,938 B2 (2004).
52. L.P. Firsova, *Radiochemistry*, 41, 272 (1999).
53. L.P. Firsova, *Radiochemistry*, 41, 279 (1999).
54. L.P. Firsova, O.I. Bogatyrev, *Radiochemistry*, 43, 188 (2001).
55. G. Shulga, R. Kalyuzhnaja, L. Mozheiko, F. Rekner, A. Zezin, V. Kabanov, *Vysokomol.Soedin.*, A24, 1516 (1982).
56. G. Shulga, *Dr habil Thesis*, Latvian State Institute of Wood Chemistry, Riga, Latvia (1998).
57. G. Telysheva, G. Shulga, *J. Agric. Engng Res.* 62, 221 (1995).
58. G. Shulga, F. Rekner, J. Varslavan, *J. Agric. Engng Res.* 78, 309 (2001).

59. T. Zheltonzhskaya, O. Demchenko, I. Rakovich, *Macromol. Symp.*, 203, 173 (2001).
60. S. Filipchenko, T. Zheltonozhskaya, O. Demchenko, T. Vitovetskaya, N. Kutsevol, L. Sadovnikov, O. Romankevich, V. Syromyatnikov, *Chem. Izv. Ecolog.*, 9, 1521 (2002).
61. O.V. Demchenko, T.B. Zheltonozhskaya, N.V. Kutsevol, V.G. Syromyatnikov, I.I. Rakovich, O.O. Romankevich, *Chem. Inz. Ekol.* 8, 463 (2001).
62. T.B. Zheltonozhskaya, N.E. Zagdanskaya, O.V. Demchenko, L.N. Momot, N.M. Permyakova, V.G. Syromyatnikov L.R. Kunitskaya, *Uspekhi Khimii (Rus. Rev. Chem.)*, 73, 811 (2004).
63. T. Zheltonozhskaya, V. Syromyatnikov, A. Eremenko, B. Permyakova, O.Akopova *Extended Abstr. of European Polymer Congress*, Moscow, Russia, CD P3, 1271 (2005).
64. N.M. Permyakova, T.B. Zheltonozhskaya, V.V. Shilov, N.E. Zagdanskaya, L.R. Kunitskaya, V.G. Syromyatnikov, L.S. Kostenko, *Theor. Exper. Chem.* 41, 382 (2005).
65. O.I. Artem'ev, M.G. Yashkarova, L.K. Orazzhanova, *Vestnik of Semey University*, 2, 52 (1998).
66. L.A. Bimendina, M.G. Yashkarova, L.K. Orazzhanova, S.E. Kudaibergenov, *Eurasian Chem. Tech. Journal*, 7, 139 (2005).
67. A.M. Kabdyrakova, L.A. Bimendina, M.G. Yashkarova, L.K. Orazzhanova, O.I. Artem'yev, A.V. Protskiy, *Abstr. of 2nd Intern. Conf. "Semipalatinsk Test Site. Radiation heritage and non-proliferation problems"*, Semipalatinsk, 2, 97 (2005).
68. S.E. Kudaibergenov, L.A. Bimendina, L.K. Orazzhanova, *Abstr. 9th Intern. Symp. "Macromolecule-Metal Complexes"*, New York, 2001.
69. S.E. Kudaibergenov, L.A. Bimendina, M.G. Yashkarova, L.K. Orazzhanova, *Abstr. 11th Intern. Symp. "Macromolecule-Metal Complexes"* Pisa, Italy, 2005.
70. S.E. Kudaibergenov, L.A.Bimendina, M.G. Yashkarova, *Research Journal of Chemistry and Environment*, 10, 2 (2006).
71. S.E. Kudaibergenov, L.A. Bimendina, M.G. Yashkarova, L.K. Orazzhanova V.B. Sigitov, *Synthesis, properties and application of novel polymeric betaines based on aminocrotonates*, Semipalatinsk: Semipalatinsk State University, 2006.
72. S.E. Kudaibergenov, L.A. Bimendina, M.G. Yashkarova, *J. Macromol. Sci. Part A: Pure & Appl. Chem.* 44, 911 (2007).
73. L.K. Orazzhanova, M.G. Yashkarova, L.A. Bimendina, S.E. Kudaibergenov, *J.Appl.Polym.Sci.*, 87, 759 (2003).
74. M.G. Yashkarova, L.K. Orazzhanova, B.S. Gaisina, L.A. Bimendina, *Materials of 2nd Intern. Conf. "Semipalatinsk Test Site. Radiation heritage and non-proliferation problems"*, Semipalatinsk, 1, 228 (2005).
75. E.A. Bekturov, L.A. Bimendina, G.K. Mamytbekov, *Complexes of water soluble polymers and hydrogels*, Almaty: Gylym, 2002.
76. M. Jiang, M. Li, M.Xiang, H. Zhou, *Adv. Polym. Sci.*, 146, 121-183 (1999).
77. E.A. Bekturov, L.A. Bimendina, S.E. Kudaibergenov, *Polymer complexes and catalysts*, Alma-Ata: Nauka, 1981.
78. B.E. Rivas, E.D. Pireira, I. Moreno-Villoslada, *Progr. Polym. Sci.*, 28, 173 (2003).

79. *Macromolecule–Metal Complexes* (Ciardelli, H., Tsuchida, E. and Wohrle, D. eds.) Heidelberg: Springer–Verlag, 1996.
80. A.B. Zezin, N.M. Kabanov, A.I. Kokorin, V.B. Rogacheva, *Vysokomol. Soedin.*, A19, 118 (1977).
81. A.B. Zezin, V.B.Rogacheva, S.P. Valueva, N.I. Nikonorova, M.F. Zansokhova, A.B. Zezin, *Rossiyskie nanotehnologii*, 1, 191 (2006).
82. S.E. Kudaibergenov, L.A. Bimendina, L.K. Orazzhanova, *Abstr. Intern. Conf.* *"Heavy metals and radionuclides in environment"*, Semipalatinsk, 1999.
83. S.E. Kudaibergenov, L.A. Bimendina, M.G. Yashkarova, *Regionalnyi Vestnik Vostoka*, 1, 110 (2007).
84. S.E. Kudaibergenov, L.A. Bimendina, M.G. Yashkarova, L.K. Orazzhanova, *Kazakhstan Patent*, # 18988 (2007).

# CHAPTER 12

# HYDROGEN-BONDED LAYER-BY-LAYER POLYMER FILMS AND CAPSULES

Veronika Kozlovskaya, Eugenia Kharlampieva, and Svetlana A. Sukhishvili[*]

*Department of Chemistry and Chemical Biology*
*Stevens Institute of Technology*
*Castle Point on Hudson, Hoboken, NJ, USA*
*[*]E-mail: ssukhish@stevens.edu*

## 1. Introduction

Hydrogen-bonding plays an important role in determining the three-dimensional structures of proteins and nucleic acids. Specific shapes adopted by these macromolecules with assistance of hydrogen-bonds control their physiological or biochemical functions.

In the case of synthetic molecules, hydrogen-bonding self-assembly has also been used to produce three-dimensional supermolecular structures.[1,2,3] Such self-assembled materials can exhibit strong responses to various environmental stimuli such as pH, temperature or solvent composition. Hydrogen-bonded complexes of synthetic polymers present another example of hydrogen-bonded self-assembly, where two initially soluble polymers are bound through multiple hydrogen bonds along the polymer backbone. Hydrogen-bonded polymer complexes in aqueous solutions have been studied for decades.[4,5,6,7] However, deposition of such complexes at surfaces through simple precipitation results in poor control over quality and thickness of the produced films.

Surface-mediated layer-by-layer (LbL) deposition results in high control over the properties of the resulting films and allows a wide range of materials to be incorporated within films. First established by Decher[8,9] and co-workers, LbL technique has been widely used over the past two decades inspiring the design of novel materials with advanced properties. The LbL procedure deposition involves alternating exposure of various surfaces to solutions of polymers able to form complexes when mixed in solution. Monolayer adsorption of a polymer is achieved at each deposition step, and the total film thickness increases with the number of steps. During this process, nanostructured polymer films with well-controlled morphology and thickness can be formed. Advantages of this technique result from the versatility of interpolymer interactions and the self-limiting nature of the amount of polymer deposited. **Figure 1** contrasts the ability of nanoscale control over polymer chain arrangements in solution and at surfaces.

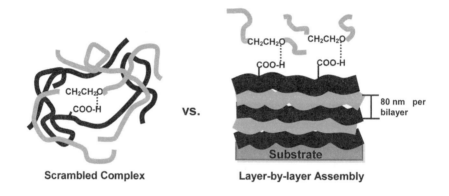

**Scrambled Complex**          **Layer-by-layer Assembly**

**Fig. 1.** Solution-based polymer complexes (left) vs. LbL films (right). In contrast with scrambled complexes in solution, the LbL film exhibits nanoscale control while conformally coating the assembly substrate. Redrawn with permission from Lutkenhaus, J.L.; Hrabak, K.D.; McEnnis, K.; Hammond, P.T. "Elastomeric flexible free-standing hydrogen-bonded nanoscale assemblies", *J. Am. Chem. Soc. 127*, 17228–17234 (2005). Copyright 2005 American Chemical Society.

There is virtually almost no limit to the species that can be incorporated within LbL films. Electroactive polymers, organic dyes, semiconductor quantum dots, electrochemically active species, inorganic colloidal nanoparticles, clay, and biologically active molecules, such as proteins, DNA, and polysaccharides can all be constituents.[10,11]

Advantageously, the LbL technique can be used to create functional coatings on substrates of almost any shape allowing great freedom in designing various dimensional systems such as hollow capsule[12] or nanotube structures.[13]

Though non-electrostatic interactions, such as hydrophobic ones, have been recognized as a significant contributor to film growth,[14,15] LbL technique has been widely applied to charged polymers of unlike charge.[10] Covalent bonding has also been realized through formation of permanent links between film constituents during deposition. Pioneering works of Rubner[16] and Zhang[17] in 1997 showed that polymer self-assembly can be driven by hydrogen-bonding interactions between adjacent layers. Since then, hydrogen-bonded LbL films have been receiving growing attention.

The use of weak polyelectrolytes allows control of the film properties by variation of the pH during film self-assembly, as well as post-deposition manipulation of film stability. Self-assembly of water soluble weak polyelectrolytes, such as polycarboxylic acids, and neutral polymers,[18,19] yields films which are easily erasable by a pH increase. The ability of hydrogen-bonded films to absorb large amounts of functional molecules such as dyes or drugs, combined with film disintegration at neutral pH values suggests future applications of such systems in drug delivery. Crosslinking of hydrogen-bonded films provides film stability at physiological pH and makes them attractive candidates for biomedical applications. Yet in other applications, such as electrochemical devices and ion-conductive membranes for fuel cells, solvent-free hydrogen-bonded membranes are used.[20]

In the present review, we focus on hydrogen-bonded films and capsules produced by LbL self-assembly of a neutral polymer and a weak polyelectrolyte in aqueous and/or organic solutions. We will discuss factors affecting the growth of hydrogen-bonded multilayers, such as the molecular weight of the polymer constituents, ionic strength, pH and temperature on the film deposition and the post-self-assembly response. Finally, recent results of response properties of cross-linked hydrogen-bonded films and capsules are overviewed.

## 2. Hydrogen-Bonded Polymer/Polymer Films

Stockton and Rubner have demonstrated that LbL film deposition in aqueous solutions can be driven primarily by hydrogen-bonding interactions.[16] Those authors showed that polyaniline can be co-self-assembled with a number of nonionic polymers such as poly(vinylpyrrolidone) (PVPON), poly(vinyl alchohol) (PVA), poly(acrylamide) (PAAM) and poly(ethylene oxide) (PEO). In this example, self-assembly of neutral polymers occurred in a wide pH range, but polyaniline was allowed to adsorb from acidic solutions where this polymer was soluble. Multilayers were also constructed from a conjugated copolymer of poly(phenylenevinylene) type containing hydroxyl groups which were capable of hydrogen-bonding with the amine groups of a co-self-assembled poly(ethyleneimine).[21]

The use of organic solvents for multilayer formation widens the range of polymers to water-insoluble and nonionic polymers which can be incorporated in the multilayers. Zhang and coworkers were exploring the

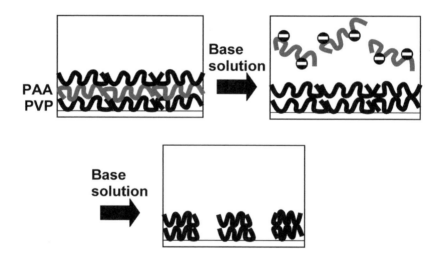

**Fig. 2.** Formation of micropores in hydrogen-bonded PAA/PVP multilayer films upon their exposure to basic pH. At the onset of film immersion, PAA dissolves into the basic solution, while PVP remains on the substrate. Redrawn with permission from Fu, Y.; Bai, Sh.; Cui, Sh.; Qiu, D.; Wang, Zh.; Zhang, X. "Hydrogen-bonding-directed layer-by-layer multilayer assembly: Reconformation yielding microporous films", *Macromolecules 35*, 9451–9458 (2002). Copyright 2002 American Chemical Society.

use of organic solvent or combinations of an organic solvent and water for construction of hydrogen-bonded films.[22,23] When a combination of poly(vinylpyridine) (PVP) with poly(acrylic acid) (PAA)[24, 25] or PVP with carboxylated dendrimers[26] was used in self-assembly, microporosity could be introduced (**Fig. 2**) within the film by exposing the film to aqueous solutions at basic pH where PAA or the dendrimer components selectively dissolved.

**Table 1**. Hydrogen-bonded multilayers in various solvents.

| # | Multilayer pair | Solvent used for deposition | Ref. |
|---|---|---|---|
| 1 | Polyaniline (PA)/polyacrylamide (PAAM) PA/poly(vinylpyrrolidone) (PVPON) PA/poly(vinyl alcohol) (PVA) PA/poly(ethylene oxide) (PEO) | water | 16 |
| 2 | Poly(acrylic acid) (PAA)/poly(4-vinylpyridine) (PVP) | water/ethanol | 17 |
| 3 | PAA/ PEO PAA/ PVPON Poly(methacrylic acid) (PMAA)/ PEO PMAA/ PVPON | water | 18,19 |
| 4 | PAA/bromo ethyl partially quaternized poly(4-vinylpyridine) | methanol | 25 |
| 5 | PAA/ PAAM | water | 27 |
| 6 | PMAA/ PAAM | water | 28 |
| 7 | PAA/ poly(N-isopropylacrylamide) (PNIPAAM) | water | 46 |
| 8 | PMAA/ poly(vinyl methyl ether) (PVME) PMAA/ poly(vinylcaprolactam) (PVCL) | water | 32 |
| 9 | PMAA/ poly(N-vinylpyrrolidone-co-NH$_2$) | water | 60,80 |
| 10 | PVP/ poly(4-vinylphenol) | ethanol or ethanol/DMF | 23 |
| 11 | PVP/ p-(hexafluoro-2-hydroxylisopropyl)-α-methylstyrene-co-styrene | chloroform | 22 |
| 12 | PVP/ carboxyl-terminated polyether dendrimer | methanol | 26, 29 |
| 13 | Amino-substituted poly(p-phenylene vinylene)/carboxyl-acid-substituted poly(p-phenylene vinylene) | dichloromethane | 30 |
| 14 | Poly[styrene-alt-(maleic acid)] (PSMA)/ PEO | water | 45 |
| 15 | Poly[(styrene sulfonic acid)-co-(maleic acid)]/ PNIPAAM | water | 31 |

Although water presents a highly competitive environment for hydrogen-bonding interactions, hydrogen-bonded self-assembly in aqueous environment has been widely used over the past decades. This is largely due to the use of aqueous media in biological and drug delivery applications. In particular, significant progress has been made towards tuning the growth of hydrogen-bonded films based on polycarboxylic acids through variation of the film constituents and deposition conditions. At acidic pH values, the polyacids are completely protonated, and hydrogen-bonded LbL deposition involves a pair of polymers which are electrically neutral. **Table 1** summarizes the data on film thickness for different polycarboxylic acid/neutral polymer systems obtained by LbL deposition on planar substrates from low ionic strength solutions at pH 2.

### 1.1 *Chemical nature of multilayer components*

In hydrogen-bonded multilayers, the bilayer thickness was found to correlate with the number of intermolecular binding points within the film. Specifically, for neutral polymers of comparable length, such as PEO-200kDa, PVME-200kDa, PNIPAAM-300kDa, and PVPON-360kDa, the highest film thicknesses were observed for loosely bound systems with weak intermolecular hydrogen-bonding such as between carboxylic groups of PMAA and ether groups of PEO or PVME.[32] For these films, as well as for other weakly associated polymer pairs, such as PMAA/PAAM and PMAA/PHEA (**Table 2**), linear growth was observed only after deposition of 6 or 8 layers. Note that when the more hydrophilic PAA is involved in the self-assembly with weakly associated polymers such as PEO and PAAM, exponential growth at large layer numbers was observed.[33,34] In addition, by switching from PMAA to the more hydrophilic PAA in self-assembly with PAAM resulted in an exponential film growth at layer numbers larger than 8, as well as in a ~2-fold larger bilayer thickness.[35] In contrast, more strongly bound systems, such as films of PMAA with PVPON, PNIPAAM or PVCL, had significantly smaller bilayer thicknesses, and showed linear growth starting from the first layer deposited. For all the polymer pairs in **Table 2,** the average bilayer thickness is given for the linear growth

**Table 2**. Average bilayer thickness ellipsometrically determined for PMAA (Mw 150 kDa) assembled with PEO (Mw 200 kDa), PVPON, (Mw 360 kDa), PAAM (Mw 5,000 kDa), PNIPAAM (Mw 300 kDa), PVME (Mw 200 kDa), PVCL (Mw 1.8 kDa), and poly(2-hydroxyethyl acrylate) (PHEA, Mw 600 kDa) in aqueous solutions at pH 2 and 23°C. Reprinted with permission from Kharlampieva, E.; Sukhishvili, S.A. "Hydrogen-bonded layer-by-layer polymer films", *Journal of Macromolecular Science, part C: Polymer Reviews 46*, 377–395 (2006). Copyright 2006 Taylor & Francis Group, LLC.

| Polymer system | Bilayer thickness, nm |
|----------------|------------------------|
| PMAA/PEO | 20 |
| PMAA/ PVPON | 4 |
| PMAA/PAAM | 7 |
| PMAA/ PNIPAAM | 6 |
| PMAA/PVME | ~ 16 |
| PMAA/PVCL | ~ 5 |
| PMAA/ PHEA | 13 |

regime. The dependence of the association strength on the nature of the hydrogen bond donor and acceptor groups is well known for polymer blends[36,37] and polymer mixtures in solution.[7] Formation of thicker multilayers results from a smaller number of interlayer binding points and, consequently, a more loopy conformation of the polymer chains in the case of binding between the carboxylic groups of PMAA with the ether groups of PEO or PVME. An analogy can be drawn with electrostatically assembled films where, too, thicker layers are reported for polyelectrolytes with lower charge density.[38]

### 1.2. *Molecular weight of polymers*

Strongly and weakly bound polymer systems also showed strikingly different molecular weight dependence of bilayer thickness. Hammond's group systematically studied the effect of molecular weight of PEO for PAA/PEO films and observed ~7-fold increase in PAA/PEO thickness reported when PEO $M_w$ increased from 1.5 to 20 kDa, and a very small further increase of thickness for a PEO $M_w$ change from 20 kDa to 4000 kDa.[33] In addition, those authors reported that films constructed with PEO of very low $M_w$ (1.5 kDa) are of a poor quality and high roughness. Using a relatively weakly bound PAA/PAAM system, Rubner and coworkers also found thicker films for PAAM of 5000 kDa than those constructed with shorter PAAM of 800 kDa.[34] Studying the effect of molecular weight of polymers in PVPON/PAA hydrogen-bonded films, Guan and co-workers [39] found that molecular weight has a significant effect on film thickness only when the molecular weight is low. When PVPON with $M_w$ of 10 kDa was used, the thickness of the resultant 30 bilayer films increased in the order of PAA-2kDa/PVPON < PAA-15kDa/PVPON < PAA-100kDa/PVPON. Compared with the films from PVPON with $M_w$ of 10 kDa, the films from PVPON with $M_w$ of 40 kDa had larger thickness, especially for the PAA-2kDa/PVPON-40 kDa film, which is almost twice as thick as the PAA-2kDA/PVPON-10 kDa film. Accordingly, films comprising polymers with strong hydrogen atom acceptors such as pyrrolidone or caprolactam rings rarely show variation of thickness with higher molecular weights. We have demonstrated this for PMAA/PVCL films composed of PVCL with different molecular weights, including a very low Mw of 1.8 kDa.[32] We also found that similar miltilayer thicknesses were produced when the $M_w$ of PVPON was varied from 55 kDa to 1,300 kDa. The absence of $M_w$ dependence of the film thickness in the abovementioned examples correlates with a high density of intermolecular hydrogen bonds in these systems, and is in good agreement with smaller thickness for an individual polymer layer.

### 1.3. *The pH of polymer solutions*

When the carboxylic groups involved in hydrogen-bonding become ionized, the hydrogen bonds between multilayer constituents are disrupted. This has severe consequences for the film stability. The weaker the hydrogen-bonding between the self-assembled polymers, the smaller is the number of hydrogen bonds which need to be dissociated for the film to dissolve. With weaker hydrogen-bonding systems, film dissolution occurs with a lower number of dissociation "defects" and at lower pH value. A significant contribution in understanding the effect of pH on formation of hydrogen-bonded films was made by Hammond and co-workers.[33] For PAA/PEO systems, film thickness was very sensitive to deposition the pH in the range ~2.8 to 3.5, where a severe decrease in the adsorbed amounts occurred. Film growth was completely prohibited at pH 3.5 due to increased ionization of the polyacid. With PMAA/PEO films, the 'modulation window' of inhibition of film growth was shifted to slightly higher pH values.[33] However, for stronger associated polymer systems, such as PMAA/PVPON, the critical pH above which film growth was prohibited, was shifted to pH 4.5.[35] A recent detailed study of the influence of pH on the hydrogen-bonded LbL assembly of PVPON and PAA by Xu and co-workers[40] revealed that at pH values above 4.0, the assembly of PVPON and PAA is almost halted and the deposition pH can greatly affect the film thickness, from several angstroms to hundreds of nanometers per dipping cycle. Importantly, they found that the critical pH value at which the hydrogen-bonded PAA/PVPON film could still be built up was significantly lower than the critical pH of disintegration of the PAA/PVPON film pre-assembled at acidic pH values. The critical disintegration pH value was found to be 5.5, while the critical buildup pH value was 4.0.

Along with exploration of higher pH values for hydrogen-bonded self-assembly, the effect of a very low deposition pH on PAA/PVPON film thickness was studied. In the pH range 0.2–1.0, a rough film was produced with an average thickness of ~48 nm per bilayer.[40] This unusually high thickness and roughness were explained by conformational changes of PAA at extremely low pH. Specifically, compaction and aggregation of PAA chains at extreme pH values was suggested.[40]

### 1.4. *Ionic strength of polymer solutions*

The effect of salt on growth of hydrogen-bonded films is expected to be complex and highly dependent on the nature of the self-assembled neutral polymers and low molecular weight salt, as well as on the ionization of the self-assembled polycarboxylic acid. In the case of PVPON, for example, strong ionic specificity is expected, since PVPON was reported to preferentially bind with large ionizable inorganic anions and cations of small size.[41] Another example includes strong chelating of $Cu^{2+}$ with carboxylic groups, which was reported to enhance the hydrogen-bonding strength between PVPON and PMAA or PAA in solution.[42] The specificity of ionic binding also affects the solubility of the components of the hydrogen-bonded multilayers. For PAA solutions, precipitation was observed at sodium chloride concentrations above 0.8 M.[43] Still another factor to be considered is that ionization of weak polyacids is enhanced in salt solutions, and reduced amounts of polymers deposited within the film can be observed for this reason.

Hammond and coworkers[33] found that the addition of monovalent and divalent salts during assembly of PAA/PEO systems resulted in an initial slight enhancement of film growth over a wide range of ionic strength, followed by a significant inhibition of polymer deposition at lithium triflate concentrations higher than 0.5 M. The mass reduction at high ionic strength was more pronounced for a divalent cation salt because of stronger competition of the salt with polymer components. In the case of the strongly associated PMAA/PVPON films, film thickness increased by only ~10% when self-assembly was performed in 0.5 M NaCl solution.[35] Overall, the effect of moderate concentrations of salts on growth of hydrogen-bonded films is smaller than that in electrostatically assembled systems. Instead of electrostatic screening and direct ion-ion interactions, resulting in a large increase of the amount of polymers deposited in electrostatically associated films,[44] the effects of salts on hydrogen-bonding self-assembly largely include dehydration and ion-dipole interactions which are highly specific to the nature of the salt and the polymer. Note, that very recently, a relatively strong effect of salt on hydrogen-bonding was reported for the case of a hydrophilic-hydrophobic copolymer, PSMA assembled with PEO.[45] In particular, deposition in 0.2 M NaCl dramatically increased the thickness (~ 2-fold) and roughness of PSMA/PEO films as compared to films deposited from 0.02 M NaCl solutions. The latter example illustrates that the hydrophobicity of one of the film components can drastically enhance the sensitivity of the amounts of polymers deposited to the concentration of the salt used during deposition.

### 1.5. *Concentration of assembly solution*

Recent results[39] showed that the concentration of the assembly solution has an effect on thickness of the produced hydrogen-bonded films. The main observation for the hydrogen-bonded PAA/PVPON films was that a higher concentration assembly solution leads to a larger amount of polymer adsorbed onto the surface and thus thicker films are produced. As a result, (PAA-100 kDa/PVPON-10 kDa)$_{20}$ films deposited from 1.0% (w/w) solutions were 55% and 40% thicker than those deposited from 0.1% and 0.3% (w/w) assembly solutions, respectively.

### 1.6. *Temperature of polymer solutions*

The effect of temperature on LbL deposition of hydrogen-bonded multilayers was first demonstrated by Caruso and coworkers for hydrogen-bonded layers containing a neutral temperature-responsive polymer, PNIPAAM.[46] Those authors reported that PAA/PNIPAAM films fabricated at 30°C showed higher thickness and much lower roughness compared to films prepared at 10 or 21°C[46] **(Fig. 3)**. Specifically, larger amounts of PNIPAAM were deposited within films when the lower critical solution temperature (LCST) of PNIPAAM (32°C) was approached. In our group, we studied the behavior of other temperature-responsive polymers, PVME (LCST ~36°C)[47] and PVCL (LCST ~35°C),[48] and found a similar trend. The increase in thickness

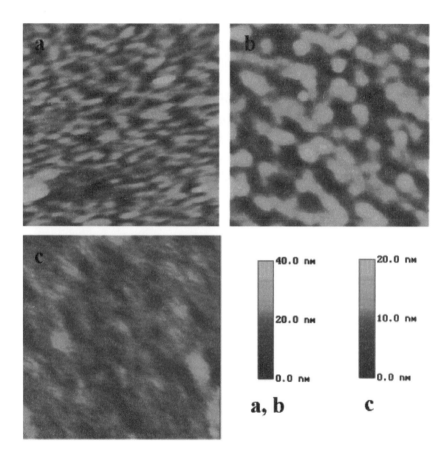

**Fig. 3.** SFM images of (PAA/PNIPAAM)$_{10}$ thin films prepared at different temperatures: (a) 10°C, (b) 21°C, and (c) 30°C. Scan area 1 micron x 1 micron. Reprinted with permission from Quinn, J.; Caruso, F. "Facile tailoring of film morphology and release properties using layer-by-layer assembly of thermoresponsive materials", *Langmuir 20*, 20–22 (2004). Copyright 2004 American Chemical Society.

when temperature was elevated to 25°C and 30°C was largely due to enhanced deposition of the temperature-sensitive polymers at higher temperatures where deterioration of the solvent quality for this polymer occurred.[35] Recently, we studied hydrogen-bonding systems that involved polymers which were not considered to be temperature-responsive due to a very high LCST, such as PAA/PEO and PAA/PVPON, and we also found an effect of increased film thickness at higher temperatures.[35] The data on temperature effects on the deposition of hydrogen-bonded films at surfaces correlated with an earlier work on hydrogen-bonded complexes in solution. For example, for PEO, whose LCST ranges from 100°C to 150°C, depending on its molecular weight,[49] no significant difference in the solvent quality is expected from 20°C to 60°C. However, solution studies of the thermodynamics of the complexation of PEO with PMAA or PAA in water revealed positive entropy of complex formation, suggesting an increase in entropy of water and an important role of hydrophobic interactions.[50] Indeed, stabilization of PAA/PEO and PMAA/PEO complexes with increased temperature has been observed.[51,52] Similar to PEO, PVPON is an amphiphilic water soluble polymer, which spreads at the air/water interface and exhibits LCST in aqueous solutions.[53] A strikingly different behavior was observed for multilayers of PAA with PAAM, which were found to produce thinner films and eventually deteriorate when self-assembled at elevated temperatures.[35] In contrast with other neutral hydrogen-bonding polymers discussed above, PAAM is more hydrophilic and forms complexes with PAA or PMAA which have UCST.[54,55] As temperature rises above UCST, intermolecular hydrogen bonds between the polyacid and PAAM are disrupted, and polymer chains swell and/or dissociate. In the case of PAA/PAAM interpenetrated networks, an UCST of 20°C to 25°C was reported.[55] The temperature-induced weakening of interpolymer interactions is consistent with hydrogen-bonding being the main contribution to PAA/PAAM binding energy, as was also found for PAA/PAAM complexes in solution.[56]

## 2. Hydrogen-Bonded Polymer/Particle Films

The self-assembly via hydrogen-bonding was also applied to fabricate micro/nanoparticle-containing thin films. Most of these methods rely on pre-assembly modification of particle surfaces with species able to form hydrogen bonds with the polymer used for LbL build-up. Gold particles of ~ 3 nm in diameter with carboxyl group or pyridine group tailored surfaces were successfully assembled in methanol solutions with either PVP or PAA, respectively[57] as shown in **Fig. 4**. Other surface-modified

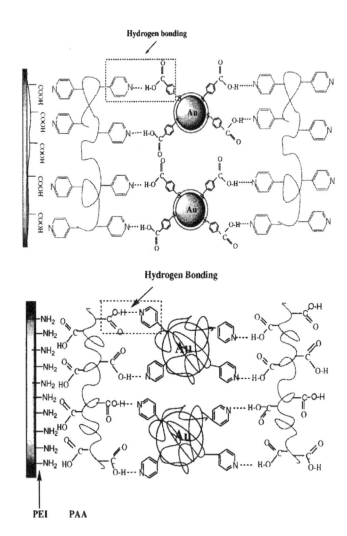

**Fig. 4.** Schematic representation of the buildup of multilayer assembly by consecutive adsorption of PVP and 4-MBA-tailored Au nanoparticles (top) and schematic representation of the buildup of multilayer assembly by consecutive adsorption of PAA and PVP-tailored Au nanoparticles (bottom). Reprinted with permission from Hao, E.; Lian, T.; "Buildup of polymer/Au nanoparticle multilayer thin films based on hydrogen bonding", *Chemistry of Materials 12*, 3392–3396 (2000). Copyright 2000 American Chemical Society.

semiconductor or metal nanoparticles, such as nanometer-sized CdSe synthesized in the presence of 4-mercaptobenzoic acid (4-MBA) in DMF/methanol, were incorporated into thin films directly through hydrogen-bonded self-assembly.[58]

## 3. Response of Hydrogen-Bonded Polymer Films to Environmental Stimuli

### 3.1. *Response of uncross-linked hydrogen-bonded films*

#### 3.1.1. *pH-Response*

A unique feature of the hydrogen-bonded self-assembly involving polycarboxylic acids to controllably dissolve at higher pH values, was first illustrated by Sukhishvili and Granick.[18,19] In contrast to electrostatically bound films, hydrogen-bonding interactions stabilize the protonated form of the polyacid and significantly suppress polyacid ionization within the multilayer as compared to polyacid ionization in solution.[19] When pH-induced ionization of the polyacid exceeds a critical value, film dissolution occurs. The critical pH of film disintegration was shown to be controlled by the strength of the hydrogen-bonding interactions within the film.[35] **Table 3** shows that the pH stability of the films improved as the strength of hydrogen-bonding and/or hydrophobicity of the film components was increased. Correspondingly, the critical pH for film disintegration was consistently higher for stronger bound systems. Specifically, the enhanced pH stability of PMAA/PVCL multilayers compared to PMAA/PVPON systems correlates with the presence of two extra methyl groups in the caprolactam ring and points to a strong contribution of hydrophobic interactions to film stabilization. Additional hydrophobic contribution to stabilization of the hydrogen-bonded self-assembly is even more evident from comparison of the pH-stability of PMAA/PVME films with the PMAA/PEO system. Specifically, more hydrocarbon moieties in the PVME chains resulted in a shift of the critical pH by about 1.6, as compared to PMAA/PEO, whose dissociation occurred at pH 4.6.

**Table 3.** Critical dissolution pH and critical ionization for hydrogen-bonded polymer systems presented in Table 2. The critical pH value was defined as the pH at which more than 90% of film did not dissolve for one hour. The critical ionization is PMAA ionization at the critical pH. Reprinted with permission from Kharlampieva, E.; Sukhishvili, S.A. "Hydrogen-bonded layer-by-layer polymer films", *Journal of Macromolecular Science, part C: Polymer Reviews 46*, 377–395 (2006). Copyright 2006 Taylor & Francis Group, LLC.

| Polymer System | Critical dissolution pH | Critical Ionization |
|---|---|---|
| PMAA/PHEA | 4.0 | ~1% |
| PMAA/PEO | 4.6 | ~1% |
| PMAA/PAAM | 5.0 | ~4% |
| PMAA/PVME | 6.0 | ~2% |
| PMAA/PNIPAAM | 6.2 | ~3% |
| PMAA/PVPON | 6.4 | ~8-10% |
| PMAA/PVCL | 6.95 | 30% |

The effect of the molecular weight on the film disintegration was also a subject of study.[19,35] It was shown that the stability of hydrogen-bonded systems was controlled by the molecular weight of the polymers. However, the difference observed was largely kinetic, and resulted in differences in the effective rate in the film disintegration for PVPON of various lengths, when the critical pH was defined through the disintegration rate of the film.[19] Specifically, using this definition, the critical pH for PMAA/PVPON systems was slightly decreased with decreasing PVPON molecular weight from 1300 kDa to 55 kDa. However, when adsorbed amounts were allowed to equilibrate at each pH value, the effect largely disappeared and films exhibited the same critical pH value for PVPON of various molecular weights.[35] In the case of PAA/PVPON films, it has been also found that the higher the polymer molecular weights, the lower their mobility and slower the erosion rate in water. Film thickness of the (PAA-100kDa/PVPON-40kDa)$_{30}$ decreased by 23% after films soaking in water for 7 days, while the lower molecular weight PAA-2kDa/PVPON-10kDa film was completely eroded after 6 hours.[39]

A similar trend was found in studies of polyelectrolyte complexes in salt solutions, where the critical salt concentration for dissociation of the polyelectrolyte chains was only weakly dependent on the molecular weight of the polyelectrolyte components in the limit of high molecular weight.[59] Indeed, the dissociation of polymeric segments is determined by the destruction of cooperative sequences of associating units, and for sufficiently long chains, the molecular weight has only a kinetic effect on chain dissociation.

The pH-response of PVPON/PMAA films was also studied using *in situ* imaging AFM.[60] Interestingly, when approaching the critical dissolution pH, PMAA/PVPON films did not swell in response to increased ionization of PMAA, but, instead, decreased in thickness. It is interesting that accumulation of extra charge within hydrogen-bonded films triggered by an increase in pH resulted in significant morphological changes within the film. **Figure 5** illustrates significant roughening of a 10-layer PMAA/PVPON film which was self-assembled at pH 2 and then exposed to pH 6. The increase in roughness from 4.4 to 8.5 nm for pH 2 and 6, respectively, is caused by microphase separation of the hydrogen-bonded components when the ionization of the polycarboxylic acid is increased. A similar increase in incompatibility, driven by enhanced ionization of a weak polyacid was reported for blends of hydrogen-bonded polymers.[61]

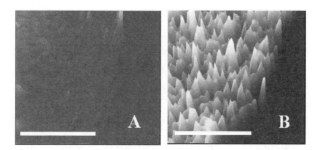

**Fig. 5.** Surface morphology of wet (PMAA/PVPON)$_5$ films deposited on silicon wafers at pH 2 and exposed to pH = 2 (panel A), and to pH = 6 (panel B) as probed by AFM (processed by SPIP software). The scale bar on both panels is 25 $\mu$m. Reprinted with permission from Kharlampieva, E.; Sukhishvili, S.A. "Hydrogen-bonded layer-by-layer polymer films", *Journal of Macromolecular Science, part C: Polymer Reviews 46*, 377–395 (2006). Copyright 2006 Taylor & Francis Group, LLC.

## 3.1.2. *Effect of salts*

As has been mentioned above, the effect of salts on the stability of hydrogen-bonded mutlilayers is expected to strongly depend not only on the total ionic strength, but also on the type of salt. Hammond and coworkers found that PAA/PEO films showed very high stability to lithium triflate salt exposure over a broad range of ionic strength at pH 2.5, with film dissolution in only 2M salt solution.[33] Note that when salt is added to hydrogen-bonded multilayer at pH values where the self-assembled polyacid is partially ionized, several additional effects should be considered. First, the addition of salt might decrease repulsions between ionized acidic groups, favoring film stabilization. At the same time the inclusion of salt results in increased osmotic pressure within the film, favoring film dissolution. For example, for strongly bound PMAA/PVPON system, we found that salt had a destabilizing effect on film stability, and the critical dissolution pH of the film was reduced from 6.4 to pH 5.0 as a result of increased PMAA ionization within the film in the presence of 0.5M NaCl.[35] Note that specific binding of ions with neutral polymer components might also affect the film dissolution transition. For example, a strong affinity of PVPON to large polarizable anions, such as $I^-$, is reported in the literature.[41]

## 3.1.3. *Effects of temperature*

Study of temperature effect on polymer films showed that hydrogen-bonded multilayers provide a good environment favoring temperature response of the self-assembled polymers. Specifically, Quinn and Caruso demonstrated enhanced dye release from PAA/PNIPAAM multilayers at elevated temperatures.[46] They also showed that the dye release rate from multilayers significantly slowed down by introducing a more hydrophobic polymer within the film.[45] A series of hydrogen-bonded LbL films, containing temperature-responsive polymers such as PVME and PVCL, exhibited a strong enhancement of dye permeability in a temperature range from 25 to 35°C.[32] The temperature responsive behavior of PVME- and PVCL-containing films was contrasted with that of films which were composed of polymers with similar functional groups, that do not demonstrate temperature sensitivity in this temperature range, such as PEO or PVPON. The fact that temperature response was clearly pronounced even for strongly bound PMAA/PVCL films, illustrates a promise of hydrogen-bonded self-assembly to produce temperature-responsive coatings for controlled release and adhesion applications.

Recently, the effect of post-assembly extreme heating on the stability of pre-assembled hydrogen-bonded PVPON/PAA films, was explored. It was shown that, at temperatures up to 120°C, water was driven out of the assembled films, but hydrogen bonds between PVPON and PAA were not destroyed.[62] The high temperature did not damage the films, and the films remained rather smooth and flat.

### 3.1.4. *Humidity response*

In contrast to the electrostatic assemblies of poly(allylamine hydrochloride)/poly(styrene sulfonate) films, with water content as high as 40–56%,[63] hydrogen-bonded PAA/PVPON films, incubated at pH = 2, have been found to have a much lower water content, of only 4%.[62] By alternating heating/humidifying steps, Yang *et al.* observed the reversible film deswelling.[62] The optical thickness, calculated from the position of the Fabry-Perot fringes, decreased upon heating and resumed upon incubation in an environment of a relative humidity ranging from 30 to 75%, with high reversibility. A linear relationship between the optical thickness of the PVPON/PAA films and ambient humidity was established. A 10%-increase in optical thickness of a 16-bilayer PAA/PVPON film was measured, when the relative humidity at 25°C was increased by 50%. This illustrates a potential use of the hydrogen-bonded PVPON/PAA films in humidity sensing.

### 3.1.5. *Incorporation of small functional molecules*

Among the most interesting properties of hydrogen-bonded films is their ability to incorporate and release small functional molecules. In one approach, a positively charged model dye was included within hydrogen-bonding films during self-assembly when added to polycarboxylic acid solution. The self-assembly pH was maintained at 4, which provided slight ionization to the polyacid. PMAA/PEO dye-containing films showed strong retention of dyes at a constant pH. Incorporated dye could be released as a result of film dissolution at a higher pH value.[19] In the second strategy, a dye, Rhodamine 6G (R6G), was loaded into preassembled PMAA/PEO multilayers in which an excess charge was created by a slight increase of the external pH.[64] It was demonstrated that the positively charged dye was efficiently bound with ionized carboxylic groups and uniformly distributed within the film. The dye loading capacity was high, ~30 wt % at pH 4.2. Interestingly, the absorbed dye was released, not only with film dissolution at higher pH values, but also in response to the suppressed ionization of self-assembled PMAA, resulting from either adsorption of a negatively charged polymer on top of the film, or lowering of the external pH.[64] Apart from pH-controlled loading and release, temperature-induced release of a Rhodamine B dye was also demonstrated by Quinn and Caruso from PAA/PNIPAAM films,[46] and from the films of PEO and a hydrophilic-hydrophobic copolymer of styrene and maleic acid.[45] In the latter case, dye retention at low temperatures was improved due to higher film hydrophobicity. Studies described above demonstrate that multilayers containing PEO are highly permeable to low molecular weight molecules such as dyes.

Finally, functional molecules, such as a positively charged dye, could be included within hydrogen-bonded films by replacement of one of the multilayer components.[35] Specifically, when R6G solutions were brought in contact with PMAA/PVPON or PMAA/PEO films at higher pH values, dye molecules replaced neutral polymers within the film, producing electrostatically-associated PMAA/R6G films (**Fig. 6**).[35] R6G remained strongly associated with PMAA chains and the film was stable at pH > 6 where PMAA was completely ionized. It was shown that for both systems, films with very high dye loading capacity of ~130–145 wt % dye/PMAA were produced. In addition, the films had a highly controlled thickness, which was templated by the individual thickness and number of polymer layers in the hydrogen-bonded self-assembly. While stable at pH > 6.5, PMAA/R6G multilayers gradually disintegrated at lower pH values, when ionization of the carboxylic groups decreased. The latter example demonstrates the potential value of functional molecule/polymer films for controlled delivery applications.[35]

## polymer replacement

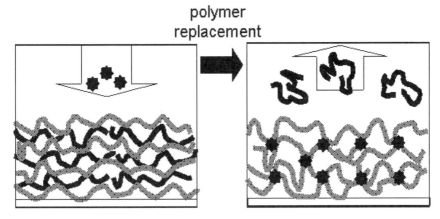

**Fig. 6.** Replacement of neutral polymers, PVPON or PEO, from hydrogen-bonded PMAA/PVPON or PMAA/PEO films with small dye molecules. Redrawn with permission from Kharlampieva, E.; Sukhishvili, S.A. "Hydrogen-bonded layer-by-layer polymer films", *Journal of Macromolecular Science, part C: Polymer Reviews 46*, 377–395 (2006). Copyright 2006 Taylor & Francis Group, LLC.

### 3.2. *Hydrogen-Bonded Capsules*

Polymeric capsules produced with the LbL deposition technique have recently attracted a great deal of research attention. The multilayer capsules are formed by alternate adsorption of interacting species onto colloidal templates. After the desired thickness is achieved, sacrificial cores are dissolved leaving behind hollow capsules (**Fig. 7**).

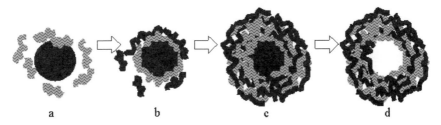

a   b   c   d

**Fig. 7.** Layer-by-layer approach to multilayer hollow capsule formation, where polymeric species are sequentially deposited onto a template particle (a, b) and a core-shell particle is formed (c). When the core is removed, hollow multilayer capsules are produced (d).

The attractive features of such capsules include: (1) the ease of control of the capsule layer thickness by simple variation of the number of deposition cycles; (2) the possibility of including macromolecules of various chemical nature within the capsule wall; and (3) the availability of internal volume within such capsules, which can be solid or filled with solvent and serve as a depot for delivery of chemicals, or as a microcompartment for chemical reactions or inclusion of catalysts.

The LbL approach for fabrication of microcapsules was introduced by the Colloids and Interface group at the Max Plank Institute (Germany) in the late 1990s.[65] The past two decades have witnessed many advances in the field of micro/nano-engineered capsules of electrostatically assembled species.[66]

Zhang with co-workers[67] were first to apply hydrogen-bonding-directed self-assembly to 3D systems and prepared multilayers of PVPON and m-methylphenol-formaldehyde resin (MPR) deposited from methanol on silica (SiO$_2$) or polystyrene (PS) particles. While dissolving PS cores in THF resulted in rupturing of the capsule walls by remains of PS cores, intact hollow capsules were successfully produced when SiO$_2$ templates were used and subsequently dissolved in HF solutions. The PVPON/MPR hollow capsules were uncharged and, therefore, had a tendency to aggregate. The capsules also had a rather rough surface. To prevent capsule aggregation and to ensure a homogeneous morphology, an additional polyelectrolyte bilayer of PAA/PAH was deposited on the top of the PVPON/MPR hydrogen-bonded layers before core dissolution.

In many biomedical or biotechnology applications of the polymeric capsules, it is preferable to use aqueous environment for polymer self-assembly. Therefore, there is a specific interest to explore hydrogen-bonded self-assembly of polymers in water. Despite the concerns that in hydrogen-bonding self-assembly, unlike electrostatic LbL, reducing polyelectrolyte's linear charge density might destabilize particle suspension, robust multilayer capsules of water soluble polymers of PMAA/PVPON or PMAA/PEO first on cadmium carbonate (CdCO$_3$) and later on SiO$_2$ particles were successfully produced[68] (**Fig. 8**). The LbL formation of hydrogen-bonded PMAA/PVPON and PMAA/PEO was monitored by zeta-potential measurements of the surface charge on

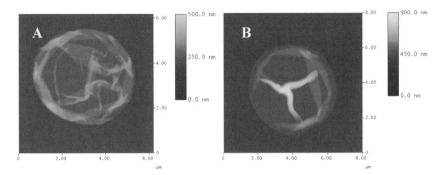

**Fig. 8.** AFM images of 4-bilayer (A) and 10-bilayer (B) hydrogen-bonded PMAA/PVPON capsules. Capsules were produced at pH 2, transferred to freshly cleaved mica surfaces, and allowed to dry for 2 hours. Elsner, N.; Kozlovskaya, V.; Sukhishvili, S. A.; Fery, A. "pH-Triggered softening of cross-linked hydrogen-bonded capsules", *Soft Matter 2*, 966–972 (2006). Reproduced by permission of The Royal Society of Chemistry.

the particles after each deposition step. Interestingly, the reversal of surface charge, a typical feature of electrostatic self-assembly of polyelectrolytes, when monitored by zeta-potential measurements of polymer-coated particles, was not observed during hydrogen-bonding self-assembly.

After assemblying a PMAA layer, the particles acquired a negative net surface charge and, starting from the neutral PVPON or PEO polymer layer, the net surface charge remained negative. In other words, the zeta-potential oscillated between negative values. The fact that the negative charge of the multilayer surface was maintained throughout the entire self-assembly process prevented particle aggregation. Similar oscillations of zeta-potential were observed by Rubner and co-workers[69] during hydrogen-bonded self-assembly of PAA/PAAM on melamine formaldehyde (MF) colloidal particles, but a larger amplitude of the zeta-potential variations was observed. The differences in the amplitude of zeta-potential oscillations, during self-assembly of hydrogen-bonded polymers in different systems, seem to be related to the thickness of the adsorbed neutral polymers. When an uncharged polymer adsorbs onto a charged surface, the surface charge is "buried" within the film. In addition, the effective slip plane, at which the electrophoretic mobility (and zeta-potential) is measured, moves away from the solid surface, due to hydrodynamic immobilization of water bound to the polymer loops. The degree of zeta-potential lowering therefore is dependent on the amount of neutral polymer adsorbed and on the structure of the adsorbed layer.

Increased hydrophobicity of a neutral polymer, however, might overwhelm the stabilizing effect of a slight net negative charge at the particle surface and promote particle coagulation. For example, while PMAA/PVPON capsules could be easily produced on silica cores at pH 2, PMAA/PVME capsules had a tendency to coagulate and produce irreversibly coagulated clusters when temperature approached ~30°C. This can be explained by the increased PVME hydrophobicity at temperatures close to its LCST of 35-36°C. In this case, to minimize the aggregation of polymer-coated particles, multilayer deposition was performed at a lower temperature (+ 10°C), which is far below the LCST of PVME, and the deposition pH was changed from 2 to 4 to increase the negative charge of the PMAA chains.[32]

### 3.2.1. *The effect of template*

The properties of hydrogen-bonded capsules depend on the templates used for capsule formation. Ideally, the core dissolution should result in complete removal of the core components, with no effect on the capsule wall.

In hydrogen-bonding self-assembly, both organic cores such as MF and its functionalized analogs, or PS, and inorganic cores such as $CdCO_3$, AgCl or $SiO_2$ were used. The drawback of using organic cores is that the oligomers formed upon dissolution are trapped within the capsule and also are entangled with the multilayer shell.[70] The use of inorganic particles is somewhat more advantageous over organic ones, as complete elimination of the core is achieved. Inorganic particles, however, have poorly defined surface charge characteristics. This uncertainty can be eliminated by pretreatment of the cores with precursor layers. For example, the use of poly (ethylene imine) (PEI) reversed the surface charge and/or enhanced the affinity of the hydrogen-bonded multilayer to the core. Interestingly, the PEI priming layer had different effects on the growth of weakly bound PMAA/PEO and strongly bound PMAA/PVPON hydrogen-bonded capsules. While successful deposition of PMAA/PVPON layers was performed on $CdCO_3$ particles, no PMAA/PEO capsules were produced when a layer of PEI was pre-adsorbed on the carbonate cores at pH=3.5. The effect was explained by a strong positive charge of the PEI-treated surface, which increased ionization of the PMAA above the threshold of 4% at which subsequent deposition of a PMAA/PEO multilayer became impossible.[71]

Different methods of removing various cores can introduce drastic changes in the capsule morphology. Yang *et al.* showed that after annealing PMAA/PVPON layers deposited on AgCl cores with a solution of KSCN in $HNO_3$ to produce hollow capsules, the capsule walls were rougher than those formed on silica cores. The high ionic strength of the solutions used for the core dissolution also resulted in capsule shrinkage and formation of holes.[72] Moreover, silver particles were formed on the capsule wall as a result of the photolysis of AgCl during the buildup of the capsules and dissolution of the AgCl cores.

Decomposition of silica particles in acidic solutions of hydrofluoric acid was found to be highly compatible with the high stability of hydrogen-bonded PMAA/PVPON polymer films in acidic media. The walls of the PMAA/PVPON capsules produced on silica cores were continuous and smooth.

Interestingly, hydrogen-bonded PMAA/PVPON microcapsules followed the shape of the 10-$\mu$m template on which they were deposited. Spherically-shaped hollow capsules were obtained using spherical non-porous silica cores, while in the case of rhombohedral cadmium carbonate templates, PMAA/PVPON hollow capsules produced were rhombohedral (**Fig. 9**).[71] However, using much smaller cubic 120 nm-sized AgCl particles as templates for PMAA/PVPON deposition, resulted in hollow capsules that lost the core shape and became spherical.

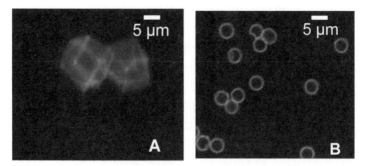

**Fig. 9.** Fluorescence microscopy image of hollow PMAA/PVPON capsules produced on rhombohedral cadmium carbonate cores (A), and confocal laser scanning microscopy image of hollow PMAA/PVPON capsules produced on spherical silica particles (B). Alexa Fluor 488 dihydrazide sodium salt was used for capsule staining in solution. Left image is reprinted with permission from Kozlovskaya, V.; Yakovlev, S.; Libera, M.; Sukhishvili, S.A. "Surface priming and the self-assembly of hydrogen-bonded multilayer capsules and films", *Macromolecules 38*, 4828-4836 (2005). Copyright 2005 American Chemical Society.

### 3.2.2. *The effect of pH*

The effect of pH on hydrogen-bonded self-assembly was studied in greater details for hydrogen-bonded films containing weak polycarboxylic acids.[19] Such films dissolve at slightly acidic or neutral pH values when ionization of the polyacid exceeds the critical value. This is described in detail above. For PMAA/PVPON films with PVPON of Mw 55 kDa, a critical pH value of 6.4 was reported. A similar pH response is also expected for hydrogen-bonded capsules as illustrated in **Fig. 10**. **Figure 11** shows a different critical pH value of 6.0 for hydrogen-bonded capsules. pH-Triggered dissolution of hollow (PMAA/PVPON)$_5$ capsules was followed by AFM after such capsules were dried on mica. These data are shown in **Fig. 11**. The capsule wall thickness started decreasing at pH 6.0, with the rate of ~24% loss of initial thickness in 1.5 hours, and the dissolution rate greatly increased at pH values only ~0.2 pH units higher. Moreover, in addition to thinning of the capsule wall, severe capsule rapture occurred upon the pH increase. Hydrogen-bonded capsules have a potential to be advantageously used as containers for fast release of the container

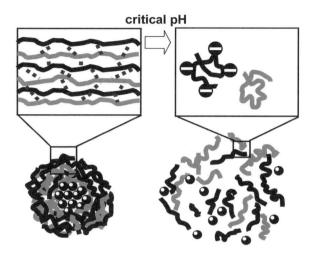

**Fig. 10.** Release of cargo from a hydrogen-bonded multilayer capsule at a critical pH value when hydrogen bonds are disrupted.

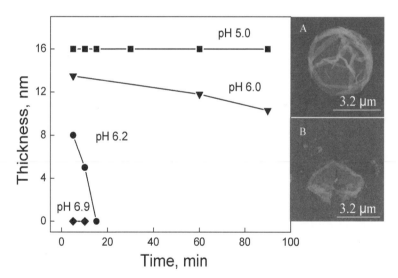

**Fig. 11.** Kinetics of dissolution of $(PMAA/PVPON)_5$ capsules at various pH values. Capsules were exposed to various pH values for a certain time, transferred to mica surfaces and dried under a stream of nitrogen. The pH values were maintained with 0.01 M phosphate buffers. AFM images of the $(PMAA/PVPON)_5$ capsule dried after exposure to pH 5.0 (A) and pH = 6.0 (B) for 90 min. Graph is reproduced from Elsner, N.; Kozlovskaya, V.; Sukhishvili, S. A.; Fery, A. "pH-Triggered softening of cross-linked hydrogen-bonded capsules", *Soft Matter 2*, 966–972 (2006). Reproduced by permission of The Royal Society of Chemistry.

contents. The pH-triggering mechanism of capsule dissolution can be used for controlled delivery applications.

### 3.2.3. *Crosslinked hydrogen-bonded capsules*

Other potential biomedical or biosensing applications of hydrogen-bonded LbL materials require, however, their stability in a wide pH range. Crosslinking as a route to stabilize multlilayers of weak polyelectrolytes has been widely used for electrostatically assembled films and capsules.[73] Several strategies have been proposed in the literature to afford hydrogen-bonding films stability at neutral and basic pH values.

To improve pH stability, there has been proposed a strategy of blending hydrogen-bonded LbL films and electrostatically assembled polymers.[74] However, this approach cannot be universally applied to polymer systems. Specifically, it has been recently shown that adsorption of a polycation can trigger replacement of the polymer chains within the hydrogen-bonded self-assembly resulting in dissociation of hydrogen-bonded stacks.[75] An alternative approach includes stabilization of hydrogen-bonded layers through chemical,[68,76] thermal or photo-cross-linking.[77] Chemically cross-linked two-component PAA/PAAM hydrogen-bonded films and capsules have recently shown much promise for biomedical applications. For example, Rubner and co-workers showed that cross-linked hydrogen-bonded capsules can be used as templates for synthesis of Ag nanoparticles,[78] imparting antibacterial properties to such capsules.

In our group, carbodiimide chemistry has been applied to covalently crosslink PMAA within PMAA/PVPON or PMAA/PEO hydrogen-bonded films or capsules.[79] Two types of cross-linked capsules were produced: (1) type I PMAA/neutral polymer capsules[80] and (2) type II PMAA capsules.[79] Both types are stable in a wide range of pH values from 2 to 12. An astounding feature of those capsules is that at high pH values there are no intermolecular interactions between the polymer components and the capsule wall resembles a crosslinked ultrathin hydrogel. Note that similar type I PAA/PAAM capsules were produced in Rubner's group.[76] Capsules of type I were produced by self-assembly of a random amino group-containing PVPON-co-NH$_2$ copolymer rather than PVPON.[80] The copolymer was synthesized by free-radical solution polymerization of N-vinylpyrrolidone and glycidyl methacrylate in dioxane using AIBN as initiator, followed by modification with ethylenediamine. There exists a significant difference between the capsules of the two types. After the copolymer was self-assembled with PMAA and crosslinked using carbodiimide chemistry, cross-linking occurred between PVPON-co-NH$_2$ and PMAA chains. Both components were retained within the film when the external pH was increased above a critical pH value of ~6.0, in spite of complete dissociation of intermolecular hydrogen bonds. Correspondingly, when pH is decreased below the critical value, the hydrogen bonds were re-formed. In the case of type II capsules, crosslinking was performed between PMAA chains within the PMAA/PVPON film. In this strategy, PVPON did not contain functional groups for crosslinking and was, therefore, released from the film at basic pH values. The wall of type II capsules thus was ultrathin PMAA hydrogel. Lowering the pH induced deswelling of such hydrogel capsule walls as a consequence of decreased PMAA ionization, but in contrast to capsules of type I, this did not result in re-formation of bonds between polymer chains. Remarkably, both types of capsules underwent reversible size changes (shown in **Fig. 12**) in response to variations in pH and/or ionic strength. The swelling/deswelling of (PMAA/PVPON-co-NH$_2$)$_7$ crosslinked capsules was reversible, but showed distinct hysteresis at pH values from 4 to 6 as shown in inset in **Fig. 12**. Specifically, a larger capsule size was detected in this region when pH was lowered, compared to that when pH was increased. Indeed, hydrogen-bonding between PVPON and PMAA chains at low pH values strongly suppresses PMAA ionization. Also dissociation of hydrogen bonds occurs when a critical ionization of PMAA is reached. When approaching the same pH region from the high pH values, in the absence of interchain association, ionization of the -COOH groups is higher than that in the pre-assembled film. Therefore, it is necessary to bring the hydrogel network to a lower pH value to induce formation of hydrogen bonds. The latter appears as the capsule size hysteresis between swelling and deswelling curves in the inset in **Fig. 12**.

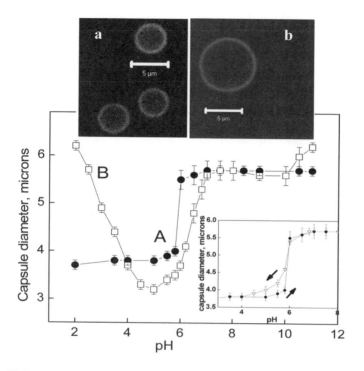

**Fig. 12.** Top: CLSM images of (PMAA/PVPON-co-NH$_2$)$_7$ type I (a) and (PMAA)$_7$ type II (b) capsules at pH 2. Bottom: pH dependence of the diameter of the (PMAA/PVPON-co-NH$_2$)$_7$ capsules (curve A, filled circles) and (PMAA)$_7$ capsules (curve B, open squares) cross-linked for 18 hours. The inset shows the hysteresis of (PMAA/PVPON-co-NH$_2$)$_7$ capsule size upon increasing (filled circles) and decreasing (open triangles) pH. The pH values were supported by 0.01 M phosphate buffer. Reprinted with permission from Kozlovskaya, V.; Sukhishvili, S.A. "pH-Controlled permeability of layered hydrogen-bonded polymer capsules", *Macromolecules 39*, 5569–5572 (2006). Copyright 2006 American Chemical Society.

Other strategies of chemical cross-linking of hydrogen-bonded layers were also explored. For example, when end-functionalized PEG or PVPON were used in a self-assembly, chemical cross-linking was also possible through the end groups of the neutral polymer, and two-component films were also produced.[81] However, through this strategy, only 50% of the PEG-dicarboxymethyl could be retained within the cross-linked multilayer film when this film was exposed to pH 7.5.

Another example is cross-linking through disulfide bonds, which was made possible through the use of poly(methacrylic acid) modified with cysteamine (PMAA-SH, with a functionalization degree of 18%) and PVPON.[82] The PMAA-SH/PVPON capsules were exposed to hydrogen peroxide solution to form S-S crosslink bonds. Interestingly, the thickness of PMAA-SH/PVPON capsules was twice as that of the PMAA/PVPON capsules, most likely due to increased hydrophobicity of the functionalized PMAA. Interestingly, the triggered deconstruction of the PMAA-SH/PVPON by a thiol-disulfide exchange reagent – dithiothreitol – at pH 7 resulted in release of the encapsulated, fluorescently labeled, protein FITC-transferrin.[82]

### 3.2.4. *Mechanical properties of hydrogen-bonded capsules*

The mechanical properties of the hydrogen-bonded PMAA/PVPON capsules at low pH values were studied using AFM single capsule force spectroscopy.[60] The value of Young's modulus for hydrogen-bonded PVPON/PMAA capsules of 610 MPa at pH 2 is comparable to that earlier obtained for electrostatically assembled microcapsules under similar conditions. Much lower Young's modulus was, however, reported by Hammond and coworkers for PAA/PEO film,[20] and the value obtained for PMAA/PVPON capsules might seem surprising. However, the more glassy mechanical response of PMAA/PVPON multilayers can be explained by stronger intermolecular hydrogen-bonding in the PMAA/PVPON system. Specifically, a large fraction (50%) of pyrrolidone units participates in interpolymer hydrogen bonding,[19] resulting in the formation of a glassy material. With non-crosslinked PMAA/PVPON and crosslinked PMAA/PVPON-co-NH$_2$ capsules, the microcapsule stiffness was characterized at pH 2 as a function of wall thickness. At this pH, both systems are expected to be dominated by hydrogen-bonding interactions. Microcapsule stiffness was shown to be proportional to the square of the wall thickness. This was in good agreement with previous observations on other microcapsule systems, and predicted by continuum mechanical models for shells.

Dramatic changes in mechanical behavior occurred when the pH was increased above 6 (**Fig. 13**), when intermolecular hydrogen bonds dissociated. Decrease in stiffness for (PMAA)$_{10}$ capsules was

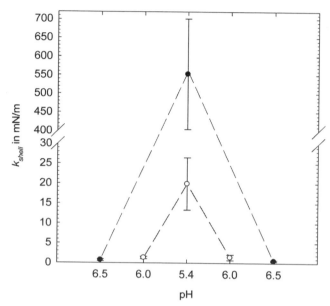

**Fig. 13.** Reversibility of the softening process for crosslinked (PMAA/PVPON-co-NH$_2$)$_7$ (open circles) and cross-linked (PMAA)$_{10}$ (filled circles) microcapsules for pH changes from 5.4 to 6 or 6.5, respectively. Elsner, N.; Kozlovskaya, V.; Sukhishvili, S. A.; Fery, A. "pH-Triggered softening of cross-linked hydrogen-bonded capsules", *Soft Matter 2*, 966–972 (2006). Reproduced by permission of The Royal Society of Chemistry.

from 550 ± 150 mN m$^{-1}$ to less than 1 mN m$^{-1}$, and from 18 ± 9 mN m$^{-1}$ to 1.4 ± 0.5 mN m$^{-1}$ for (PMAA/PVPON-co-NH$_2$)$_7$ capsules. The values of Young's modulus at pH > 6 are orders of magnitude lower than those reported in the literature for polymeric electrostatically assembled multilayer systems. It is remarkable that the deformation properties of the capsules change in a very narrow range of pH between 5.4 and 6. These changes are in sharp contrast to the electrostatically stabilized multilayers, where an increase in salt concentration only gradually changes the elastic modulus. The low stiffness of crosslinked hydrogen-bonded capsules at high pH is explained by the absence of intermolecular associations in the capsule walls. Indeed, the capsule wall resembles ultrathin hydrogel at high pH. The mechanical properties of a capsule can be designed to be switchable in response to environmental stimuli, and hydrogen-bonding interactions are very promising for such a design. For drastic changes in the mechanical properties of microcapsules, dissociation of intramolecular interactions within the multilayer should occur. In case of electrostatically assembled multilayers, dissociation of ionic polymer–polymer pairs can be achieved via an increase in salt concentration or pH, but extreme conditions are required. For many hydrogen-bonded systems, softening will occur at a moderate pH range, which is essential for many applications.

### 3.2.5. *pH-Controlled permeability of hydrogen-bonded capsules*

Various approaches have been developed for loading and subsequent re-leasing of substances from the microcapsule interior.[83,84] The pH-depend-ent permeability of the cross-linked wall of hydrogen-bonded capsules is promising in this regard. In this strategy, changing pH switches the capsules from the open state when they are permeable for species with a certain molecular weight, to the closed state when the capsules become impermeable to the same molecules (**Fig. 14**).

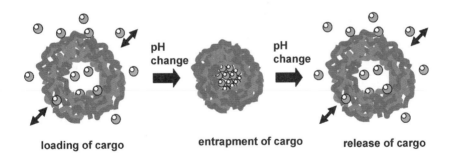

loading of cargo            entrapment of cargo            release of cargo

**Fig. 14.** Encapsulation and release of cargo to and from pH-sensitive capsules.

Specifically, when LbL deposition involved poly(carboxylic acids),[79] cross-linking resulted in capsules with hydrogel-like walls. As shown in **Figs. 12** and **14**, the capsule size and the corresponding mesh size of the hydrogel wall dramatically changed in response to pH variation. This was used to encapsulate FITC-labeled dextrans which could freely permeate the capsule wall at high pH. The dextrans were trapped inside the capsule after pH was changed to decrease the hydrogel mesh size and shrink the capsules. Subsequent release of the dextrans was achieved when pH was switched back (**Fig. 15**).

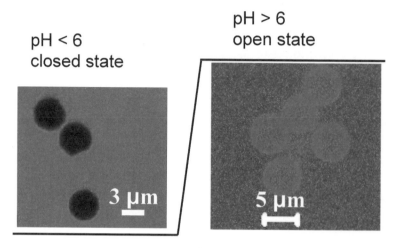

**Fig. 15.** pH-Control of permeability of (PMAA/PVPON-co-NH$_2$)$_7$ type I capsules to FITC labeled dextran with Mw 4 kDA at pH 4 (left) and pH 7 (right).

As the permeability of the PMAA capsule wall was dependent on the mesh size of the hydrogel network, variation of the cross-linking density allowed adjusting the mesh size of the hydrogel capsule walls in a controlled way. For example, the 5-hour cross-linked PMAA capsules were readily permeable to FITC-labeled dextrans from 4 kDA to 500 kDa at pH 8, as shown in Table 4, while an increase in cross-linking time afforded selective permeability of the same dextran molecules. As a result, the 18-hour cross-linked PMAA capsules became impermeable to the dextrans of 500 kDa at pH 8.

Interestingly, retention of the neutral component in two-component pH-responsive (PMAA/PVPON-co-NH$_2$) capsules, drastically affected the pH-dependent permeability of the capsules.[80] This is evident from comparison of the permeability of cross-linked one- and two-component capsules to Cascade Blue (CB) dye molecules and FITC-dextrans at various pH values. As shown in **Table 4**, there was a striking difference in the permeability of (PMAA/PVPON-co-NH$_2$)$_7$ capsules, which were able to form hydrogen bonds at acidic pH, and the (PMAA)$_7$ hydrogel capsules, in which such hydrogen bonds could not be formed. The (PMAA)$_7$ capsules demonstrated high permeability of the capsule wall to macromolecules. However, an increased cross-linking density of one-component hydrogel PMAA capsules did not allow controlled permeation of oligomeric molecules with Mw 4 kDa (**Table 4**). In a sharp contrast, the (PVPON-co-NH$_2$/PMAA)$_7$ system allowed permeability control for (a) oligomeric compounds at low pH through hydrogen-bonding and (b) larger macromolecules at high pH through dissociation of the hydrogen-bonded segments of the cross-linked polymer network.

**Table 4.** Permeability of cross-linked (PMAA/PVPON-co-NH$_2$)$_7$ and (PMAA)$_7$, as well as non-cross-linked (PMAA/PVPON)$_7$ capsules to FITC-labeled dextrans of different molecular weights and CB dye at pH 4.6 and 8a. Reprinted with permission from Kozlovskaya, V.; Sukhishvili, S.A. "pH-Controlled permeability of layered hydrogen-bonded polymer capsules", *Macromolecules 39*, 5569–5572 (2006). Copyright 2006 American Chemical Society.

| Permeant | CB dye | | | | | | FITC-labeled dextrans | | | |
|---|---|---|---|---|---|---|---|---|---|---|
| M$_w$ | 596 Da | | 4 kDa | | 70 kDa | | 150 kDa | | 500 kDa | |
| pH | 4.6 | 8 | 4.6 | 8 | 4.6 | 8 | 4.6 | 8 | 4.6 | 8 |
| (PVPON-NH$_2$/PMAA)$_7$ | O | O | X | O | X | X | X | X | X | X |
| (PVPON/PMAA)$_7$ | X | -[b] | X | - | X | - | X | - | X | - |
| (PMAA)$_7$ 5 h crosslinked | O | O | O | O | O | O | O | O | X | O |
| (PMAA)$_7$ 18 h crosslinked | O | O | O | O | X | O | X | O | X | X |

[a] Symbols X and O indicate capsule impermeability and permeability, respectively, to a specific permeant molecule, after 20 min of observation. [b] The permeability of hydrogen-bonded non-cross-linked (PMAA/PVPON)$_7$ capsules cannot be measured at pH 8 due to their dissolution at pH 6.

## 4. Conclusions

Hydrogen-bonded LbL films are receiving growing attention. Studies of the pH-response of hydrogen-bonded films containing water-soluble non-ionic polymers and weak polyacids are now complemented by knowledge on the effects of ionic species and temperature on film growth and post-assembly film response. Gaining deeper insights in fundamental aspects of self-assembly, will allow the design of materials based on hydrogen-bonded self-assembly with new useful properties. Today, several applications for hydrogen-bonded films have already been proposed and are being explored. The possibility to manipulate film properties by simple changes in solution pH, temperature or ionic strength make hydrogen-bonded films and capsules attractive candidates for controlled release applications. For example, both non-crosslinked and crosslinked films and capsules derived from hydrogen-bonded self-assembly, show environmental response to pH, and disintegration of the non-crosslinked films at higher pH values or enhanced swelling of the crosslinked films or capsules, can be used for drug release. The ease of removal of hydrogen-bonded films in aqueous solutions can also be used to create surface patterns after crosslinking of hydrogen-bonded films using ink-jet printing or photolithography techniques. Crosslinked hydrogen-bonded layers have demonstrated an exceptional resistance toward mammalian cell adhesion.[69] Hydrogen-bonded multilayers have been also rendered antibacterial properties after they have been used as matrices for *in situ* synthesis of Ag nanoparticles.[85] Still another range of applications includes the use of free-standing hydrogen-bonded films in sensor and device applications, such as the recently suggested use of PAA/PEO films as fuel-cell membranes.[33]

When deposited onto particulates, hydrogen-bonded LbL self-assembly produces capsules which can be designed to drastically change their size and mechanical properties at neutral pH values. The capsule containers derived from hydrogen-bonded multilayers allow selective control of permeation of macromolecules based on their molecular weight. Such capsules present a unique platform for encapsulation and release of a wide range of chemicals based on precise tuning of the capsule wall mesh size triggered by environmental stimuli.

## Acknowledgment

This work was partially supported by the National Science Foundation (Award DMR-0513197).

# References

1. A. W. Bosman, R. P. Sijbesma and E. W. Meijer, *Materials Today*, 4, 34 (2004).
2. W.H. Binder, *Monatshefte für Chemie*, 136, 1 (2005).
3. L. Häggman, C. Lindblad, H. Oskarsson, A. S. Ullström and I. Persson, *J. Am. Chem. Soc.*, 125, 3631 (2003).
4. F. E. Bailey, R. D. Lindberg and R. W. Callard, *J. Polym. Sci., Part A*, 2(2), 845 (1964).
5. T. Ikawa, K. Abe, K. Honda and E. Tsuchida, *J. Polym. Sci.*, 13(7), 1505 (1975).
6. E. A. Bekturov and L. A. Bimendina, *Adv. Polym. Sci.*, 41, 99 (1981).
7. E. Tsuchida and K. Abe, *Adv. Polym. Sci.*, 45, 1 (1982).
8. J. D. Hong, G. Decher and J. Schmitt, *Thin Solid Films*, 831, 210 (1992).
9. G. Decher, *Science*, 277, 1232 (1997).
10. P. Bertrand, A. Jonas, A. Laschewsky and R. Legras, *Macromol. Rapid Commun.*, 21, 319 (2000).
11. K. Ariga, J. P. Hill and Q. Ji, *Phys. Chem. Chem. Phys.*, 9, 2319 (2007).
12. F. Caruso, R. A. Caruso and H. Möhwald, *Science*, 282, 1111 (1998).
13. S. Hou, C. C. Harrell, L. Trofin, P. Kohli and C. R. Martin. *J. Am. Chem. Soc.*, 126, 5674 (2004).
14. N. A. Kotov, *NanoStructured Materials*, 12, 789 (1999).
15. S. L. Clark and P. L. Hammond, *Langmuir*, 16, 10206 (2000).
16. W. Stockton and M. Rubner, *Macromolecules*, 30, 2717 (1997).
17. L. Wang, Z. Q. Wang, X. Zhang, J. C. Shen, L. F. Chi and H. Fucks, *Macromol. Rapid Commun.*, 18, 509 (1997).
18. S. A. Sukhishvili and S. Granick, *J. Am. Chem. Soc.*, 122, 9550 (2000).
19. S. A. Sukhishvili and S. Granick, *Macromolecules*, 35, 301 (2002).
20. J. L. Lutkenhaus, K. D. Hrabak, K. McEnnis and P. T. Hammond, *J. Am. Chem. Soc.*, 127, 17228 (2005).
21. I. Benjamin, H. Hong, Y. Avny, D. Davidov and R. Neumann, *J. Mater. Chem.*, 8, 919 (1998).
22. L. Wang, Y. Fu, Sh. Cui, Zh. Wang, X. Zhang, M. Jiang, L. Chi and H. Fuchs, *Langmuir*, 16, 10490 (2000).
23. H. Zhang, Zh. Wang, Y. Zhang and X. Zhang, *Langmuir*, 20, 9366 (2004).
24. Y. Fu, Sh. Bai, Sh. Cui, D. Qiu, Zh. Wang and X. Zhang, *Macromolecules*, 35, 9451 (2002).
25. Sh. Bai, Zh. Wang, X. Zhang and B. Wang, *Langmuir*, 20, 11828 (2004).
26. H. Zhang, F. Yu, D. Wang, L. Wang, Zh. Wang and X. Zhang, *Langmuir*, 19, 8497 (2003).
27. S. Y. Yang and M. F. Rubner, *J. Am. Chem. Soc.*, 124, 2100 (2002).
28. S. Y. Yang, J. D. Mendelsohn and M. F. Rubner, *Biomacromolecules*, 4, 987 (2003).

29. F. Huo, H. Xu, L. Zhang, Y. Fu, Z. Wang and X. Zhang, *Chem. Commun.*, 874 (2003).
30. Z. Liang, M. Cabarcos, D. Allara and Q. Wang, *Adv. Mater.*, 16, 823 (2004).
31. J. F. Quinn and F. Caruso, *Adv. Func. Mater.*, 16, 1179 (2006).
32. E. Kharlampieva, V. Kozlovskaya, J. Tyutina and S. A. Sukhishvili, *Macromolecules*, 38, 10523 (2005).
33. D. DeLongchamp and P. Hammond, *Langmuir*, 20, 5403 (2004).
34. G. Decher and J. B. Schlenoff, in *Multilayer thin films. Sequential assembly of nanocomposite materials*, (Wiley-VCH, Weinheim, 2003) p. 151
35. E. Kharlampieva and S. A. Sukhishvili, *Journal of Macromolecular Science, Part C: Polymer Reviews*, 46, 377 (2006).
36. M. M. Coleman, A. M. Lichkus and P. C. Painter, *Macromolecules*, 22, 586 (1989).
37. Sh. W. Kuo, Sh. Ch. Chan and F. Ch. Chang, *Macromolecules*, 36, 6653 (2003).
38. K. Glinel, A. Moussa, A. M. Jonas and A. Laschewsky, *Langmuir*, 18, 1408 (2002).
39. Y. Guan, S. Yang, Y. Zhang, J. Xu, C. Han and N. A. Kotov, *J. Phys. Chem. B*, 110, 13484 (2006).
40. S. Yang, Y. Zhang, X. Zhang and J. Xu, *Soft Matter*, 3, 463 (2007).
41. M. Takano, K. Ogata, S. Kawauchi, M. Satoh and J. Komiyama, *J. Polymer Gels and Networks*, 6, 217 (1998).
42. R. Subramanian and P. Natarajan, *J. Polym. Sci.: Poly. Chem. Ed.*, 22, 437 (1984).
43. Sh. Jin, M. Liu, Sh. Chen and Y. Chen, *European Polymer J.*, 41, 2406 (2005).
44. V. Izumrudov, E. Kharlampieva and S. A. Sukhishvili, *Macromolecules*, 37, 8400 (2004).
45. J. F. Quinn and F. Caruso, *Macromolecules*, 38, 3414 (2005).
46. J. Quinn, and F. Caruso, *Langmuir*, 20, 20 (2004).
47. N. Yanul, Yu. Kirsh and E. Anufrieva, *Thermal Analysis and Calorimetry*, 62, 7 (2000).
48. Y. Maeda, *Langmuir*, 17, 1737 (2001).
49. M. Rackaitis, K. Strawhecker and E. *J. Polym. Sci. B: Polym. Phys.*, 40, 2339 (2002).
50. I. M. Papisov, V. Yu. Baranovskii, Ye. I. Sergieva, A. D. Antipina and V. A. Kabanov, *Polymer Sci. USSR* , A16, 1311 (1974).
51. Y. Osada and M. Sato, *J. Polym. Sci. Part B: Polym. Lett.*, 14, 129 (1976).
52. V. Khutoryanskiy, Z. Nurkeeva, G. Mun and A. Dubolazov, *J. Appl. Polymer Sci.*, 93, 1946 (2004).
53. L. Gargalló, A. Leiva, L. Alegría, B. Miranda, A. Gonźalez and D. Radic, *J. Macromol. Sci, Physics*, B43, 913 (2004).
54. K. Hosoya, Y. Watabe, T. Kubo, N. Hoshino, N. Tanaka, T. Sano and K. Kaya, *J. Chromatography A*, 1030, 237 (2004).

55. L. Y.Chu, Y. Li, J. H. Zhu and W. M. Chen, *Angew. Chem. Int. Ed.*, 44, 2124 (2005).
56. G. Staikos, K. Karayanni and Y. Mylonas, *Macromol. Chem. Phys.*, 198, 2905 (1997).
57. E. Hao and T. Lian, *Chem. Mater.*, 12, 3392 (2000).
58. E. Hao and T. Lian, *Langmuir*, 21, 7879 (2000).
59. V. A. Izumrudov and M. V. Zhiryakova, *Macromol. Chem. Phys.*, 200, 2533 (1999).
60. N. Elsner, V. Kozlovskaya, S. A. Sukhishvili and A. Fery, *Soft Matter*, 2, 966 (2006).
61 . V. V. Khutoryanskiy, A. V. Dubolazov, Z. S. Nurkeeva and G. A. Mun, *Langmuir*, 20(9), 3785 (2004).
62. S. Yang, Y. Zhang, Y. Guan, S. Tan, J. Xu, S. Cheng and X. Zhang, *Soft Matter*, 2, 699 (2006).
63. R. Steitz, R. Leiner, R. Siebrecht and R. V. Klitzing, *Colloids Surf. A*, 163, 63 (2000).
64. E. Kharlampieva and S. A. Sukhishvili, *Langmuir*, 20, 9677 (2004).
65. G. B. Sukhorukov, E. Donath, S. Davis, H. Lichtenfeld, F. Caruso, V. I. Popov and H. Möhwald, *Polym. Advn. Technol.*, 9, 759 (1998).
66. G. B. Sukhorukov and H. Möhwald, *Trends in Biotechnology*, 25, 93 (2007).
67. Y. Zhang, Y. Guan, S. Yang, J. Xu and C. C. Han, *Adv. Mater.*, 15, 835 (2003).
68. V. Kozlovskaya, S. Ok, A. Sousa, M. Libera and S. A. Sukhishvili, *Macromolecules*, 36, 8590 (2003).
69. S. Y. Yang, D. Lee, R. E. Cohen and M. F. Rubner, *Langmuir*, 20, 5978 (2004).
70. S. Moya, L. Dähne, S. Leporatti, E. Donath and H. Möhwald, *Colloids Surf. A*, 183, 27 (2001).
71. V. Kozlovskaya, S. Yakovlev, M. Libera and S. A. Sukhishvili, *Macromolecules*, 38, 4828 (2005).
72. S. Yang, Y. Zhang, G. Yuan, X. Zhang and J. Xu, *Macromolecules*, 37, 10059 (2004).
73. S. A. Sukhishvili, *Curr. Opin. Coll. Int. Sci.*, 10, 37 (2005).
74. J. Cho and F. Caruso, *Macromolecules*, 36, 2845 (2003).
75. E. Kharlampieva and S. A. Sukhishvili, *Langmuir*, 20, 10712 (2004).
76. S. Y. Yang, D. Lee, R. E. Cohen and M. Rubner, *Langmuir*, 20, 5978 (2004).
77. Y. Y. Sung and M. F. Rubner, *J. Am. Chem. Soc.*, 124, 2100 (2002).
78. D. Lee, M. F. Rubner and R. E. Cohen, *Chem Mater.*, 17, 1099 (2005).
79. V. Kozlovskaya, E. Kharlampieva, M. L. Mansfield and S. A. Sukhishvili, *Chem. Mater.*, 18, 328 (2006).
80. V. Kozlovskaya and S. A. Sukhishvili, *Macromolecules*, 39, 5569 (2006).
81. V. Kozlovskaya, E. Kharlampieva and S. A. Sukhishvili, *Polymeric Materials: Science and Engineering*, 93, 86 (2005).

82 . A. N. Zelikin, J. F. Quinn and F. Caruso, *Biomacromolecules,* 7, 27 (2006).

83. I. L. Radchenko, G. B. Sukhorukov and H. Möhwald, *Colloids Surf., A,* 202, 127 (2002).

84. D. V. Volodkin, N. I. Larionova and G. B. Sukhorukov, *Biomacromolecules,* 5, 1962 (2004).

85. D. Lee, R. E. Cohen and M. F. Rubner, *Langmuir,* 21, 9651 (2005).

# INDEX